Lecture Notes of the Institute for Computer Sciences, Social Informatics and Telecommunications Engineering 184

More information about this series at http://www.springer.com/series/8197

Yifeng Zhou · Thomas Kunz (Eds.)

Ad Hoc Networks

8th International Conference, ADHOCNETS 2016
Ottawa, Canada, September 26–27, 2016
Revised Selected Papers

 Springer

Editors
Yifeng Zhou
Ottawa, ON
Canada

Thomas Kunz
Carleton University
Ottawa
Canada

ISSN 1867-8211   ISSN 1867-822X (electronic)
Lecture Notes of the Institute for Computer Sciences, Social Informatics
and Telecommunications Engineering
ISBN 978-3-319-51203-7   ISBN 978-3-319-51204-4 (eBook)
DOI 10.1007/978-3-319-51204-4

Library of Congress Control Number: 2016960291

Printed on acid-free paper

This Springer imprint is published by Springer Nature
The registered company is Springer International Publishing AG
The registered company address is: Gewerbestrasse 11, 6330 Cham, Switzerland

Preface

The EAI International Conference on Ad Hoc Networks (AdHocNets) is a major annual international event in the ad hoc networking community. This year's AdHoc-Nets conference was held in Ottawa, Ontario, Canada. The aim of the AdHocNets conferences is to provide a forum to bring together researchers from academia and industry as well as government to meet and exchange ideas and discuss recent research work on all aspects of ad hoc networking.

Over the last decade, many efforts have been devoted to this area, producing enormous contributions addressing the fundamental issues of ad hoc networking including modeling, protocol and algorithm design, and security architectures and mechanisms etc. Recently, the implementation of ad hoc networking has been growing substantially, and we believe that it is time to invest greater research efforts into addressing the challenges related to real and commercial applications, and bring the huge potentials of ad hoc networking and years of research and development beyond experiments and laboratories.

The conference included a number of general sessions with regular papers as well as invited papers from renowned researchers in the field. The invited papers provide visions, trends, challenges, and opportunities in the area of ad hoc networking and emerging applications. The conference also featured two workshops on ad hoc network security and vulnerability, and convergence of wireless directional network systems and software defined networking, respectively. This volume of LNICST includes all the technical papers presented at AdHocNets 2016. It is our hope that the proceedings will be a useful and timely reference for researchers in their effort to understand the real-world challenges for ad hoc networking, and to develop innovative solutions in addressing these challenges.

September 2016

Yifeng Zhou
Thomas Kunz
Stefan Fischer
Zhangdui Zhong
Marc St-Hilaire

Organization

Steering Committee

Chair

Imrich Chlamtac   Create-Net, Italy

Members

Shiwen Mao   Auburn University, USA
Jun Zheng   Southern University, China

Organizing Committee

General Chair

Yifeng Zhou   Communications Research Centre, Canada

Technical Program Co-chairs

Thomas Kunz   Carleton University, Canada
Stefan Fischer   University of Lübeck, Germany
Zhangdui Zhong   Beijing Jiaotong University, China

Publicity and Social Media Co-chairs

Mike Tanguay   National Defence, Canada
Xiaohong Jiang   Future University Hakodate, Japan
Mouna Rekik   Inria, France
Changle Li   Xidian University, China

Publication Chair

Marc St-Hilaire   Carleton University, Canada

Workshops Co-chairs

John Matyjas   U.S. Air Force Research Laboratory, USA
Kamesh Namuduri   University of North Texas, USA

Web Co-chairs

Qianshu Wang   University of Waterloo, Canada
Tianhao Zhou   University of Michigan, USA

Local Chair

Jiali Shang   Canada Remote Sensing Application Development,
                           Canada

Conference Managers

Sinziana Vieriu   European Alliance for Innovation
Lenka Koczova   European Alliance for Innovation

Venue Manager

Ivana Allen   European Alliance for Innovation

Technical Program Committee

A. Sprintson   Texas A&M University, USA
A. Nayak   University of Ottawa, Canada
A. Soomro   Air Force Research Laboratory, USA
A. Gallais   University of Strasbourg, France
A. Guillen-Perez   Universidad Politecnica de Cartagena, Spain
A. Sen   Arizona State University, USA
B. Xiao   East China Normal University, China
C.H. Lung   Carleton University, Canada
D. Brown   DRDC, Canada
H. Tang   DRDC Ottawa, Canada
H. Rutagemwa   Communications Research Centre
I. Korpeoglu   Bilkent University, Turkey
J. Pan   University of Victoria, Canada
K. Namuduri   University of North Texas, USA
K. Wu   University of Victoria, Canada
M. Tsukada   University of Tokyo, Japan
M. Kellett   DRDC Ottawa, Canada
M. Salmanian   DRDC, Canada
M. Sookhak   Carleton University, Canada
M. Li   DRDC Ottawa, Canada
M. Ni   Beijing Jiaotong University, China
N. Mitton   Inria Lille - Nord Europe, France
N. Mastronarde   University at Buffalo, USA
O. Yang   University of Ottawa, Canada
O. Farooq   University College Cork, Ireland
P.A. Ward   University of Waterloo, Canada
P.K. Sahu   University of Montreal, Canada
R. Gravina   University of Calabria, Italy
R. Liscano   University of Ontario Institute of Technology, Canada
R. Sanchez-Iborra   Universidad Politecnica de Cartagena, Spain
R. Song   DRDC Ottawa, Canada

S. Bilen	Pennsylvania State University, USA
S. Chessa	University of Pisa, Italy
S. Fischer	University of Lübeck, Germany
S. Kumar	San Diego State University, USA
S. Krco	DunavNET
S. Papavassiliou	National Technical University of Athens, Greece
S. Pudlewski	Air Force Research Laboratory, USA
T. Burnop	Air Force Research Laboratory, USA
T. Kunz	Carleton University, Canada
W. Li	Beijing University of Posts and Telecommunications, China
Y. Gadallah	American University in Cairo, Egypt
Y. Sugaya	Tohoku University, Japan
Y. Zhou	Communications Research Centre Canada
Z. Zhong	Beijing Jiaotong University, China

Contents

Modelling and Analysis

Protocols

Workshop on Practical ad hoc Network Security and Vulnerability

**Workshop on Convergence of Airborne Networking, Wireless
Directional Communication Systems, and Software Defined Networking**

Wireless Sensor Networks

Management of Surveillance Underwater Acoustic Networks

Michel Barbeau$^{(\boxtimes)}$, Zach Renaud, and Wenqian Wang

School of Computer Science, Carleton University Ottawa, Ontario K1S 5B6, Canada
barbeau@scs.carleton.ca, {ZachRenaud,wenqianwang}@cmail.carleton.ca

Abstract. A Surveillance Underwater Acoustic Network (SUAN) is a sensor network specialized in the detection of sea surface or subsurface physical intruders, e.g., seagoing vessels. Network management provides the ability to remotely monitor and update the state of SUAN nodes. It is a crucial feature because of the difficulty of physical access once they have been deployed in sea or underwater. We explore three network management approaches: out-of-band, in-band and bio-inspired. Out-of-band management assumes the availability of high-speed wireless channels for the transport of management messages. The acoustic bandwidth of the SUANs is not directly used. In-band management uses the low date rate and short range underwater acoustic communication paths. Network management traffic is mixed together with data traffic. The bio-inspired approach does not require management traffic. Learning-by-imitation is used to transfer the settings node-to-node. It is useful in cases where it is really hard to convey information using messages because of harsh conditions.

Keywords: Surveillance underwater acoustic network · Sensor network · Sensor network management · Underwater acoustic communications · Network management · Routing · Bio-inspired network management

1 Introduction

A Surveillance Underwater Acoustic Network (SUAN) is a sensor network specialized in the detection of sea surface or subsurface physical intruders. Physical intruders are seagoing vessels. They are detected using acoustic and magnetic sensors. Transgressors produce acoustic noise and magnetic perturbations. The acoustic detection range is in the order of kilometers. The magnetic detection range is in the order of a fraction of a kilometer. When a presence is detected, alarms are produced. They are forwarded to a sink. SUANs involve deployment of underwater sensors, communication nodes and gateway buoys, playing the roles of sinks. Acoustic waves are used for underwater communications.

Once they are deployed in sea or underwater, physical access to SUAN nodes becomes difficult or literately impossible. Network management provides the ability to remotely monitor and update the state of nodes. It comprises getters

© ICST Institute for Computer Sciences, Social Informatics and Telecommunications Engineering 2017
Y. Zhou and T. Kunz (Eds.): ADHOCNETS 2016, LNICST 184, pp. 3–14, 2017.
DOI: 10.1007/978-3-319-51204-4_1

(e.g., data rate selection), setters (e.g., interface shutdown) and data models of managed information. It aims at automation and scalability. Classical network management approaches include remote login, Web interface, management protocol, such as Simple Network Management Protocol (SNMP) [10], and software-defined network.

We have explored out-of and in-band SUAN management. Out-of-band management assumes the availability of high-speed wireless channels for the transport of messages. The acoustic bandwidth of the SUANs is not directly used. Gateway buoys have this capability. To reach seabed sensors, in-band management is required. The acoustic bandwidth of the SUANs is used, that is, the low data rate and short range underwater acoustic communication paths. Management and data traffic are mixed together.

Our out-of and in-band network management work is integrated in our GNU Radio [1] Location-free Link State Routing (LLSR) protocol implementation [6, 7,27,33]. For out-of-band management, we use the popular and well supported SNMP. In every node, an agent, in conjunction with a translation and state management layer, handles incoming requests that retrieve or modify the state of variables. The state of a node can be accessed and modified through a myriad of available client side tools. On the hand, SNMP is a protocol designed for the Internet. It generates too much overhead with respect to the capacity of underwater communication channels.

For in-band management, there are two main issues. Firstly, given the low bandwidth of underwater acoustic paths, management messages need to be small. Secondly, in sensor networks, the flow of traffic is optimized for the sensor-to-sink direction. Management messages are expected to flow in the opposite direction. We create small management messages. We leverage the protocol elements of LLSR and augment it with a simple strategy for the transport of management traffic from sinks to sensors.

There are instances where underwater communications are very difficult. Solely few bytes can be exchanged. For such cases, we investigate a non-classical approach. We propose bio-inspired SUAN management. Every node learns its settings by imitating its neighbors. That approach does not generate traffic overhead.

Background and related work are reviewed in Sect. 2. Out-of-band, in-band and bio-inspired SUAN management are respectively discussed further in Sects. 3, 4 and 5. We conclude with Sect. 6.

2   Background and Related Work

The SUAN concept has been defined by Benmohamed et al. [8], Otnes et al. [25] and Rice et al. [28]. According to Otnes et al. [25], the architecture of a SUAN consists of seabed sensors, communication nodes and gateway buoys. In a SUAN, at least one gateway plays the role of sink. It forwards data, collected by sensors and relayed by communication nodes, to a fusion center. Underwater communications (sensors to communication nodes to gateways) are done using acoustic waves. Gateway-to-fusion center communications are done using electromagnetic waves. With respect to underwater communications, low data rates (in hundreds or thousands bits per second) and short ranges (e.g., 1000 m) are assumed. Small message payloads (e.g., 300 bytes) and small message rates (e.g., lower than one message per second per sensor) are also assumed. Intruder detection strategies that favor low traffic generation are desirable. Hence, issues specific to SUANs include the design of light weight communication protocols and placement of detection data processing functions (e.g., local processing of alarms in sensors versus global processing in fusion centers). The data fusion aspect is investigated in Braca et al. [9]. Underwater communication performance taking into account the various impairments is studied in Huang et al. [17]. The work comprises simulation and sea trial results. The main observation is that bit and packet error rates can be very high, even over short distances.

SUANs use underwater acoustic waves, a mechanical phenomenon. In contrast, terrestrial and space wireless communications are based on an electromagnetic phenomenon. Hence, acoustic waves are generated, propagate and are detected according to rules that differ from the ones of electromagnetic waves. Acoustic waves are produced by mechanical vibrations, a vibrator. Because of the elasticity of water, the resulting acoustic pressure propagates undersea. The detection of the acoustic pressure is done using a hydrophone.

Our research activities comprise an experimental facet. We adopted a software-defined approach. It is ideal for the academic environment because of its open character, flexibility and possibility to software-define low level protocols, i.e., at the physical and link layers. Modems, and accompanying protocols, are implemented in software using GNU Radio. They run in the Linux environment on a mini PC. To interface with the acoustic world, a recording studio-quality sound card is used. The vibrator and hydrophone plug in. In lieu of vibrator, the strength of the signal is raised with an audio power amplifier. It feeds a portable underwater speaker. Both the hydrophone and speaker have long cables that allow operation in depths. Other hardware options are described in the software-defined underwater communication projects of Demirors et al. [12] and Dol et al. [13].

The design of link layer and network layer communication protocols specific to SUANs has been investigated. MAC protocol performance issues are discussed in Otnes et al. [25] and Guerra et al. [16]. In particular, the time correlation between messages from different sensors and their impact in medium access are studied. MAC protocol and power control issues are addressed by Karlidere and Cayirci [19]. The DFLOOD routing protocol for SUANs has been defined by Otnes and Haavik [24], with improvements by Komulainen and Nilsson [21] and simulation work by Austad [4]. Security and clustering are discussed by Islam et al. [18]. The design of jamming resistant routing is investigated by Goetz et al. [15]. A delay tolerant network approach is introduced by Azad et al. [5]. Much of the protocols in commercial hardware are proprietary and vendor specific. JANUS is multiple-access acoustic protocol being standardized by the North Atlantic Treaty Organization (NATO) [2]. The physical layer uses

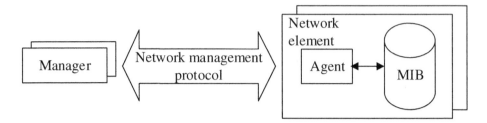

Fig. 1. Generic network management architecture.

frequency hopping transmission and binary frequency shift keying modulation. The link layer follows the carrier sense multiple access with collision avoidance protocol.

Figure 1 shows a generic architecture that most of the network management systems actualize in one way or another. Left side, there is one or several *managers* on which management applications are running. Right side, there are managed *network elements*. On each of them is running a server called the *agent*. The managers and agent communicate together using a *network management protocol*. The protocol is used to transport get and set requests, from the managers to the agents, as well as responses and alarms, in the reverse direction. In each network element there are manageable resources. Each of them is abstracted as a *managed object*. They are all collectively represented in a structure called the *Management Information Base* (MIB). A MIB is a collection of identified managed objects, i.e., constants and variables. They have standard formats. Hence, they look the same across different network elements. A manager controls a network element by inspecting and updating managed objects, using the management protocol. An agent may spontaneously notify a manager when a managed object reaches a condition by sending an alarm.

SNMP [10] is a well established Internet management framework, following the model of Fig. 1. For out-of-band SUAN management purposes, that is, sink node management, the use of SNMP is a good choice. One reason is the availability of a large base of SNMP software tools. For the aim of in-band SUAN management, i.e., submersed nodes, use of SNMP faces the problem of limited underwater acoustic bandwidth. Indeed, SNMP messages are not short. SNMP involves three layers of encapsulation (UDP, IP and data link). These three layers prefix each SNMP message with at least 50 bytes (16 + 20 + 14) of headers. The size of a SNMP message is variable. Depending on the version and exact operation being performed, a SNMP message easily consists of a few tens of bytes.

To the best of our knowledge, the management of SUANs has received little attention in the scientific literature. On the other hand, the management of wireless sensor networks has been a research topic [11,14,23,30]. Hereafter, two examples are reviewed. Management Architecture for Wireless Sensor Network (MANNA), by Ruiz et al. [29], is a policy-based system. It collects information, maps it into a network model, executes management functions and provides management services. The policies define management functions that are executed when conditions are met. Management is done by analyzing and updating the MIBs through the MANNA protocol. The sensor nodes are organized into clusters. Each node sends its status to its cluster head. The head is responsible for executing local management functions. It aggregates management data received from sensor nodes, which is forwarded to a base station. Several cluster heads, in a hierarchical architecture, can work together to achieve global network management.

The Sensor Network Management System (SNMS), by Tolle and Culler [32], is an interactive tool for monitoring the health of sensor networks. It has two main functions: query-based health data collection and event logging. Query functions retrieve and oversee the parameters of each node. The event-driven logging function set the parameters of the nodes. The nodes report data according to the values of dynamically configured thresholds. SNMS supports collection and dissemination of traffic patterns.

3 Out-of-Band Network Management

We investigated out-of-band management of SUANs [27]. For the purpose of compatibility with our work on software-defined communications and LLSR [7], it has been implemented in the GNU Radio environment. Given a communication application, a GNU Radio component is embedded to support remote management. A Python based SNMP agent is used above the NET-SNMP library in conjunction with a translation and state management layer to handle requests that retrieve or modify the state of variables within LLSR. SNMP requires the presence of a MIB in order to handle communication between a managed application and an agent. Creating this object database is a crucial task for adding SNMP support where none currently exists. The strength of this approach is that the state of LLSR can be remotely accessed and modified using off-the-shelf client side tools. The downside is the amount of traffic overhead that is required.

4 In-Band Network Management

LLSR forwards packets, produced by sensor nodes and relayed by communication nodes, in the direction of the sink. It is a routing model typical of sensor networks. For the purposes of forwarding, the next hop is selected according to a link-state metric. The sink announces its presence using broadcast beacon packets. Each of them contains the sink address, a hop count (zero, initially) and a numerical value reflecting the quality of the path to it, which for instance may be a degree of redundancy. Near surface underwater nodes, that are the sink one-hop neighbors, receive these beacon packets. They store the three items of information extracted from beacon packets into a neighbor node table. These nodes are now considered connected to the sink. In turn, they produce beacon packets. This procedure is performed repeatedly, hop-by-hop, until leaf sensor nodes are reached and become also connected to the sink through the communication nodes. When the network is fully connected, each node periodically broadcasts beacon packets. Network connectivity is maintained. From its neighbor node table, each node selects a next hop node, for packet forwarding.

To integrate in-band network management into LLSR, the sink node becomes responsible of forwarding management messages down to sensors and collecting the management data they produce. Each node has a SNMP-inspired agent to handle the management requests. There is also a MIB mapping abstract managed objects to actual resources. LLSR has been extended with four main protocol elements. Firstly, the management messages that are generated by the sink node reach their destination using network flooding. Secondly, the delivery of management packets to their final destination is confirmed. In third place, management response messages may be returned to the sink. Finally, management messages are authenticated and integrity checked.

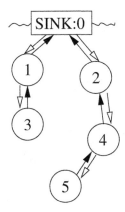
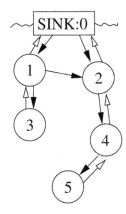

Fig. 2. Data packet forwarding.

Fig. 3. Management message forwarding.

Figure 2 illustrates the flow of data traffic. Black-end arrows indicate data packets. Their sources are sensors, node 5 in this example. Figure 3 shows the flow of management traffic. Black end arrows indicate management messages. Their source is the sink. In both figures, white-end arrows are acknowledgments. For data traffic, the data collected by each node is forwarded upward to the sink (node 0). For management message routing, Fig. 3 illustrates network flooding. Each node repeats once every management packet it receives. In Fig. 3, nodes 0 and 1 both deliver the same management message to node 2. At node 2, when it is received for the first time, the message number and arrival time are recorded in a table. An acknowledgment is sent to node 0. When the same management message is received for the second time, a table lookup performed by message number succeeds. The management message is ignored.

PROTO ID	PKT SRC	MGMT TRACK	MGMT ORG	VALUE	DEST	OPT	OID	HASH VALUE

Fig. 4. Management packet structure.

In each node, there is an agent for handling management messages. The agent has a MIB for mapping abstract managed objects to network parameters. The handling of a management message results into a response upward forwarded to the sink. The management packet structure is shown in Fig. 4. Each field is one byte, for a total of nine bytes. Management messages are short because of the low data rate of acoustic communications. PROTO ID is the identifier of the management protocol. PKT SRC is the address of the sender. MGMT TRACK is the tracking number of the management packet. MGMT ORG is the address of the management message origin sink. DEST is the address of the destination node. The default of field VALUE is zero. It is used to store a value accompanying a set operation. OPT indicates the type of management operation: a set (1) or a get (0). OID is the identifier of the managed object on which the operation is performed. HASH VALUE is an authentication field. The sink node shares a unique secret key with each network node. It is used to generate the value of the authentication field.

5   Bio-inspired Network Management

The use of the bio-inspired approach for distributed decision making in wireless networks has been discussed by Barbarossa and Scutari [3]. Hereafter, we use the approach for SUAN management purposes. With respect to out-of or in-band management, the bio-inspired approach does not generate traffic overhead. We leverage *learning-by-imitation*. It is used in robotics [22,31]. In the context of SUANs, a managed node acquires and updates its configuration by observing and replicating the behavior of other nodes. The network configuration parameters are transferred node-to-node without the use of network management messages.

Learning-by-imitation makes sense in networks in general because nodes must consistently apply a similar behavior. Epitomes are protocol elements such as the transmit probability and request-clear to send exchange in data link multiple access with collision avoidance.

Learning-by-imitation involves a teacher-observer relation. Figures 5 and 6 illustrate two possibilities. In both figures, an arrow represents a teacher-observer relationship, which is enabled when two nodes are within communication range. In Fig. 5, it is address-based. Each node has a numerical address. Node zero is the sink. There is a teacher-observer relationship when the address of the former is lower than the one of the latter. In Fig. 6, it is according to depth. There is a teacher-observer relationship when the depth of the former is lower than the one of the latter. Depth information needs to be available. Each node needs the ability to determine its depth and to learn the depths of neighbors. Note the

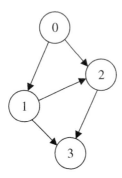
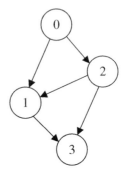

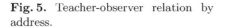

Fig. 5. Teacher-observer relation by address.

Fig. 6. Teacher-observer relation by depth.

inversion of the relationship between nodes one and two. In both cases, the goal is to establish a teacher-observer relation tree rooted at the sink. Establishment of a hierarchy is a typical animal behavior.

Figure 7 shows an example. It is about the beacon broadcast period. It is a parameter of LLSR. During the SUAN operation, the teacher periodically sends beacons. An observer collects traffic traces of the teacher. The observer measures beacon-to-beacon time differences. In Fig. 7, the x axis represents observation points. The y axis represents time values, in seconds. For each point, the actual period applied by the teacher is shown as a diamond, in seconds. For the first 20 points, the actual beacon period is one second. For the next 20 points, it is three seconds. For the last 20 points, it is two seconds. Because of various factors, such as processing time, the observations are noisy. In this example, zero mean white Gaussian noise and a signal-to-noise ratio of 22 dB are assumed. The noisy observations are shown as hollow circles. The observer uses the theory of system identification to estimate the value of the parameter from the observations [20]. System identification is about the construction of mathematical models of dynamic systems using collected observations. More particularly, on-line estimation algorithms have the ability to determine the parameters in reaction to the availability of data during the operation of a system. In Fig. 7, it is a problem of *time-varying parameter tracking*. The *recursive least-squares* algorithm is used [20]. The estimate resulting from each observation point is shown as a filled circle. The observer uses the current estimate to set the value of its own beacon broadcast period.

For comparison purposes, Fig. 7 plots as yellow circles the calculation of beacon periods from observations using a moving average equation. Figure 8 shows the Root Mean Square Error (RMS), as a function of the observation point, of the estimates using the recursive least-squares algorithm (diamonds) and moving average (circles). The RMS of the former is always substantially better than the one of the latter.

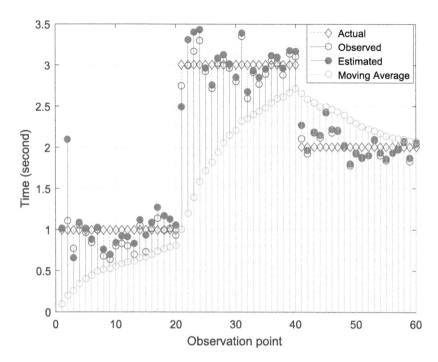

Fig. 7. Time-varying parameter tracking using the recursive least squares algorithm. (Color figure online)

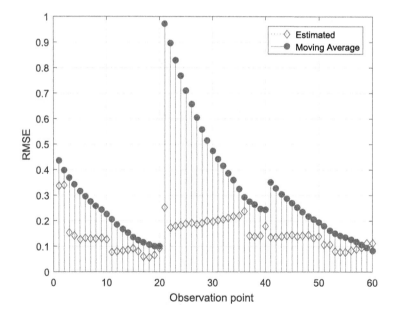

Fig. 8. Root mean square error (RMSE).

6   Conclusion

SUAN is an emerging area. It has been present in the literature only for a few years. Most of the projects, that have been published, discuss deployments of handful numbers of nodes. Much larger systems are expected in the future, as the technology will be perfected. The Internet type of communication protocols work solely for elements of SUANs with classical wireless access. SUAN specific protocols are required for submerged node communications. Interoperability between components from different sources is desirable. However, little standardization work has been accomplished so far.

Building SUANs is an important problem. Planning mechanisms enabling fine tuning their operation after their deployment is equally important. Physical access to SUAN elements may become very challenging as they may have to be recovered from the seabed. We have explored out-of-band, in-band and bio-inspired SUAN management. Out-of-band management is applicable to nodes that are directly accessible over a wireless channel, i.e., gateway buoys which are not entirely submersed. The approach leverages the vast amount of available SNMP tools. The work essentially amounts to defining and implementing a MIB, which we have demonstrated for the LLSR protocol. The disadvantage is the high traffic overhead for this context. For submersed communication nodes and sensors, we have developed an in-band management approach. The underwater acoustic bandwidth is used and shared with normal data traffic. Because of the low data rates and unreliable links, a light weight solution is highly desirable. Another aspect of the problem is the fact that the nature of the traffic in sensor networks is asymmetric, sensor to sinks. Sensor network routing protocols are designed for that traffic model. The solution we propose comprises short messages and leverages the protocol elements defined in the host routing protocol, which is LLSR in the example discussed in this paper. Our ideas have bee implemented, simulated and tested [6,7,27,33]. We have established the foundations of bio-inspired SUAN management. Nodes adopt the behavior of neighbors higher in a hierarchy. They learn and imitate their behavior. No management traffic required. Further work is needed to find strategies for learning the settings for various kinds of managed objects.

References

1. GNU radio - the free and open software radio ecosystem. gnuradio.org. Accessed June 2016
2. JANUS wiki. www.januswiki.org. Accessed June 2016
3. Abdolee, R., Champagne, B., Sayed, A.H.: Diffusion adaptation over multi-agent networks with wireless link impairments. IEEE Trans. Mob. Comput. 15(6), 1362–1376 (2016)
4. Austad, H.: Simulation of subsea communication network. Master's thesis, University of Oslo, Department of Informatics, Norway (2003)
5. Azad, S., Casari, P., Zorzi, M.: Coastal patrol and surveillance networks using AUVs and delay-tolerant networking. In: OCEANS, pp. 1–8, May 2012

6. Barbeau, M.: Location-free Link State Routing (LLSR) implementation for GNU Radio (2016). http://www.github.com/michelbarbeau/gr-llsr.git. Accessed Apr 2016
7. Barbeau, M., Blouin, S., Cervera, G., Garcia-Alfaro, J., Kranakis, E.: Location-free link state routing for underwater acoustic sensor networks. In: 28th Annual IEEE Canadian Conference on Electrical and Computer Engineering (CCECE), Halifax, Nova Scotia, Canada, May 2015
8. Benmohamed, L., Chimento, P., Doshi, B., Henrick, B., Wang, I.J.: Sensor network design for underwater surveillance. In: IEEE Military Communications Conference (MILCOM), pp. 1–7, October 2006
9. Braca, P., Goldhahn, R., Ferri, G., LePage, K.: Distributed information fusion in multistatic sensor networks for underwater surveillance. IEEE Sens. J. **16**(11), 4003–4014 (2016)
10. Case, J., Fedor, M., Schoffstall, M., Davin, J.: Simple network management protocol (SNMP), RFC 1157. rfc-editor.org/rfc/rfc1157.txt
11. de Oliveira, B.T., Margi, C.B., Gabriel, L.B.: TinySDN: enabling multiple controllers for software-defined wireless sensor networks. In: IEEE Latin-America Conference on Communications (LATINCOM), pp. 1–6, November 2014
12. Demirors, E., Sklivanitis, G., Melodia, T., Batalama, S.N., Pados, D.A.: Software-defined underwater acoustic networks: toward a high-rate real-time reconfigurable modem. IEEE Commun. Mag. **53**(11), 64–71 (2015)
13. Dol, H., Casari, P., van der Zwan, T.: Software-defined open-architecture modems: historical review and the NILUS approach. In: Underwater Communications and Networking (UComms), pp. 1–5, September 2014
14. de Gante, A., Aslan, M., Matrawy, A.: Smart wireless sensor network management based on software-defined networking. In: 27th Biennial Symposium on Communications (QBSC), pp. 71–75, June 2014
15. Goetz, M., Azad, S., Casari, P., Nissen, I., Zorzi, M.: Jamming-resistant multi-path routing for reliable intruder detection in underwater networks. In: Proceedings of the Sixth ACM International Workshop on Underwater Networks (WUWNet), p. 10 (2011). Observation of strains. Infect Dis. Ther. **3**(1), pp. 35–43, pp. 1–10. ACM, New York, NY, USA
16. Guerra, F., Casari, P., Zorzi, M.: MAC protocols for monitoring and event detection in underwater networks employing a FH-BFSK physical layer. In: Proceedings of IACM UAM (2009)
17. Huang, J., Barbeau, M., Blouin, S., Hamm, C., Taillefer, M.: Simulation and modeling of hydro acoustic communication channels with wide band attenuation and ambient noise. Int. J. Parallel, Emergent Distrib. Syst. dx.doi.org/10.1080/17445760.2016.1169420 (2016)
18. Islam, M.R., Azad, S., Morshed, M.M.: A secure communication suite for cluster-based underwater surveillance networks. In: International Conference on Electrical Engineering and Information Communication Technology (ICEEICT), pp. 1–5, April 2014
19. Karlidere, T., Cayirci, E.: A MAC protocol for tactical underwater surveillance networks. In: IEEE Military Communications Conference (MILCOM), pp. 1–7, October 2006
20. Keesman, K.J.: System Identification - An Introduction. Springer, London (2001)
21. Komulainen, A., Nilsson, J.: Capacity improvements for reduced flooding using distance to sink information in underwater networks. In: Underwater Communications and Networking (UComms), pp. 1–5, September 2014

22. Lopes, M., Santos-Victor, J.: A developmental roadmap for learning by imitation in robots. IEEE Trans. Syst. Man Cybern. Part B Cybern. **37**(2), 308–321 (2007)
23. Luo, T., Tan, H.P., Quek, T.Q.S.: Sensor OpenFlow: enabling software-defined wireless sensor networks. IEEE Commun. Lett. **16**(11), 1896–1899 (2012)
24. Otnes, R., Haavik, S.: Duplicate reduction with adaptive backoff for a flooding-based underwater network protocol. In: OCEANS - Bergen, 2013 MTS/IEEE, pp. 1–6, June 2013
25. Otnes, R., Voldhaug, J.E., Haavik, S.: On communication requirements in underwater surveillance networks. In: OCEANS 2008 - MTS/IEEE Kobe Techno-Ocean, pp. 1–7, April 2008
26. Porter, M.B.: The KRAKEN normal mode program (draft). http://www.oalib. hlsresearch.com/Modes/AcousticsToolbox/manual/kraken.html. Accessed 20 Apr 2015
27. Renaud, Z.: Network management for software defined radio applications. Honours project report, School of Computer Science, Carleton University. http://www. github.com/michelbarbeau/gr-llsr/blob/master/docs/ZachRenaud.pdf. Accessed Apr 2016
28. Rice, J., Wilson, G., Barlett, M., Smith, J., Chen, T., Fletcher, C., Creber, B., Rasheed, Z., Taylor, G., Haering, N.: Maritime surveillance in the intracoastal waterway using networked underwater acoustic sensors integrated with a regional command center. In: International Waterside Security Conference (WSS), pp. 1–6, November 2010
29. Ruiz, L.B., Nogueira, J.M., Loureiro, A.A.F.: MANNA: a management architecture for wireless sensor networks. IEEE Commun. Mag. **41**(2), 116–125 (2003)
30. Sayyed, R., Kundu, S., Warty, C., Nema, S.: Resource optimization using software defined networking for smart grid wireless sensor network. In: 3rd International Conference on Eco-friendly Computing and Communication Systems (ICECCS), pp. 200–205, December 2014
31. Schaal, S.: Is imitation learning the route to humanoid robots? Trends Cogn. Sci. **3**(6), 233–242 (1999)
32. Tolle, G., Culler, D.: Design of an application-cooperative management system for wireless sensor networks. In: Proceedings of the Second European Workshop on Wireless Sensor Network, pp. 121–132, January 2005
33. Wang, W.: Performance management of hydroacoustic surveillance networks. Master of computer science project report, school of computer science, Carleton University. http://www.github.com/michelbarbeau/gr-llsr/blob/master/ docs/WenqianWang.pdf. Accessed June 2016

Relative Localization for Small Wireless Sensor Networks

Yifeng Zhou[1(✉)] and Franklin Wong[2]

[1] Communications Research Centre Canada, Nepean, Canada
yifeng.zhou@canada.ca
[2] Defence Research and Development Canada, Ottawa, Canada
franklin.wong@drdc-rddc.gc.ca

Abstract. In this paper, we investigate relative localization techniques based on internode distance measurements for small wireless networks. High precision ranging is assumed, which is achieved by using technologies such as ultra-wide band (UWB) ranging. A number of approaches are formulated and compared for relative location estimation, which include the Linear Least Squares (LLS) approach, the Maximum Likelihood Estimation (MLE) approach, the Map Registration Approach (MAP), the Multidimensional Scaling (MDS) approach and the enhanced MDS approaches. Finally, computer simulations are used to compare the performances and effectiveness of these techniques, and conclusions are drawn on the suitability of the relative localization techniques for small networks.

Keywords: Wireless sensor networks · Localization · Ranging · Ultra-wide band (UWB) · Least squares (LS) · Maximum likelihood method (MLE) · Multidimensional scaling (MDS)

1 Introduction

Localization refers to the process of estimating the locations of objects based on various types of measurements and the use of a number of anchors. Anchors are simply objects that know their coordinates *a priori*. Localization is a prerequisite for many military operations where location information must be known *a priori* in order to monitor the environment, gather data measurements, track objects to make right decisions. Although GPS can be used for providing coordinates, it requires line-of-sight (LOS) conditions to satellites, and does not work reliably in urban and indoor environments. In addition, GPS is subject to jamming. In the last two decades, many localization techniques have been developed for wireless sensor network applications [1,2]. In general, localization can be relative or global. Relative localization provides relative coordinates that are defined without reference to an external coordinate system while global localization provides coordinates that are defined in the form of specific geographic coordinates such as latitude and longitude. Relative coordinates can be derived

© ICST Institute for Computer Sciences, Social Informatics and Telecommunications Engineering 2017
Y. Zhou and T. Kunz (Eds.): ADHOCNETS 2016, LNICST 184, pp. 15–26, 2017.
DOI: 10.1007/978-3-319-51204-4_2

from corresponding global coordinates. Relative coordinates are not unique, and are arbitrary rigid transformations of their global coordinates.

In this study, a number of relative localization approaches are formulated and discussed, which include the Linear Least Squares (LLS) approach, the Maximum Likelihood Estimation (MLE) approach, the Map Registration Approach (MAP), the Multidimensional Scaling (MDS) approach and the enhanced MDS approaches. All approaches are based on internode distance measurements that are assumed to be provided by the ultra-wide band (UWB) ranging technology. UWB radios employ very short pulse waveforms with energy spread over a wide swath of the frequency spectrum. Due to the inherently fine temporal resolution of UWB, arriving multi-path components can be sharply timed at a receiver to provide accurate time of arrival estimates, and thus the internode distance measurements. The LLS and the MLE method are based on the multilateration technique, which is seen to be one of the most popular localization techniques [1,3]. The MDS and MAP approaches are based on the approaches in [4–6], respectively. They use internode distance measurements to provide relative coordinates. MDS requires the full knowledge of the Euclidean distance matrix of the nodes, which is usually not available in practice due to the limited ranging capability. Unavailable distance measurements need to be approximated, which may introduce large localization errors. The MAP approach is a more elaborated approach that is proposed to counter this difficulty by dividing the network into many small sub-groups with adjacent groups sharing common nodes, constructing local maps for the all sub-group, and merging them into a global map. The MAP approach is able to alleviate the problems due to using the shortest path distances for remote sensor nodes.

The rest of the paper is organized as follows. In Sect. 2, the LLS and MLE methods are formulated. The procedures for determining the anchors are discussed in detail in this section. In Sect. 3, the MDS and the MAP method are discussed and formulated in the context of relative localization. In Sect. 4, the performance of various approaches are evaluated using computer simulations. A number of different application scenarios are simulated, which include fully and partially connected networks. Finally, conclusions are drawn on the suitability of various approaches for relative localization for small networks.

2   The MLE and LLS Methods

The linear least squares (LLS) method and the maximum likelihood estimation (MLE) method, in general, have two steps. The first step is to estimate three node locations. These nodes will be used as anchors. The second step is to iteratively estimate the locations of the rest of the nodes.

First, an arbitrary node is selected, denoted by s_1, as the first anchor and define it as the origin of the coordinate system. Secondly, a neighbour node of s_1, denoted by s_2, is selected as the second anchor. Define the line connecting s_1 and s_2 as the x-axis. The coordinates of s_2 are given by $(d_{12}, 0)$, where d_{12} is the measured distance between s_1 and s_2. Select a third node, s_3, that is a neighbour

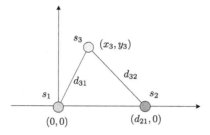

Fig. 1. Geometry of the first three anchor nodes.

to both s_1 and s_2 with distances d_{31} and d_{32}, respectively. The geometry of the first three anchor nodes are shown in Fig. 1. If the distances d_{21}, d_{31} and d_{32} satisfy the triangle inequality relationship, the coordinates of s_3 can be obtained as the intersections of two circles with centers at s_1 and s_2 and radii of d_{31} and d_{32}, respectively. They are given as [7]

$$x_3 = \frac{d_{21}^2 + d_{31}^2 - d_{32}^2}{2d_{21}}, \quad y_3 = \pm\sqrt{d_{31}^2 - x_3^2}. \tag{1}$$

In (1), the positive root is selected for y_3. Note the selection is arbitrary and will not affect the performance of relative localization. When the triangle inequality is not satisfied due to distance measurement errors, the two circles will not intersect. In this case, the coordinates of s_3 can be estimated using the following nonlinear least squares solution

$$\min_{\{x_3,y_3\}} \left(\sqrt{x_3^2 + y_3^2} - d_{31}\right)^2 + \left(\sqrt{(x_3 - d_{21})^2 + y_3^2} - d_{32}\right)^2. \tag{2}$$

Note that (2) is a nonlinear optimization problem and an analytical solution does not exist. Numerical techniques are required to solve for minimizing x_3 and y_3.

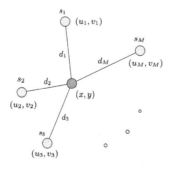

Fig. 2. Geometric of anchor nodes and the node to be localized.

2.1   Formulation of the MLE and LLS Methods

The second step of MLE and LLS is to estimate the locations of the rest of the nodes based on the nodes with known locations. Figure 2 shows the geometric configuration of anchors and the node to be localized. In the figure, M anchors are used, and their coordinates and measured distances to the node to be localized are denoted by $\{u_m, v_m, d_m\}$, for $m = 1, 2, \ldots, M$, respectively. The multilateration approach is to estimate the coordinates (x, y) given $\{u_m, v_m, d_m; m = 1, 2, \ldots, M\}$. The maximum likelihood method minimizes the following sum of squared errors between the measured distances and hypothetical ones based on the unknown sensor node location

$$\min_{x,y} \sum_m \left[\sqrt{(x - u_m)^2 + (y - v_m)^2} - d_m \right]^2. \tag{3}$$

Under the assumption that $\{d_m; m = 1, 2, \ldots, M\}$ contain additive measurement errors that are an independent, identically distributed (i.i.d.) Gaussian process with zero mean, (3) can be shown to be equivalent to the maximum likelihood estimator [8]. We refer to the formulation (3) as the maximum likelihood estimator (MLE). Since (3) is a nonlinear minimization problem, a closed-form solution does not exist, and numerical techniques are typically the resort. As mentioned before, numerical optimization techniques are subject to convergence difficulties and always suffer from the local minimum problem.

In practice, the least squares problem is often formulated in the squared distance domain to simplify the solution

$$\min_{x,y} \sum_m \left[(x - u_m)^2 + (y - v_m)^2 - d_m^2 \right]^2. \tag{4}$$

It can be shown that (4) is equivalent to solving the following least squares problem

$$B\underline{z} - \frac{r_x^2}{2} \cdot \mathbf{1} = \underline{\eta}, \tag{5}$$

where $\underline{z} = [x, y]^T$, $\mathbf{1}$ denotes an all one vector of length M,

$$\underline{\eta} = -\frac{1}{2} \begin{bmatrix} d_1^2 - r_1^2 \\ d_2^2 - r_2^2 \\ \vdots \\ d_M^2 - r_M^2 \end{bmatrix}, \quad B = \begin{bmatrix} u_1 & v_1 \\ u_2 & v_2 \\ \vdots & \vdots \\ u_M & v_M \end{bmatrix}. \tag{6}$$

and $r_x^2 = x^2 + y^2$ and $r_m^2 = u_m^2 + v_m^2$. The nonlinear term r_x^2 can be eliminated from the equation by the use of projection operations. Define $P_1^\perp$ as the orthogonal projection onto the null subspace of $\mathbf{1}$. By multiplying both sides of (5), we can obtain the following equation

$$A\underline{z} = \underline{b}, \tag{7}$$

where $A = P_1^\perp B$ and $\underline{b} = P_1^\perp \underline{\eta}$. Equation (7) is linear in $\underline{z}$ and has a closed-form least squares (LS) solution [9].

3   The MDS and MAP Methods

The MDS method was first proposed for solving the problem of sensor localization by Shang *et al.* [4,10], where either connectivity information or distance measurements between neighbor nodes were used for localization. It is based on the application of the popular multidimensional scaling (MDS) technique in statistics. It is a data analysis technique that can be used to represent a set of data as a configuration of points in some Euclidean spaces based on their similarity measures. The distances of the resulting configuration of points resemble the original similarities. There are many types of MDS techniques, including metric MDS and nonmetric MDS, replicated MDS, weighted MDS, deterministic and probabilistic MDS [11]. The classical MDS method is more attractive than the others because it has analytical solutions that can be obtained via eigendecomposition of a transform of the Euclidean distance matrix. In [12], the authors proposed an iterative MDS algorithm that uses a multivariate optimization for location estimation. The iterative MDS is similar to the least squares refinement step in [10]. The iterative MDS approach is less tractable than the classical MDS solution because it involves complex computations and suffers from global convergence problems. In general, the MDS technique is relatively resilient to distance errors due to the over-determined nature of the solution. However, MDS requires full knowledge of the Euclidean distance matrix of the sensor nodes, which is usually not available in practice due to the limited transmission range of beacons or ranging modules on each sensor node. A commonly used approach is to approximate the distances between nodes that are separated further than the transmission range by their shortest path distances. The shortest path distances can be computed using shortest path algorithms such as Dijkstra's [13] or Floyd's [14]. The approximation of the Euclidean distance matrix introduces sensor localization errors, especially when the shortest paths do not correspond well with the Euclidean distance in sparse networks or networks of irregular topology. Refer to [4,10] for the details of the MDS method.

3.1   The MAP Approach

The MAP approach refers to the map registration approach proposed by Zhou *et al.* [5,15]. It is known that the MDS approach requires that the Euclidean distance matrix for all nodes be known, which may not be always available in practice due to the limited ranging distance of the nodes. When two nodes are out of their transmission range, the distance between them cannot be directly obtained, and needs to be estimated. In MDS, the unavailable internode distances are typically approximated by its shortest path distance. A shortest path distance corresponds well to the corresponding Euclidean distance in a network of regular topology or a densely distributed network of nodes. In a sparse network or a network of nodes of irregular topology, however, a shortest path distance may not match its Euclidean distance and the use of the approximated distance matrix will result in degraded localization performance [4,5]. A more elaborate approach is to divide the network into many small sub-groups of nodes, where adjacent groups share common nodes. For each sub-group of nodes, a local map with relative coordinates of the nodes, is built using some localization techniques (*e.g.*, MDS). The local maps are then merged into a global map based on the common nodes. In [4], an incremental greedy algorithm was proposed for merging the local maps in a sequential manner. Each time a local map that has the maximal number of common nodes with the core map is selected and merged with the core map. The incremental greedy approach is locally optimal since it only explores the commonalities of the shared nodes in two maps. In practice, the common nodes are often shared by more than two local maps. In some cases, adjacent local maps may not have a sufficient number of common nodes.

The MAP approach was introduced to counter the problems of the sequential approach. Instead of using a sequential pairwise approach for merging local maps, the MAP approach constructs the global map at a global level. An affine transformation is defined for each local map to transform it to a global map. The set of optimal affine transformations are determined simultaneously by considering all available nodes that are shared by various local maps. The discrepancy is represented by the sum of the squared distances of all nodes to their respective geometric centers in the global map. Assume that a network consists of N nodes. For each node, the local map is assumed to contain its neighbor nodes within k-hops. Define a neighbor vector $\underline{c}_i$ of length N for the ith node. The nth component of $\underline{c}_i$ is given by 1 or 0 depending on whether the nth node is a neighbor node or not. Define a neighbor matrix $C = [\underline{c}_1, \underline{c}_2, \ldots, \underline{c}_N]$. For the ith local map, define an orthogonal matrix $U_i \in \mathcal{R}^{2 \times 2}$ and a row vector $T_i \in \mathcal{R}^{1 \times 2}$ to represent rotation/reflection (or a combination) and translation, respectively. Define $U \in \mathcal{R}^{2N \times 2}$ and $T \in \mathcal{R}^{N \times 2}$ as

$$U = [U_1; U_2; \ldots; U_N] \text{ and } T = [T_1; T_2; \ldots; T_N], \tag{8}$$

respectively. Let $\mathbf{z}_{ij} \in \mathcal{R}^{1 \times 2}$ denote the local coordinates of the ith sensor node in the jth local map. If the ith sensor node is not in the jth local map, then $\mathbf{z}_{ij} = \mathbf{0}$. Define a data matrix $Z_{ij} \in \mathcal{R}^{N \times 2}$, where the jth row of Z_{ij} is $\mathbf{z}_{ij}$. If the ith node is not in the jth local map, then, Z_{ij} is an all-zero matrix. Let $C_i = diag(\underline{c}_i)$ be a diagonal matrix of $N \times N$, where $diag$ puts the elements of $\underline{c}_i$ on its diagonal. For the ith local map, we construct a data matrix X_i

$$X_i = [Z_{i1}, Z_{i2} \ldots, Z_{iN}]. \tag{9}$$

Let Y_i denote an affine transform of X_i given by

$$Y_i = X_i U + C_i T. \tag{10}$$

All Y_i are in a same coordinate system that is referred to as the *global coordinate system*. The global coordinates of the sensor nodes form the *global map*. In MAP, the optimal U is obtained from the following optimization problem [5]

$$\min_{U} tr\{U^T \Sigma U\}, \tag{11}$$

subject to the constraint that U_i is an orthogonal matrix for $i = 1, 2, \ldots, N$. Denote $\mathcal{M}$ as the manifold that consists of all $U = [U_1, U_2, \ldots, U_N]$ and each U_i

is an orthogonal matrix of 2×2. Then, the constraint implies that the optimal U is in the manifold $\mathcal{M}$. In (11), tr denotes the trace of a square matrix, and

$$\Sigma = \sum_i X_i^T P_i^\perp X_i - A_s^T B_s^{-1} A_s \tag{12}$$

$$B_s = \sum_i \tilde{C}_i^T P_i^\perp \tilde{C}_i, \quad A_s = \sum_i \tilde{C}_i^T P_i^\perp X_i, \quad P_i^\perp = I - \frac{1}{N_i} \underline{c}_i \underline{c}_i^T, \tag{13}$$

and $\tilde{C}_i$ is C_i with its first column removed. The translation matrix T is related to the optimal U by $T = [\mathbf{0}; -B_s^{-1} A_s U]$.

The optimization problem (11) involves highly nonlinear criterion function, and analytic solutions are not known to exist. In [5], a gradient projection algorithm is developed for finding the optimal transforms for transforming local maps to a global map. The algorithm is developed based on a general idea by Jennrich in [16,17] and is particularly suitable to the constrained optimization problem of coordinate transformation. The algorithm is iterative, and has the advantages of faster convergence and computationally more efficient than many general numerical optimization techniques [18] for nonlinear programming. The detailed discussion of the GP algorithm can be found in [15].

4 Simulations and Performance Analysis

In this section, we use computer simulations to demonstrate the effectiveness and performance of the proposed relative localization techniques. MLE uses the LLS solution as the initial estimates in each iteration after the initial node selection process. For the MDS method, Dijkstra's algorithm [13] is used to compute the shortest paths to approximate the unavailable internode distances. For MAP, a local maps is constructed for each node, which consists of all direct neighbor nodes within its maximum ranging distance. The root mean square errors (RMSE) of the location estimates are used as a performance metric. In order to compute meaningful RMSEs for relative location estimates, all relative estimates are aligned to best conform to its ground truth node location.

The nodes are assumed to be uniformly distributed in a square area of $100\,\mathrm{m}$ by $100\,\mathrm{m}$. All nodes are assumed to have a common maximum ranging distance that can be configured. The maximum ranging distance determines whether the internode distance measurement between a pair of nodes is available or not. The network is assumed to be connected, *i.e.*, each of the nodes of the network is connected to each other either one-hop or *via* multiple hops in terms of internode ranging. The algebraic connectivity of a network is used to check whether the network is connected or not [19]. The distance measurement errors are assumed to be additive and uniformly distributed. The uniform distributed model ensures that the errors are bounded, and leads to the more conservative estimates of uncertainty than the Gaussian error model. Let $\tilde{d}_{ij}$ and d_{ij} denote the actual and measured distances between the ith and the jth node, respectively. Then, the measured distance is given by $d_{ij} = \tilde{d}_{ij} + \epsilon_{ij}$, where ϵ_{ij} is simulated to be uniformly distributed in $[-\sigma\tilde{d}_{ij}, \sigma\tilde{d}_{ij}]$ and $\sigma \in [0,1]$.

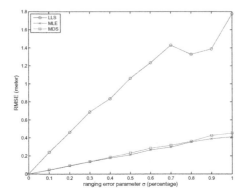

Fig. 3. Variation of RMSE for LLS, MLE and MDS *versus* σ in a fully connected network of $N = 5$ nodes.

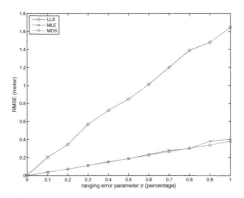

Fig. 4. Variation of RMSE for LLS, MLE and MDS *versus* σ in a fully connected network of $N = 10$ nodes.

Fully Connected Network. By a fully connected network, we mean that the maximum ranging distance for all nodes in the network is sufficiently large such that each node is able to measure its distances to all other nodes in the network. For a fully connected network, distance measurements between all pairs of nodes are available. Figures 3 and 4 show the variation of RMSE for the MLE, LLS and MDS estimates versus the ranging error parameter σ for fully connected networks with 5 and 10 nodes, respectively. The parameter σ is written in the form of percentage. In the simulations, σ varies from 0 to 0.01 (or 1%), and for each value of σ, 1000 tests are repeated to obtain the averaged RMSE results. The averaged internode distances for the networks of 5 and 10 nodes are calculated as 52.39 and 52.25 m, respectively. When $\sigma = 0.01$, it would translate into an averaged internode distance measurement error range of about ± 0.5 m. For each test, all nodes are randomly re-deployed and their random internode distance measurement errors re-generated. Thus, the RMSE results are averaged over both node distribution and random distance measurement errors. For a fully connected network, since all pairs of nodes are within the maximum ranging distance, the Euclidean distance matrix is completely available, and MAP become equivalent to MDS. Thus, only LLS, MLE and MDS are evaluated for fully connected networks. As discussed before, the LLS solutions are sensitive to geometric distribution of the nodes. In order to isolate the impact of node distribution on localization, a condition number of 400 is used to avoid scenarios that would result in ill-conditioned data matrix. In Figs. 3 and 4, it can be observed that the RMSEs of the LLS, MLE and MDS estimates increase as σ increases. The MLE and MDS estimates have similar performance, and both outperform the LLS estimates significantly, especially as σ increases. All approaches have similar performance for networks with 5 and 10 nodes.

Partially Connected Networks. In a partially connected networks, a node may not be able to measure the distances to all other nodes in the network due to the limited ranging distance of the nodes. For the MDS type approaches,

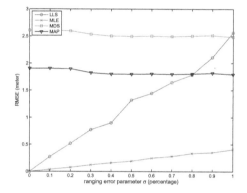

Fig. 5. Variation of RMSE for LLS, MLE, MDS, and MAP *versus* σ in a partially connected network with a maximum ranging distance of 80 m.

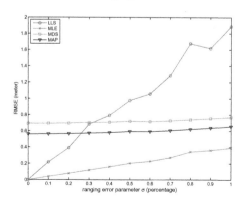

Fig. 6. Variation of RMSE for LLS, MLE, MDS, and MAP *versus* σ in a partially connected network with a maximum ranging distance of 100 m.

this means that the unavailable internode distances will need to be estimated using their corresponding shortest path distances. The shortest path distances are approximate of the Euclidean distances. Partially connected networks with 10 node are simulated. Three scenarios are simulated with the maximum ranging distances being set to 80, 100 and 120, respectively. Figures 5, 6 and 7 show the variations of RMSEs for LLS, MLE, MDS, and MAP *versus* the ranging error parameter σ in those three scenarios. Connectivity level is defined, which is computed as the averaged number of nodes that a node can measure distance to. Connectivity level increases as the maximum ranging distance is increased. For the three scenarios with maximum ranging distances of 80, 100, and 120 m, the connectivity levels are 7.72, 8.77 and 8.99, respectively. In the simulations, σ varies from 0 to 0.01 (or 1% in terms of percentage). For each value of σ, 1000 tests are repeated to obtain the averaged results. In each test, nodes are re-deployed and random ranging errors are re-produced. Similarly, a condition number of 500 is used to avoid the ill-conditioned data matrix for LLS. In Figs. 5, 6 and 7, the top figures show the variations of RMSE of the LLS, MLE, MDS and MAP estimates *versus* σ. In all three scenarios, all approaches in the simulation study show similar performance patterns. The RMSEs of the LLS and MLE estimates increase as σ increases while the RMSEs of the MDS and MAP estimates are relatively constant over the tested range of σ. The accuracy of the MDS and MAP estimates is dominated by the connectivity level of the network rather than the assumed relatively small ranging errors. In all three scenarios, MLE performs the best. The performance of MDS and MAP improves as the network connectivity improves, as can be observed from Figs. 5, 6 and 7. As shown in Fig. 7, MDS and MAP perform as well as MLE when the connectivity level is 8.99 except for small values of σ. LLS outperforms MDS and MAP for small values of σ, and is outperformed by MDS and MAP as σ increases. The demarcation point for LLS moves down as the network connectivity increases as observed from Figs. 5, 6 and 7. It is observed that MDS and MAP produce large

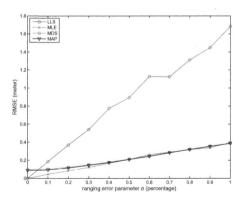

Fig. 7. Variation of RMSE for LLS, MLE, MDS, and MAP *versus* σ in a partially connected network with a maximum ranging distance of 120 m.

RMSEs when the network has low connectivity levels, and improve as the connectivity improves. Although the RMSE for MLE and LLS is less affected by the network connectivity level, MLE and LLS may run into problems in the iteration process due to the problem of insufficient number of anchors for localization in the case of low network connectivity levels.

5 Conclusions

In this paper, the MLE, LLS, MDS, and MAP methods have been formulated for estimating the relative locations of a set of node based on their internode distance measurements. Their performances were discussed and analyzed using computer simulations. Fully and partially connected networks were simulated in the study. From the simulation results, MLE and MAP, among all proposed approaches, are considered the viable solutions to relative localization for small wireless sensor networks. MLE is able to provide the superior localization performance in both fully and partially connected network scenarios. Simulation results showed that, when LLS was used to provide the initial estimates, MLE has converged to the desired optimal estimates almost every time. For partially connected networks with low connectivity, however, MLE may suffer from the problem of not having sufficient numbers of anchors in iterating across the entire network. The performance of MAP is close to that of MLE in fully connected networks and partially connected works with moderate and high connectivity levels, and deteriorates as the network connectivity decreases. In addition, MAP has the advantage of always being able to provide a localization solution in spite of the network connectivity level, although the localization accuracy may be low as in the case of networks with low connectivity levels.

References

1. Savvides, A., Han, C.C., Srivastava, M.B.: Dynamic fine-grained localization in ad hoc networks of sensors. In: Proceedings of the 7th Annual ACM/IEEE International Conference on Mobile Computing and Networking (MobiCom 2001), Rome, Italy, pp. 166–179, July 2001
2. Mao, G., Fidan, B., Anderson, B.D.O.: Wireless sensor network localization techniques. Comput. Netw. **51**(10), 2529–2553 (2007)
3. Savvides, A., Park, H., Srivastava, M.B.: The bits and flops of the N-hop multilateration primitive for node localization problems. In: Proceedings of the First ACM International Workshop on Wireless Sensor Networks and Applications, Atlanta, Georgia, USA, pp. 112–121, September 2002
4. Shang, Y., Ruml, W.: Improved MDS-based localization. In: Proceedings of the IEEE INFOCOM 2004, The 23rd Annual Joint Conference of the IEEE Computer and Communications Societies, Hong Kong, China, March 2004
5. Zhou, Y., Lamont, L.: An optimal local map registration technique for wireless sensor network localization problems. In: Proceedings of the 11th International Conference on Information Fusion (FUSION 2008), Cologne, Germany, 30 June–03 July 2008

6. Zhou, Y., Lamont, L.: A mobile beacon based localization approach for wireless sensor network applications. In: Proceedings of the Fifth International Conference on Sensor Technologies and Applications (SENSORCOMM), Nice, France, August 2011

7. Savarese, C., Rabaey, J.M., Beutel, J.: Location in distributed ad-hoc wireless sensor networks. In: Proceedings of the 2001 IEEE International Conference on Acoustics, Speech, and Signal Processing, Salt Lake City, UT, vol. 4, pp. 2037–2040, May 2001

8. Mendel, J.M.: Lessons in Digital Estimation Theory. Prentice Hall, Englewood Cliffs (1987)

9. Golub, G.H., Van Loan, C.F.: Matrix Computation, 3rd edn. The Johns Hopkins University Press, London (1996)

10. Shang, Y., Ruml, W., Zhang, Y., Fromherz, M.: Localization from connectivity in sensor networks. IEEE Trans. Parallel Distrib. Syst. **15**(11), 961–974 (2004)

11. Borg, I., Groenen, P.: Modern Multidimensional Scaling, Theory and Applications. Springer, New York (1997)

12. Ji, X., Zha, H.: Sensor positioning in wireless ad hoc networks using multidimensional scaling. In: Proceedings of the IEEE 23rd Annual Joint Conference of the IEEE Computer and Communications Societies (INFOCOM 2004), Hong Kong, China, vol. 4, pp. 2652–2661, March 2004

13. Dijkstra, E.W.: A note on two problems in connection with graphs. Numer. Math. **1**, 269–271 (1959)

14. Warshall, S.: A theorem on Boolean matrices. J. ACM **9**(1), 11–12 (1962)

15. Zhou, Y., Lamont, L.: Optimal local map registration technique for wireless sensor network localization problems. In: Mukhopadhyay, S.C., Leung, H. (eds.) Advances in Wireless Sensors and Sensors Networks. LNEE, vol. 64, pp. 177–198. Springer, Heidelberg (2010)

16. Jennrich, R.I.: A simple general procedure for orthogonal rotation. Psychometrika **66**(2), 289–306 (2001)

17. Jennrich, R.I.: A simple general method for oblique rotation. Psychometrika **67**(1), 7–19 (2002)

18. Dennisand, J.E., Schnabel, R.B.: Numerical Methods for Unconstrained Optimization, Nonlinear Equations. Prentice-Hall, Englewood Cliffs (1983)

19. Chung, F.R.K.: Spectra Graph Theory. American Mathematical Society, Providence (1997)

Performance Study of the IEEE 802.15.6 Slotted Aloha Mechanism with Power Control in a Multiuser Environment

Luis Orozco-Barbosa[✉]

Albacete Research Institute of Informatics,
Universidad de Castilla La Mancha, 02071 Albacete, Spain
luis.orozco@uclm.es

Abstract. Recent developments in wireless systems and sensor technologies have spurred the interest on the design and development of Wireless Body Area Networks (WBANs). The performance of such systems will very much rely on protocols capable of dealing with the stringent QoS requirements of the end applications. Towards this end, numerous studies are being carried out aiming to evaluate the IEEE 802.15.6 MAC protocol. In this paper, we undertake the development of a simulation tool enabling the study of the IEEE 802.15.6 slotted-Aloha MAC with power control. Our study particularly focuses on a multiuser environment where the monitoring devices of a number of users will communicate with a central hub. Such scenario will typically describe the operating conditions of a nursing unit.

Keywords: Wireless Body Area Networks · Performance evaluation · IEEE 802.15.6 · Simulation

1 Introduction

Recent developments in the area of wireless communications and sensors have enabled the design and deployment of applications in a wide variety of fields: entertainment, industry, agriculture, medicine, healthcare, among others [1]. Due to the increasing cost of healthcare and ageing population, the use of Wireless Body Area Networks (WBANs) should enable ambulatory care and continuous monitoring of patients.

Even though many Wireless Personal Area Networks (WPANs) technologies are already available in the market, such as Bluetooth, Zigbee and Wi-Fi, it has been recognized that they do not meet the requirements of WBANs. In particular, major challenges to wireless communications designers are (1) the definition of the proper protocol architecture addressing the Quality of Service (QoS) requirements of the various medical parameters, (2) an adequate characterization of the wireless channel temporal variations providing a better insight on the major challenges facing the design of effective protocol mechanisms and (3) energy-efficient protocol mechanisms [2]. The first issue is being addressed by the IEEE 802.15.6 Standard developed by the Task Group 6 (TG6) of the IEEE 802.15 Working Group [3]. The IEEE 802.15.6 standard defines the MAC layer mechanisms supporting three different physical layers, namely,

© ICST Institute for Computer Sciences, Social Informatics and Telecommunications Engineering 2017
Y. Zhou and T. Kunz (Eds.): ADHOCNETS 2016, LNICST 184, pp. 27–37, 2017.
DOI: 10.1007/978-3-319-51204-4_3

Narrowband (NB), Ultra-Wideband (UWB), and Human Body Communications (HBC) layers [4]. Among its main features, the IEEE 802.15.6 MAC protocol incorporates eight different user priorities for accessing the medium in an aim to meet the QoS requirements of the various medical applications and vital signals communications to be supported.

Regarding the characterization of the wireless channels within the context of WBANs, experimental efforts are being carried out on the operating conditions of WBANs [5]. It is clear that the fact of placing the wireless nodes in and around the human body, see Fig. 1, requires an analysis of the impact over the channel behavior introduced by the human body and the relative displacement of the various wireless nodes.

Nowadays BSN research still faces many key technical challenges. One of the

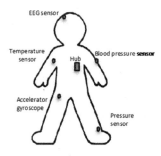

Fig. 1. Body Area Network

ultimate goals of many current studies is the development of wireless channel models to be used in the performance analysis of the proposed protocol mechanisms of the IEEE 802.15.6 MAC protocol [6].

In this paper, we undertake the study of the IEEE 802.15.6 slotted-Aloha MAC protocol. First, we examine the priority mechanism and conflict access resolution algorithm implemented by the MAC protocol. Taking as a baseline, a simple Markov Chain model recently reported in the literature, we then evaluate the performance of the proposed mechanisms having an enhanced capture effect. We evaluate the impact of using a simple power control scheme over the performance of the IEEE 802.15.6. Our study and simulator development efforts set the basis for further studies using power control, wireless channel models and experimental traces reported in the literature and/or gathered in our labs.

In the following this document is organized as follows. Section 2 overviews some recent works in the area of channel modeling and power control. Section 3 describes the IEEE 802.15.6 slotted-Aloha mechanism. Section 4 reviews a simple model of the IEEE 802.15.6 slotted-Aloha MAC protocol, the use of power control mechanism and major efforts towards the development of representative channel wireless models within the scope of WBANs. We also provide some results under different scenarios. Section 5 draws some conclusions and outlines our future research plans.

2 WBANs Power Control and Channel Modelling for MAC Performance

Numerous research initiatives around the world are actively working on the characterization of the wireless channel for WBANs. It is widely accepted that WBANs introduce a very different environment with respect to the ones characterizing other wireless technologies. Many studies are then conducting field trials with the main goal of developing channel models from the data being obtained [5]. Within these efforts, the study of the stability and or temporal variation WBANs are particular relevant to the development of tools for evaluating MAC protocols. In [5], the authors have investigated the wireless channel temporal variations. In a first set of simulations, they vary the transmission power while fixing the average path loss to the coordinator. As expected, their findings report that the number of unacknowledged frames increases exponentially increases as the transmission power is decreased. Their results further confirm that the retransmissions do not prove effective on improving the frame transmission success rate.

In [9], the authors characterize the stability of a WBAN narrowband based on real-time measurements of the time domain channel impulse response (CIR) at frequencies near the 900- and 2400-MHz industrial, scientific, and medical (ISM) bands. Their findings confirm that body movement has considerable impact on the stability of the channel. This is numerically confirmed by the length of the coherence time. Furthermore, they conclude that there is a greater temporal stability at the lower frequency, namely 900 MHz.

Based on the results reported in [5, 9], it is clear that a power control strategy may play a major role on improving the performance of the MAC protocol. In fact, several efforts on evaluating the use of transmission power control in the context of contention-based MAC protocols have been reported in the literature [10–13]. Most of these studies focus on the slotted Aloha mechanism aiming to improve its performance through the capture effect. In [10], the capture event success probability is improved by controlling the probability of choosing the high power transmission level. In [12], the authors develop a wireless random access protocol with multiple power levels. Under the proposed protocol, each node contends following a random access mechanism incorporating a random transmission power value selected among a given set. Following a game theory approach, the channel throughput is optimized taking into account the network load. In [13], the authors develop a three-state discrete-time Markov chain model of a contention-based MAC protocol including the conditional capture probability. Their work included the development of an experimental platform validating their results. In [14], a model and experimental evaluation of the capture effect in an IEEE 802.15.4 is also presented. All these studies show the great benefits of incorporating the use of multiple power levels into the operation of contention-based mechanism.

3 IEEE 802.15.16 Slotted Aloha Mechanism

In this section, we briefly describe the slotted-Aloha MAC mechanism of the IEEE 802.15.6 standard. The IEEE 802.15.6 slotted-Aloha MAC mechanism defines eight different User Priorities; UPi, $i = 0, \ldots, 7$. A parameter, Contention Probability (CP) is used to implement the prioritized access of differing users, where $CP \in [CP_{max}, CP_{min}]$. The values of CP_{max} and CP_{min} are set differently for each UP_i as shown in Table 1.

Table 1. Contention probability thresholds for slotted Aloha access

Priority	Traffic type	CP_{max}	CP_{min}
0	Background (BK)	1/8	1/16
1	Best effort (BE)	1/8	3/32
2	Excellent effort (EE)	1/4	3/32
3	Video (VI)	1/4	1/8
4	Voice (VO)	3/8	1/8
5	Medical data or network control	3/8	3/16
6	High-priority medical data	1/2	3/16
7	Emergency	1	1/4

A sensor node contends to transmit its packet as follows:

1. A node generates a random number z from the interval [0, 1].
2. The CP value is set to CP_{max} for a new packet.
3. A node obtains a contended allocation by the current Aloha slot if $z \leq CP$.
4. If the transmission fails, the CP value is halved for an even number of retransmissions, and is fixed to the previous value for an odd number of retransmissions. If halving CP makes the new CP smaller than CP_{min}, the CP value is set to CP_{min}.

The CP value does not change when it gets settled with CP_{min} at the m-th retransmission. The value of m is given by:

$$m = 2 \left\lceil \log_2 \left(\frac{CP_{max}}{CP_{min}} \right) \right\rceil. \tag{1}$$

4 Protocol Models and Simulation

4.1 The IEEE 802.15.6 Slotted-Aloha MAC Protocol

In this section, we first follow the work reported in [7]. The authors introduce an analytical model to evaluate the saturation throughput of the IEEE 802.15.6 slotted Aloha mechanism. In their analysis, the authors assume a single hop star topology consisting of a central hub and a fixed number of nodes. Each and every node always has a packet available for transmission under an ideal channel scenario. The nodes implement the IEEE 802.15.6 slotted-Aloha MAC protocol described in Sect. 3.

The authors introduce a simple Discrete Time Markov Chain (DTMC) model. The DTMC is obtained from the state transition of the IEEE 802.15.6 slotted-Aloha MAC protocol. The key element of the model is that it captures the decreasing probability of transmission stated in the IEEE 802.15.6 standard and the division of the time into sequence of well-defined packets of fixed length.

In the DTMC, $\{0, 1, 2, \ldots, m\}$ are the states through which a node goes based on the contention probability. For the tagged node being in state k, the slot-wise event probabilities are denoted as follows:

- $p(k, i)$: remains idle in the current slot without attempting to transmit
- $p(k, s)$: transmits successfully in the current slot
- $p(k, c)$: transmits in the current slot that ends up in collision

The one-state transition probabilities in the DTMC are:

$$
\begin{cases}
Pr(0|0) = p(0, s) + p(0, i) \\
Pr(k+1|k) = p(k, c) & k \in [0, m-1] \\
Pr(0|k) = P(k, s) & k \in [1, m] \\
Pr(k|k) = p(k, i) & k \in [1, m-1] \\
Pr(m|m) = p(m, c) + p(m, i)
\end{cases}
\tag{2}
$$

Under the saturation case scenario, the collision probability for the node i is:

$$
\gamma_i = 1 - (1 - \tau_i)^{(n_i - 1)} \prod_{j=0, j \neq i}^{7} (1 - \tau_j)^{n_j}.
\tag{3}
$$

From the analysis reported in [7], the probability that node i transmits in a slot is given by:

$$
\tau_i = \frac{CP_{\max(i)}(1 - 2\gamma_i^2)}{(1 - \gamma_i^2)\{1 - (\sqrt{2}\gamma_i)^{m_i}\} + (1 - 2\gamma_i^2)(\sqrt{2}\gamma_i)^{m_i}}.
\tag{4}
$$

The pair of non-linear Eqs. (3) and (4) are solved by means of Equilibrium Point Analysis (EPA) [8]. Taking this model as a starting point, and as a first step to evaluate the priority and backoff mechanisms of the IEEE 802.15.6 MAC protocol, we have developed a simulation tool. This first set of results complement the finding reported in [7]. In an aim to gain insight into the effectiveness of the priority mechanism, we consider a network consisting of one Class 5 sensor node while increasing the number of Class 0 nodes.

Figure 2 shows the normalized saturation throughput for the two different priority classes, Class 0 and Class 5 and the channel. The figure depicts both the results derived from the model and simulation results. Each simulation run was set for 106 slots. There is a good match between the results obtained from the model and simulation. As already stated, the purpose of this first part of the study has been centered on the development and validation of a simulation tool allowing us to further explore the performance of the MAC protocol: identify the impact/use of the priority mechanism

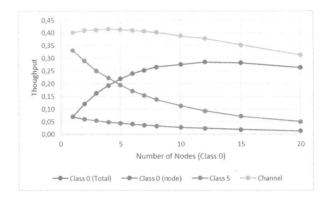

Fig. 2. Throughput – IEEE 802.15.6 Standards

and back-off mechanism. Furthermore, in order to get a more accurate evaluation tool reflecting the control mechanisms of the IEEE 802.15.6 slotted Aloha mechanism, we have set the maximum retransmission parameters to 10 as specified in the standard. In this case, a frame exceeding this threshold is discarded.

Figure 3 shows the frame loss probability for both classes being considered. From the results depicted in Figs. 2 and 3, it is clear that the higher-priority traffic, Class 5 is heavily penalized as the load generated by the low-priority class increases. Figure 2 shows that the throughput of a Class 5 node decreases at a much faster pace than the throughput of a Class 0 node as the number of nodes of Class 0 increases. The results clearly show that the priority mechanism does not provide QoS guarantees to the high-priority traffic.

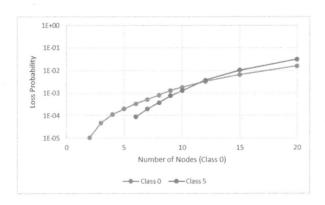

Fig. 3. Frame loss probability

In order to gain further insight into the performance of the backoff mechanism of the IEEE 802.15.6 MAC standard, we consider the case of halving the CP value as soon as a transmission fails.

Figure 4 depicts the throughput for both classes and the channel. The results show a slight degradation when a reduced number of Class 0 nodes, up to five, are present. This clearly shows that the backoff over reacts in this case. However for a network configuration between five and 10 nodes, the throughput improves: in this case the backoff proves effective. Figures 1 and 3 show very similar results, as the number of Class 0 nodes increases. This is due to the fact that as the number of active nodes increases, the backoff mechanism quickly fixes the CP parameters to its minimum value.

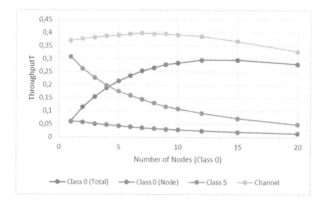

Fig. 4. Throughput – modified

4.2   Simulation Scenarios and Results

Following the results reported in Sect. 3A and the literature review, we propose to supplement the IEEE 802.15.6 MAC protocol by incorporating into it the use of multiple power levels selected randomly by the nodes. Under the proposed scheme, a node being ready to transmit will randomly select its transmission power. The proposed mechanism should not only prove effective on limiting the adverse condition of the channel, but it will also be effective in producing a capture effect in the presence of a limited number of simultaneous transmissions. The latter effect has lately attracted the attention of researchers due to the increasing number of wireless devices and emerging wireless technologies [11, 12]. Remember that our main aim is to explore the performance of the IEEE 802.15.6 in scenarios where a large number of devices are collocated, such as in a nursing home.

We undertake the study of the proposed scheme under two main scenarios. In order to get a better insight on the impact of the retransmission probability used by the different priorities, we study two classes: Class 0 and Class 5. We assume that a node can transmit using one of two power levels selected randomly. Similar to the study reported in [11], we assume that all nodes experience the same channel gain. The transmission of the target node i will be successful in the presence of simultaneous transmission if the following three conditions are met.

1. *Node i* transmits with the highest power level, denoted by P_i.
2. All other nodes transmit with power levels lower than P_i.
3. The instantaneous SINR of tagged *Node i* is higher than the target SINR at the hub (receiver).

The number and values of the power levels to use will very much depend on the actual technology being used. As for the actual random selection scheme of the power level, the results in [12] have shown that uniform random scheme proves to be effective. Condition c is derived from the channel model. From the results reported in [5, 12, 14], the actual values will depend very much on the end-user environment.

In our simulations, we explore the performance of the two classes of traffic. Since our main aim is to explore the role of the contention probability defined by the IEEE 802.15.6 and the power control mechanism, we consider two scenarios: a network exclusively comprising only Class 0 (Class 5) nodes and a network consisting of one Class 5 and one or more Class 0 nodes. For both scenarios, we consider the use of two power levels, p_{high} and p_{low}, randomly selected following a uniform random distribution. A transmission is successful in a given slot if (1) the aforementioned conditions are met or (2) in the case that only one node transmits.

Figures 5 and 6 show the results obtained for both classes and the two MAC protocols, namely, the IEEE 802.15.6 standard MAC and the proposed power-enhanced MAC protocol. Besides the fact that the capture effect considerable improves the network performance, it is evident that the use of a higher retransmission probability, Class 5, has a negative impact as the number of competing nodes increases. In fact, in one of our previous works using game theory principles, we have shown that the nodes should reduce their retransmission probability as the load increases [12].

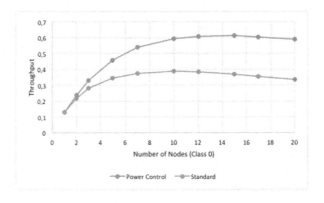

Fig. 5. Throughput - Class 0

Figures 7 and 8 depict the results for a network consisting of a single high-priority node (Class 5) and a given number of low-priority nodes (Class 0). In the first case, Fig. 7, the nodes of both classes randomly select one of the two available transmission powers with equal probability, i.e., $p_{high} = p_{low} = 0.5$. The results clearly show an improvement on the throughput of both classes. However, similar to the results

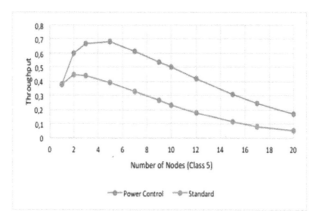

Fig. 6. Throughput - Class 5

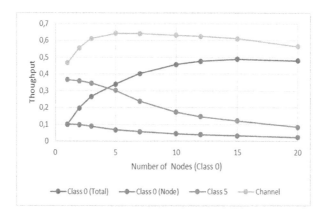

Fig. 7. Throughput – symmetric

depicted in Fig. 1, the throughput of Class 5 gets penalized as the number of Class 0 nodes increases.

Figure 8 show the results for both classes when the probability of selecting the high power transmission level has been set to 0.9 for the Class 5 and 0.1 for Class 0. In this case, the Class 5 throughput exhibits better results. However, the overall throughput degrades as the network load increases. The results also show that for a typical WBANs consisting of five to seven nodes with one high-priority node, the setting of the power levels is rather straightforward, i.e., $p_{high} = p_{low} = 0.5$ for all classes.

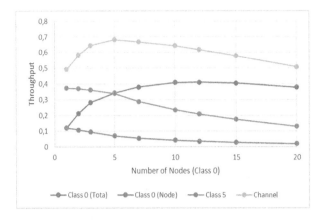

Fig. 8. Throughput – asymmetric

5 Conclusions and Future Work

In this work, we have first reviewed the principles of operation of the IEEE 802.15.6 slotted-Aloha MAC protocol. Taking as a basis the model recently reported in [7], we have validated our simulation tool. We have then explored the performance of the MAC protocol making use of a simple power control scheme. Some preliminary results have been reported focusing on the network parameters. Our future work plans comprise: the inclusion of channel models, the evaluation of a multiple-class network, and the QoS guarantees as perceived by the end-user application.

Acknowledgement. This work has been supported by MINECO (Spain) under grant TIN2012-38341-C04.

References

1. Imran, M., Ullah, S., Yasar, A.U.H., Vasilakos, A., Hussain, S. Enabling: Technologies for next-generation sensor networks: prospects, issues, solutions, and emerging trends. Int. J. Distrib. Sens. Netw. **2015** (2015). doi:10.1155/2015/634268. Article ID 634268
2. Rashwand, S., Misic, J., Khazaei, H.: Performance analysis of IEEE 802.15.6 under saturation condition and error-prone channel. In: Proceedings of IEEE WCNC 2011, pp. 1167–1172 (2011)
3. Part 15.6: Wireless Body Area Networks, IEEE Standard for Local and Metropolitan Area networks. IEEE Std. 802.15.6 (2012)
4. Ullah, S., Mohaisen, M., Alnuem, M.A.: A review of IEEE 802.15.6 MAC, PHY and security specifications. Int. J. Distrib. Sens. Netw. **2013** (2013). doi:10.1155/2013/950704. Article ID 950704
5. Boulis, A., Tselishchev, Y., Libman, L., Smith, D., Hanlen, L.: Impact of wireless channel temporal variation on MAC design for body area networks. ACM Trans. Embed. Comput. Syst. (TECS) **11**(S2) (2012). Article 51

6. Tselishchev, Y., Libman, L., Boulis, A.: Reducing transmission losses in body area networks using variable TDMA scheduling. In: Proceedings of IEEE International Symposium on World of Wireless, Mobile and Multimedia Networks (WoWMoM) (2011). 10 pages
7. Chowdhury, S., Ashrafuzzaman, K., Kwak, K.S.: Saturation throughput analysis of IEEE 802.15.6 slotted Aloha in heterogeneous conditions. IEEE Wirel. Commun. Lett. 3(3), 257–260 (2014)
8. Woodward, M.E.: Communication and computer networks – modelling with discrete-time queues. IEEE Computer Society Press (1994)
9. Zhang, J., Smith, D.B., Hanlen, L.W., Miniutti, D., Rodda, D., Gilbert, B.: Stability of narrowband dynamic body area channel. IEEE Antennas Wirel. Propag. Lett. 8, 53–56 (2009)
10. Takanashi, H., Kayama, H., Iizuka, M., Morikura, M.: Enhanced capture effect for slotted ALOHA employing transmission power control corresponding to offered traffic. In: Proceedings of IEEE International Conference on Communications (ICC 1998), Vol. 3, pp. 1622–1626 (1998)
11. Karouit, A., Sabir, E., Ramirez-Mireles, F., Orozco-Barbosa, L., Haqiq, A.: A team study of a multiple-power wireless random channel access mechanism with capture effect. Math. Prob. Eng. 13. doi:10.1155/2013/187630
12. Tahir, M., Mazumder, S.K.: Markov chain model for performance analysis of transmitter power control in contention-based wireless MAC protocol. Telecommun. Syst. (2008). doi:10.1007/s11235-008-9103-3
13. Sarker, J.H., Hassan, M., Halme, S.J.: Power level selection schemes to improve throughput and stability of slotted Aloha under heavy load. Comput. Commun. 25(18), 1719–1726 (2002)
14. Gezer, C., Buratti, C., Verdone, R.: Capture effect in IEEE 802.15.4 networks: modelling and experimentation. In: IEEE International Symposium on Wireless Pervasive Computing, IEEE ISWCP 2010, Modena, Italy (2010)

New Selection Strategies of Actor's Substitute in DARA for Connectivity Restoration in WSANs

Riadh Saada$^{(\boxtimes)}$, Yoann Pigné, and Damien Olivier

Normandie Univ, UNIHAVRE, LITIS, 76600 Le Havre, France
{riadh.saada,yoann.pigne,damien.olivier}@univ-lehavre.fr

Abstract. Wireless Sensor and Actor Networks are used in many dangerous applications. When performing their tasks, actors may fail due to harsh environments in which they are deployed. Some actors are cut-vertices within network. Their loss breaks its connectivity and disrupts its operation accordingly. Therefore, restoring network connectivity is crucial. DARA is among the most popular connectivity restoration schemes. It performs multi-actor relocations in order to replace a failed cut-vertex by one of its neighbors, based on the lowest degree. In this paper, we propose new selection strategies of actor's substitute in DARA, in which the substitute selection is based on the nature of the links with neighbors rather than on the degree. Our approaches improve the performance of DARA by reducing the number of relocated actors in the recovery process by 24% on average compared to its original selection strategy. The proposed strategies are validated through simulation experiments.

Keywords: Topology management · Fault tolerance · Connectivity restoration · Wireless sensor and actor networks

1   Introduction

Wireless Sensor and Actor Networks (WSANs) [1] are used to replace or assist humans in many hazardous situations such as fire extinguishing and rescue of victims within hostile or unknown environments. A WSAN is composed of two categories of elements: sensors and actors. Sensors are usually small devices characterized by low cost and limited resources in computation, communication and energy. They are present within the network in abundance. The main duty of sensors is to probe their surroundings by collecting data about the supervised area and to report it to one or several actors, which react when necessary by performing appropriate tasks. Actors are more powerful devices. They could be mobile, they usually have advanced computation and communication capabilities, as well as a significant onboard energy, thus, they could be relatively more expensive. This is why their number within networks is fewer compared to sensors.

Sensors and actors can be subject to failure for many reasons. For instance, they can undergo external attacks from the fact that they are deployed unattended in harsh environments. They can also fail due to energy depletion or

© ICST Institute for Computer Sciences, Social Informatics and Telecommunications Engineering 2017
Y. Zhou and T. Kunz (Eds.): ADHOCNETS 2016, LNICST 184, pp. 38–49, 2017.
DOI: 10.1007/978-3-319-51204-4_4

simply because of internal malfunctions [2]. Upon failing, an actor loses all its communication links. In most applications, actors must operate in a structured and collaborative manner for better efficiency in tasks' realization. Therefore, they need to interact with each other in order to share information and coordinate their actions. Thus, it is mandatory that they stay all the time reachable from each other. In other words, it is necessary to maintain a connected inter-actor network. An actor can be a cut-vertex in the inter-actor network topology. The failure of such actor splits the network into two or many disjoint partitions and affects its connectivity. The loss of network connectivity will have a detrimental impact on its performance. Indeed, actors belonging to different sub-networks will no longer be able to communicate, and thus, information exchange between them as well as their actions' coordination will be interrupted. As a result, the overall network operation will be severely disrupted.

To deal with this situation, the inter-actor network must integrate mechanisms of resilience, so that it continues to perform its tasks normally even if some actors fail. Many contemporary fault tolerance techniques in WSANs use topology management methods [3]. One of the most popular of them is DARA [4] (Distributed Actor Recovery Algorithm). DARA is a distributed connectivity restoration technique which performs coordinated multi-actor relocation in order to replace a failed actor by a healthy one into the inter-actor network. The connectivity restoration in DARA is thus a self-healing process, exploiting the mobility of operational actors within the network. The main objective of DARA is to reduce the number of involved actors in the recovery process. For this aim, a failed actor's substitute is chosen based on the lowest degree (least number of neighbors). In this paper, we propose new selection strategies of failed actor's substitute in DARA based on the nature of links with its neighbors rather than their number. Our selection schemes improve the performance of DARA in reducing the number of relocated actors during the recovery process by 24% on average, compared to its original selection strategy. We validated our approaches through simulation experiments.

The rest of paper is organized as follows: next section describes the system model. In Sect. 3, we review related works. An analysis of DARA is given in Sect. 4. In Sect. 5, we identify a shortcoming in DARA's selection strategy and propose solutions in order to remedy it. In Sect. 6, we evaluate our proposed approaches through simulation experiments. Section 7 concludes the paper.

2   System Modeling

As mentioned earlier, a WSAN is composed of two types of elements: sensors and actors. Sensors probe their surroundings and send regularly their sensed data to actors. On the other hand, actors collect information from sensors in their neighborhood and collaborate between them to perform one or several tasks. Figure 1 illustrates an example of WSAN. The system model includes the following assumptions:

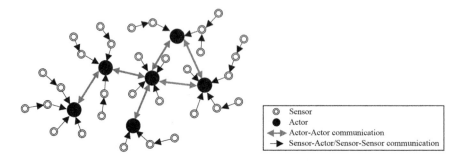

Fig. 1. Representation of a WSAN with a connected inter-actor network.

1. Each actor is identified by a unique identifier. A_i denotes the actor identified by identifier i. The set of all actors within the network is designated by $\mathcal{A}$.
2. Actors are mobile.
3. Actors are able to recognize their positions using localization techniques like GPS.
4. The communication range of an actor corresponds to the maximum Euclidean distance that its radio can reach. We denote it by r.
5. The radio range of all actors is identical, finite and significantly smaller than the dimensions of the deployment area.
6. Two actors can communicate if they are at range from each other.
7. Communications between actors are symmetric.
8. Actors are randomly deployed in an environment of interest and form a connected inter-actor network after a discovery step.

The following definitions are used in this paper:

Definition 1 (One-Hop Neighbors). *Let $A_i \in \mathcal{A}$.*
1-Hop-Neighbors(A_i) or simply Neighbors(A_i) is the set of actors that are directly reachable from A_i.

Definition 2 (Two-Hop Neighbors). *Let $A_i \in \mathcal{A}$.*
2-Hop-Neighbors(A_i) is the set of actors that are reachable from A_i through A_f, where $A_f \in Neighbors(A_i)$.

Definition 3 (Adjacent Siblings). *Let $A_i, A_f \in \mathcal{A}$ such as $A_f \in Neighbors(A_i)$.*
Adjacent-Siblings(A_i, A_f) is the set of actors that are neighbors of both A_i and A_f. Mathematically: Adjacent-Siblings(A_i, A_f) = $\{A_k - A_k \in Neighbors(A_i) \land A_k \in Neighbors(A_f)\}$

Definition 4 (Dependents). *Let $A_i, A_f \in \mathcal{A}$ such as $A_f \in Neighbors(A_i)$.*
Dependents(A_i, A_f) is the set of actors that are neighbors of A_i (without A_f) but not neighbors of A_f. Mathematically: Dependents(A_i, A_f) = $\{A_k - A_k \in Neighbors(A_i) \land A_k \notin Neighbors(A_f) \land A_k \neq A_f\}$. It is easy to see that: Dependents(A_i, A_f) = Neighbors(A_i) − Adjacent-Siblings(A_i, A_f) − $\{A_f\}$.

3   Related Works

Failure tolerance in WSANs is divided in two large categories: tolerating failure of one actor at a time and tolerating failure of multiple actors simultaneously. Our work lies within the first category. We focus on actors' failures because sensors are supposed available in abundance, as previously mentioned. When a cut-vertex actor fails, affecting network connectivity, the most obvious solution is to replace it by a new one. This solution can take a lot of time and can be dangerous if the operation is performed by humans within harsh environments. Contemporary connectivity-centric fault tolerance methods for WSANs are based on network topology management. In the literature, they are grouped into three classes: proactive techniques, reactive techniques and hybrid techniques.

Proactive techniques strive to anticipate the failure by taking some precautions at setup, allowing the network to continue operating normally in case of losing one or more actors. For example, the authors in [5] deploy a k-connected topology in which the failure of $k - 1$ actors does not break the network connectivity.

Reactive techniques aim to perform network connectivity restoration as soon as a cut-vertex failure is detected, in real time, and in a distributed manner for most algorithms. The recovery process involves available healthy actors within the network. These actors being mobile, the idea is to reposition them to the appropriate locations in order to restructure the network's topology, so that it becomes connected again. Reactive techniques are more suitable than proactive ones for failure tolerance in WSANs. Indeed, WSANs are asynchronous by nature and may be dynamic. Therefore, the recovery process must be a network self-healing using adaptive schemes [3]. This category contains a wide variety of algorithms. DARA [4] is a reference algorithm within the research community working on this field. Many proposed approaches are compared against it for assessment. DARA performs a coordinated multi-actor relocation in order to replace a failed cut-vertex. To do this, it only requires to maintain an updated list of one-hop and two-hop neighbors, its mechanism is detailed in the next section. RIM [6] adopts another strategy: when an actor fails (cut-vertex or not), all its direct neighbors move toward its position in order to establish a connectivity between them. The process is repeated in cascade as long as the movement of neighbors causes further broken links. RIM shrinks the network topology inward, affecting network coverage. Compared to DARA, it involves much more actors in the recovery process and generates overhead messaging. However, its advantage is in splitting the load between actors. Indeed, it has been proven in [6] that each actor travels a maximum distance of $r/2$. VCR [7] and LDMR [8] are variants of RIM. Some algorithms consider a secondary objective in addition to the principal one that consists in restoring network connectivity. In return, they introduce additional assumptions or consume more resources. LeDiR [9] has as a secondary objective, the preservation of shortest-path length between any pair of actors using a shortest-path routing table. C^3R [10] has the auxiliary aim of maintaining network coverage: when a cut-vertex fails, its direct neighbors coordinate to establish a schedule in order to replace it in turns, during a time interval. Excessive movements of actors consume a lot of energy, therefore, this solution is considered temporary. RACE [11] is an interesting recent work based on DARA. It restores network connectivity while minimizing its coverage loss. For that, it needs additional information about actors' criticality, which is provided by the method developed in [12].

Hybrid techniques are a compromise between the two previous categories. They anticipate the failures by assigning backups to cut-vertices at setup in a proactive manner. However, the recovery process is triggered when a cut-vertex fails like reactive approaches. PADRA [13] and DCRS [14] operate this way.

For a complete state of the art on the subject, we recommend the reader a very interesting survey available in [3].

4   DARA Analysis

DARA [4] is a distributed connectivity restoration Algorithm. As a previous knowledge, it only requires that actors are aware of their one-hop and two-hop neighbors. For this, they have to maintain updated one-hop and two-hop tables. The tables must contain degrees, positions and identifiers (IDs) of neighbors.

Actors periodically report their presence to direct neighbors by sending heartbeat messages. When an actor fails and can no longer communicate, its direct neighbors detect the failure by missing the heartbeat messages. If the failed actor is a cut-vertex, DARA is launched locally on each of its neighbors.

The main idea of DARA is to replace a failed cut-vertex by an appropriate actor among its direct neighbors. When a cut-vertex actor fails, its direct neighbors can no longer communicate between them. However, they know each other thanks to their two-hop tables. These neighbors are all considered as potential candidates to replace the failed actor A_f. Nevertheless, they must elect the most suitable of them (the Best Candidate, BC) to restore the connectivity by moving to the position of A_f. The potential candidates have the same information on each of them, so, they will come to the same result. The selection strategy of BC in DARA is based on the following criteria:

1. *Least actor degree*: The authors assume that moving an actor having few neighbors minimizes the number of involved actors in the recovery process (we will show in the next section that this is not always true).
2. *Closest proximity to A_f*: in the case of actors which have the same degree, DARA favors the nearest one to A_f in order to minimize the traveled distance.
3. *Highest actor ID*: in the case of actors having the same degree and are equidistant to A_f, DARA decides between them by picking the one with biggest ID.

Figure 2a illustrates an inter-actor network in which a cut-vertex A_f fails, dividing the network topology in two disconnected subnetworks. A_1, A_2, A_3 and A_4 initiate the recovery process as soon as they detect the failure. They execute DARA simultaneously in order to determine BC. A_2 has the highest degree and is then excluded. A_1, A_3 and A_4 have the same degree. A_4 is the farthest among them from A_f. So, it is also excluded. A_1 and A_3 have the same degree and are equidistant to A_f. Thus, A_3 is selected as BC based on the highest ID and moves to the position of A_f as shown in Fig. 2b. Upon its arrival to destination, it broadcasts a RECOVERED message.

When BC leaves its position, it may cause another network partitioning as depicted in Fig. 2b. In this case, its neighbors must perform DARA again, exactly as if BC had failed, in order to replace it by the most suitable of them. The process is repeated recursively until full connectivity restoration. However, this time, the recovery process does not concern all the neighbors of BC. Indeed, BC can have two types of neighbors according to their previous relations with A_f: the siblings (actors in the set Adjacent-Siblings(BC,A_f), see Definition 3) and the dependents (actors in the set Dependents(BC,A_f), see Definition 4). Siblings preserve direct links with BC after relocation (for example A_2 in Fig. 2b). Dependants can either preserve indirect links with BC (such as A_8 in Fig. 2b) or find them detached from it (like A_9 in Fig. 2b).

Dependants can check whether they conserve indirect links with BC after relocation or not, as explained in the following. Before leaving its position, BC sends a MOVING message to all its dependent neighbors, in order to inform them about its departure. BC integrates also in this message, the list of its sibling with A_f. When a dependent receives the message, it verifies if it is related to one of BC's siblings or not, using its neighborhood tables. If such link exists, the dependent remains connected to BC after relocation, otherwise, it concludes that it is detached from it based on limited two-hop knowledge.

BC's neighbors concerned by the cascade are its detached dependents, so that they reestablish their broken links with it. In our previous example, A_9 moves to the position of A_3 and completes the recovery process, as shown in Fig. 2c. For more details about DARA's mechanism, please refer to article [4].

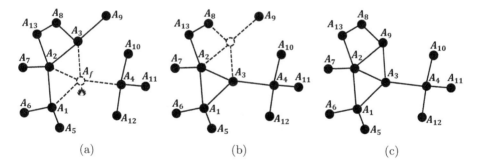

(a)   (b)   (c)

Fig. 2. An execution example of DARA. (a) represents a failure of A_f. In (b), A_3 replaces A_f. In (c), A_9 replaces A_3 and completes the recovery process.

5   Our Approach

The first selection criterion of A_f's substitute in DARA is based on the least degree of its neighbors. Abbasi et al. in [4] have motivated this choice by the

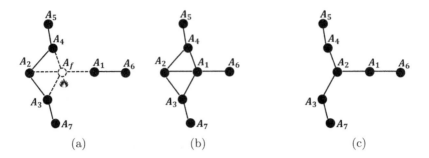

Fig. 3. An example in which DARA does not select the best substitute for A_f. (a) represents the failure of A_f. In (b), A_1 and A_6 move according to selection strategy of DARA. Alternatively in (c), the displacement of A_2 is sufficient to restore connectivity in (a).

fact that it limits the number of cascaded relocations in the recovery process. We will show through the next example that this is not always true.

In Fig. 3a, the cut-vertex A_f fails. In Fig. 3b, A_1 replaces it and A_6 replaces A_1 according to selection strategy of DARA. In Fig. 3c, the selection of A_2 as BC rather than A_1 limits the number of involved actors in recovery process, even if it has higher degree. Moreover, in our previous analysis of DARA, we have shown that neighbors of BC that may cause cascaded relocations are its dependents, when they are not related to any of its siblings with A_f. Based on these observations, we realized that it is not the candidates' degree that influences the cascades but the nature of links with their neighbors (dependency links). Indeed, in Fig. 3a, A_1 has one dependent while A_2 does not.

In order to reduce the number of involved actors in DARA, we propose three new selection strategies of a failed actor's substitute. Our selection strategies focus on the nature of links with neighbors rather than their number. They are presented in the following:

- **Strategy 1:** select the candidate with least number of dependents. This choice is motivated by the fact that dependents are the actors that may cause cascades as mentioned earlier. Therefore, we propose to minimize their number. In case of a tie, break it with distance and highest ID like in DARA.
- **Strategy 2:** select the candidate with highest number of siblings. We believe that the higher the number of siblings, the more likely dependents are related to one of them, and thus, stop relocations. In case of a tie, break it with distance and highest ID like in DARA.
- **Strategy 3:** is a compromise between the two previous ones. Here, we favor the candidate which number of dependents is null, if exists, because no cascade will be trigged in this case. Otherwise, we pick the candidate with highest number of siblings. In case of a tie, break it with distance and highest ID like in DARA.

To use our selection strategies, each potential candidate A_p needs to calculate the number of its dependent neighbors and the one of its siblings. It must also do the same for the other candidates which are its two-hop neighbors through A_f. For this purpose, we propose a distributed algorithm that does not require any other information than the one available in the tables of DARA. Our algorithm begins by counting the number of siblings. Then, it infers the number of dependents from formula (1), which is easy to demonstrate.

$$Number\ of\ Dependents\ =\ Degree - Number\ of\ Siblings\ -\ 1 \qquad (1)$$

Indication: *the neighbors' set of a candidate is composed of its siblings with A_f, its non siblings with A_f (its dependents) and A_f.*

The calculation of siblings' number by the potential candidate A_p is based on the four following rules:

- **Rule 1 (initialization):** A_p initializes its number of siblings as well as those of other candidates to zero.
- **Rule 2:** if A_p is neighbor of another candidate $A_{p'}$, it means that both are siblings with A_f. Therefore, A_p increases by 1, its number of siblings and the one of $A_{p'}$ (For example, A_2 and A_4 in Fig. 3a).
- **Rule 3:** if two candidates A_{p_1} and A_{p_2} other than A_p are neighbors, it means that both are siblings with A_f. If A_p is adjacent to at least one of them, it can infer the link between them by consulting its two-hop table. Thus, A_p increases by 1 the number of siblings of A_{p_1} and the one of A_{p_2} (for example, A_3 can detect the link between A_2 and A_4 in Fig. 3a).
- **Rule 4:** if two candidates A_{p_1} and A_{p_2} other than A_p are neighbors, it means that both are siblings with A_f. If A_p is not adjacent to any of them, it needs a three-hop vision to detect their link, which is not provided by its tables. However, A_p is aware of the positions of A_{p_1} and A_{p_2} through its two-hop table. Thus, A_p can verify whether they are neighbors or not by calculating the distance between them. If Distance$(A_{p_1}, A_{p_2}) \leq r$ (r is the communication range) then A_{p_1} and A_{p_2} are neighbors. Therefore, A_p increases by 1 their respective siblings' numbers (For example, A_1 detects the link between A_2 and A_4 in Fig. 3a).

6   Validation

The effectiveness of our approaches is validated through simulations. In this section, we describe the simulation environment and discuss the obtained results.

6.1   Simulation Environment

To evaluate our approaches, we developed a simulation environment similar to that presented in the original article of DARA [4]. For this, we used the JAVA programming language and the GraphStream[1] library. We have randomly generated connected inter-actor networks with different density levels, in an area of 1000 m × 600 m. The test consists in varying the number of actors within the network topology from 20 to 100. Their communication range is fixed at 100 m. Each of these simulation steps is run for 1000 different network topologies in which we treat the failures of all cut-vertices independently. Average measures on these 1000 topologies are then reported. Our selection schemes are assessed and compared with the original version of DARA based on the following metrics:

(1) *Number of relocated actors:* it is the number of actors involved in the recovery process. This metric evaluates the extent of the connectivity restoration process.
(2) *Total traveled distance:* it is the distance traveled by all actors during the recovery process. This metric assesses the efficiency of the different connectivity restoration strategies. It is considered as an indicator of energy consumption and network reconstruction time.
(3) *Number of exchanged messages:* this metric counts all sent messages during the recovery process. It is also an indicator of energy consumption.

6.2 Discussion

Our simulation results are depicted in Figs. 4, 5 and 6. For the original version of DARA, we got the same curve shapes as those presented in the articles [4,6], which shows the consistency of our results compared to previous works. As in the articles [4,6], increasing the network density reduces the number of cascades in the recovery process, because high connectivity degree promotes the presence of siblings in the network's topology rather than dependents. We observe the same behavior in all our proposed approaches as shown in Fig. 4. Table 1 compares the performance of the different strategies with the original version of DARA.

As previously explained, the dependents are those that may cause cascades, so we aim in our first selection strategy to minimize their number. As expected, this choice reduces the number of relocated actors by 14% compared to the original version of DARA (see Table 1). It also decreases the total traveled distance during the recovery process by 21%, and the number of exchanged messages by 4%. Decreasing the total traveled distance in recovery process improves the network reconstruction time allowing it to resume operations more quickly. It also reduces the energy consumption and extends the network lifetime accordingly. Decreasing the number of exchanged messages allows also to save energy.

Our second strategy which aims to maximize the number of siblings in the selection of BC obtains better results than the first one in terms of actors' relocation (-17%) and total traveled distance (-26%, see Table 1). The reason is: maximizing the number of siblings increases the probability that dependents are related to one of them, and thus, avoid cascades. However, this strategy increases the number of exchanged messages by 87% compared to the original version of

---

[1] http://graphstream-project.org/.

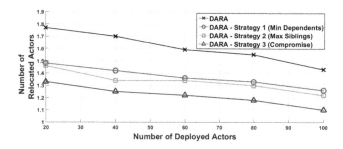

Fig. 4. Number of relocated actors with varying number of actors ($r = 100\,\text{m}$).

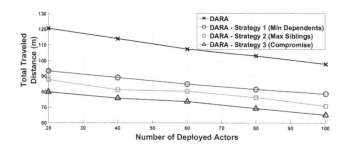

Fig. 5. Total traveled distance with varying number of actors ($r = 100\,\text{m}$).

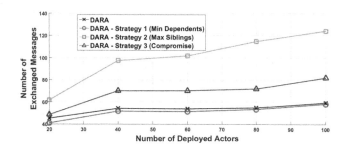

Fig. 6. Number of exchanged messages with varying number of actors ($r = 100\,\text{m}$).

Table 1. Performance analysis of our strategies compared to the original version of DARA.

Metrics	Strategy 1 (Min dependents)	Strategy 2 (Max siblings)	Strategy 3 (Compromise)
Number of relocated actors	-14.80%	-17.16%	-24.37%
Total traveled distance	-21.26%	-26.94%	-32.98%
Number of exchanged messages	-4.36%	$+87.11\%$	$+28.12\%$

DARA for the following reason: BC is selected based on the highest number of siblings. After displacement of BC, its siblings must send messages to their own neighbors in order to update their two-hop tables. Furthermore, the number of siblings increases accordingly with the network density, which increases the number of sent messages compared to the original version of DARA and the first selection strategy where BC is chosen based on a minimization criterion.

Our third and last selection strategy favors the candidate with no dependents, if exists. Otherwise, it maximizes the number of siblings in the selection of BC. Indeed, the second strategy only focuses on the number of siblings and ignores the case when a candidate has no dependents. Nevertheless, this case is very important because it stops the cascades and ends the recovery process immediately. This is what motivated the idea of the third selection strategy which outperforms all previous ones in terms of cascades limitation (-24%), as well as total traveled distance (-32%), while alleviating the number of messages introduced by the second strategy up to 28% more than the original version of DARA. We know that for relatively powerful actors, the cost of sending a message in terms of energy is negligible compared to the cost of their physical displacement. Therefore, this strategy can be a good compromise for energy preservation in WSANs.

7   Conclusion

DARA [4] is one of the most popular connectivity restoration schemes in WSANs which does some topology control in order to replace a failed cut-vertex by a healthy actor into the network. In this paper, we identified a shortcoming in its substitute selection strategy and proposed new ones that improve its performance. Our approaches was evaluated and compared with the original version of DARA through simulation experiments. As future works, we plan to apply our selection strategies on RACE [11] algorithm which relies on DARA's mechanism in order to restore network connectivity while minimizing its coverage loss. We believe that our approaches are able to improve its performance. We also plan to do real experimentations of our methods on mobile robot networks in collaboration with robotics researchers in our laboratory.

Acknowledgment. The project is co-financed by the European Union with the European regional development fund (ERDF) and by the Haute-Normandie Regional Council.

References

1. Akyildiz, I.F., Kasimoglu, I.H.: Wireless sensor and actor networks: research challenges. Ad hoc Netw. **2**(4), 351–367 (2004)
2. Akyildiz, I.F., Su, W., Sankarasubramaniam, Y., Cayirci, E.: Wireless sensor networks: a survey. Comput. Netw. **38**(4), 393–422 (2002)

3. Younis, M., Senturk, I.F., Akkaya, K., Lee, S., Senel, F.: Topology management techniques for tolerating node failures in wireless sensor networks: a survey. Comput. Netw. **58**, 254–283 (2014)
4. Abbasi, A.A., Younis, M., Akkaya, K.: Movement-assisted connectivity restoration in wireless sensor and actor networks. IEEE Trans. Parallel Distrib. Syst. **20**(9), 1366–1379 (2009)
5. Han, X., Cao, X., Lloyd, E.L., Shen, C.C.: Fault-tolerant relay node placement in heterogeneous wireless sensor networks. IEEE Trans. Mob. Comput. **9**(5), 643–656 (2010)
6. Younis, M.F., Lee, S., Abbasi, A.A.: A localized algorithm for restoring internode connectivity in networks of moveable sensors. IEEE Trans. Comput. **59**(12), 1669–1682 (2010)
7. Imran, M., Younis, M., Said, A.M., Hasbullah, H.: Volunteer-instigated connectivity restoration algorithm for wireless sensor and actor networks. In: 2010 IEEE International Conference on Wireless Communications, Networking and Information Security (WCNIS), pp. 679–683. IEEE (2010)
8. Alfadhly, A., Baroudi, U., Younis, M.: Least distance movement recovery approach for large scale wireless sensor and actor networks. In: 2011 7th International Wireless Communications and Mobile Computing Conference (IWCMC), pp. 2058–2063. IEEE (2011)
9. Abbasi, A.A., Younis, M.F., Baroudi, U.A.: Recovering from a node failure in wireless sensor-actor networks with minimal topology changes. IEEE Trans. Veh. Technol. **62**(1), 256–271 (2013)
10. Tamboli, N., Younis, M.: Coverage-aware connectivity restoration in mobile sensor networks. J. Netw. Comput. Appl. **33**(4), 363–374 (2010)
11. Haider, N., Imran, M., Younis, M., Saad, N., Guizani, M.: A novel mechanism for restoring actor connected coverage in wireless sensor and actor networks. In: 2015 IEEE International Conference on Communications (ICC), pp. 6383–6388. IEEE (2015)
12. Imran, M., Alnuem, M.A., Fayed, M.S., Alamri, A.: Localized algorithm for segregation of critical/non-critical nodes in mobile ad hoc and sensor networks. Procedia Comput. Sci. **19**, 1167–1172 (2013)
13. Akkaya, K., Senel, F., Thimmapuram, A., Uludag, S.: Distributed recovery from network partitioning in movable sensor/actor networks via controlled mobility. IEEE Trans. Comput. **59**(2), 258–271 (2010)
14. Guizhen, M., Yang, Y., Xuesong, Q., Zhipeng, G., He, L., Xiangyue, X.: Distributed connectivity restoration strategy for movable sensor networks. Chin. Commun. **11**(13), 156–163 (2014)

Asymmetric Multi-way Ranging for Resource-Limited Nodes

Erik H.A. Duisterwinkel[1,2(✉)], Niels A.H. Puts[2], and Heinrich J. Wörtche[1,2]

[1] INCAS3, Assen, The Netherlands
mail@erikduisterwinkel.nl
[2] Eindhoven University of Technology, Eindhoven, The Netherlands

Abstract. Cooperative localization in WSN is used in applications where individual nodes cannot determine their location based on external contact, like e.g. GPS. The applications we focus on are the exploration and mapping of flooded cavities that are otherwise inaccessible or difficult-to-access, e.g. underground (oil-) reservoirs or industrial tanks for e.g. mixing. High levels of miniaturization are required for the nodes to traverse these cavities; nodes will have to be stripped down to a bare minimum. Ultrasound time-of-flight is used as radio communication is infeasible. Network topology is highly unpredictable and fast changing.

We present an asymmetric multi-way ranging protocol for these highly resource-limited, miniaturized, autonomous nodes. The specific set of constraints imposed by these applications, like the use of ultrasound, high latency, low data-rates, and non-static network topology is far-reaching and has not been studied before. Simulations of the protocol show trade-off's that can be made between ranging latency, signal overlap and overall energy budget.

Keywords: WSN · Sensor swarm · Multi-way ranging · MWR · Resource-limited

1 Introduction

Underground cavities like (oil-)reservoirs, mines and geothermal sources, and industrial infrastructure like, pipelines, mixing tanks and reactors are systems which have in common that they are hard to access for in situ measurements of system structure, dynamics, conditions and integrity. A straight forward approach which has been proposed and investigated in [1,2] is based on directly injecting large quantities of miniaturized sensor systems ('sensor motes') into the flooded system[1], let them go with the flow in order to penetrate and to explore the system as visualized in Fig. 1.

For these sensor nodes to pass through the environment and explore it without disturbing it or interfering with the dynamics, the nodes need to be scaled

---

[1] In this paper, *mote* and *node* will be used interchangeably, as well as *system* and *environment*.

© ICST Institute for Computer Sciences, Social Informatics and Telecommunications Engineering 2017
Y. Zhou and T. Kunz (Eds.): ADHOCNETS 2016, LNICST 184, pp. 50–63, 2017.
DOI: 10.1007/978-3-319-51204-4_5

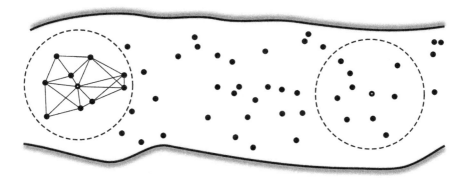

Fig. 1. Swarm of exploring sensor nodes forming a network within an enclosed environment. Nodes perform ranging transactions to neighbouring nodes within their communication range for localization and further analysis offline.

down to the centimeter or millimeter scale, depending on the application. This highly limits the resources, like energy, processing and memory, that can be taken on board the nodes. Furthermore, antennas with those dimensions will only efficiently produce radio signals with wavelengths that have an extreme high attenuation in the liquids in these environments, effectively blocking all radio communication. However, ultrasound transducers at these scales do provide larger communication ranges in these environments, but yields other problems for stable and fast communication between dynamic nodes in enclosed environments [3]. Instead of relying on communication of data, measured data is stored in memory and made available for offline analysis after retrieving the nodes from the environment.

A crucial requirement is to obtain knowledge of the positions of the nodes while traversing the environment. Structural information can be extracted from this and sensor measurement of relevant parameters (e.g. temperature, pressure, salinity) can be visualized on a map. However, during operations, neither a distributed system of anchor points nor external beacons will be available. The concept of cooperative localization [4] can be used; nodes perform measurements like time-of-flight (TOF), angle-or-arrival (AOA) or received signal strength (RSS), to gain knowledge about the position of nodes relative to neighbouring nodes within communication range. This paper introduces a *ranging protocol* to determine distances between nodes using round-trip TOF that can be used for *localization algorithms* like in [5–7], but under the specific constraints that are found in these applications.

Besides the limitation on the nodes resources, localization is further hindered by the fact that network topology – the nodes' positions and their (sparse) connectivity to neighbouring nodes – is non-static and highly unpredictable. As it is not known where neighbouring nodes are positioned, omni-directional ultrasound is used for ranging measurements. This can be achieved using e.g. tube-shaped transducers as in [8].

Ultrasound is often used for ranging applications as the propagation speed is 10^5 times lower than that of radio, therefore allowing for larger timing errors. However, the low propagation speed in combination with the significantly lower data transmission rates (typically 2–40 kb/s or even lower in more challenging environments), introduces challenges that are less often seen in radio communication [3]. Latency in the ranging transactions makes that the movement of the nodes becomes significant in the distance determination. The low data-rates, in combination with the enclosed environment and non-static topology makes signal overlap a significant hinder.

In this paper we present a novel asymmetric multi-way ranging protocol, in which trade-offs are made between the energy budget, the ranging latency and the signal overlap to optimally use the on board resources for obtaining nodes positions in offline analysis for the above mentioned applications. Depending on the application or the state that nodes are in, these trade-off's can be adjusted to address the specific situation as good as possible.

The specific challenges in developing the ranging protocol for these applications are addressed in Sect. 2. In Sect. 3 the design of a ranging protocol is described that attempts to balance between all the parameters involved. In order to assess the suitability of the protocol for these applications, the protocol is simulated in a network simulator as described in Sect. 4. Important performance metrics that assess the specific goals are shown in the results Sect. 5. Discussion and future work can be found in Sect. 6, the conclusion in Sect. 7.

2   Protocol Design Challenges

Traditional ranging protocols consist of three phases: a scanning phase, a ranging phase and a reporting phase [9]. In this paper, an attempt is being made to maximally reduce energy costs of ranging in the specific application cases described above.

Reporting of ranging measurements to neighbouring nodes is not performed as this would require extra node resources and data communication is challenging in this applications. Nodes only store measurements in their own memory.

The ranging-phase can be performed in a variety of methods. We chose for the concept of multi-way ranging (MWR), initially proposed in [10] as N-Way Time Transfer. It exploits the omnidirectional transmission and reception by using all received signals for determining distances between nodes, rather than only the signal between sender and one addressed receiver in e.g. two-way ranging (TWR) methods [11]. Therefore, the total amount of messages needed to complete a full ranging cycle using MWR scales linearly with the number of nodes, instead of quadratically in TWR methods. It significantly reduces the energy required for performing the ranging procedure.

Control of Ranging Sequence. As the network topology is non-static and connectivity sparse and fast-changing, it is not known which neighbouring nodes are within communication range. The simple sequence of events in traditional

MWR [10], where node $i+1$ transmits a ranging signal after node i, cannot be easily controlled in these applications. An alternative method is proposed in Sect. 3 using a master-slave system.

Ranging Latency. The ranging latency of a single ranging transaction takes up to 3 ms when nodes are 1 m apart (twice the propagation time, the message length and the processing time). A full ranging cycle within a large swarm, with all its individual ranging transactions, can easily take 100 ms. Depending on the movement of the nodes, a large latency significantly challenges the localization algorithm as the measured inter-node distances cannot be considered quasi-static. The latency should therefore be kept as low as possible.

Furthermore, from an energy perspective it is beneficial to reduce latency such that nodes are longer in a low-energy sleep state instead of an active listening/decoding state.

Signal Overlap. The low data-rates in combination with small inter-node distances in enclosed environments cause a significant amount of potential signal overlap. Signal overlap should be prevented as much as possible as it requires more energy and processing to filter and distinguish signals. The ranging protocol in Sect. 3 uses a time-divided communication scheme for determination of the distances to allow for reduction of signal overlap. The amount of bits transmitted should also be kept at a minimum to keep the message length as short as possible to reduce signal overlap.

Scanning Phase. As it is not known which nodes are within communication range, often in ranging protocols, a separate scanning phase is initiated before the ranging phase. In this phase, nodes determine which neighbouring nodes are within communication range to determine which nodes to perform ranging measurements to.

Such an additional scanning phase adds to the energy budget. In this paper, the scanning phase is omitted and solved by addressing all nodes by a non-unique calling identifier. It causes a trade-off between ranging latency and signal overlap.

3   Protocol

This section introduces a modified version of the regular multi-way ranging protocol to deal with the specific limitations in the usage of ultrasound in a non-static network topology with highly resource-limited nodes. It will also address the challenge of finding proper trade-off's between e.g. ranging latency and signal overlap. It is important to notice that these trade-off's can be adjusted based on the specific environments or the specific situation that nodes are in.

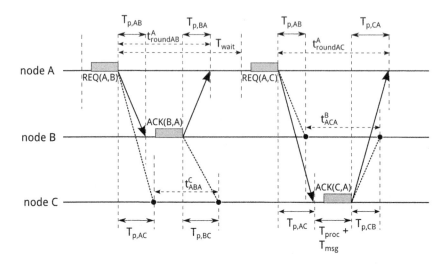

Fig. 2. One ranging cycle of master node A and slave nodes B and C. The master node transmits request (REQ) signals and slave nodes respond with and acknowledgement (ACK) signal if it is addressed to them. Knowledge of timing information $t^A_{roundAB}$, $t^A_{roundAC}$, t^C_{ABA} and t^B_{ACA} and the fixed value of the processing time T_{proc} is sufficient to determine the propagation times $T_{p,AB}$, $T_{p,AC}$ and $T_{p,BC}$ between the nodes.

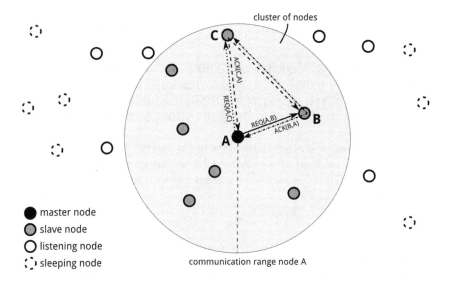

Fig. 3. Master node A initiates the ranging process to node B and C. The nodes within communication range become slave node and respond to REQ signals with an ACK signal. All nodes within the respective communication range receive the signals and store them: in offline analysis they can be used to determine round-trip TOF between nodes. The nodes outside the cluster will only receive ACK signals. In this figure, not all signals (arrows) are drawn.

3.1   Asymmetric Multi-way Ranging

Instead of traditional, 'symmetric', MWR as introduced in [10], here the ranging procedure is controlled by *master nodes* that send request (REQ) signals to its neighbouring nodes that then become *slave nodes* and respond with an acknowledgement (ACK) signal. The communication scheme that is used is illustrated in Fig. 2. The *cluster of nodes* that is formed by this master node and the slave nodes is illustrated in Fig. 3.

In a ranging transaction between master node A and slave node B, the timestamps of transmission and reception at node A provides knowledge about the round-trip TOF between nodes A and B, denoted as $t^A_{roundAB}$. The node's internal processing time T_{proc} and signal message time T_{msg} are known beforehand and are fixed, therefore, the round-trip propagation time, $T_{p,AB} + T_{p,BA}$, between A and B can be estimated. After this ranging transactions, master node A performs a similar transaction to node C and the other nodes within the cluster.

Since the nodes A, B and C are within each others communication range, also the nodes that are not addressed in the ranging transactions receive the signals. The time difference t^C_{ABA} between the arrival of REQ(A,B) and ACK(B,A) at node C and the time difference t^B_{ACA} between the arrival of REQ(A,C) and ACK(C,A) at node B, can be used to calculate the propagation time between nodes B and C using:

$$T_{p,BC} + T_{p,CB} = t^C_{ABA} + t^B_{ACA} - 2(T_{proc} + T_{msg}) \tag{1}$$

3.2   Picking the Master Node

The role of master node is being alternated between all nodes in the network. The advantage of this is that the power consumption is distributed evenly over all nodes (master nodes transmit more signals) and clusters are more distributed over the swarm.

Within the time frame of one sample T_{sample}, in which a complete ranging cycle is completed for all nodes, the role of master node is chosen randomly. This is performed by having all nodes at the beginning of a sample chose a random delay time T_D. Nodes become master when their sample timer t_s, that is set to zero at the beginning of a sample, trespasses $t_s > T_D$. Nodes become slave node when before t_s reaches T_D, it receives any REQ signal from a master node. The master node initiates the ranging transactions as described above, thereby forming a cluster of nodes as in Fig. 3.

As seen in Fig. 1, throughout the entire network, several of these clusters are formed in which ranging transactions are performed. Every sample, these clusters change based on which nodes have become master node.

3.3   Scanning the Slave Nodes

Within one ranging cycle, the master node should send a request to all slave nodes in the cluster, but it is beforehand not known which nodes are within communication range. Regular scanning techniques depend on the availability of sufficient bandwidth, processing power or time to perform broadcasting.

In this work we propose for the master node to initiate the ranging transactions to all possible hardware addresses. But as the total amount of nodes in the network can be very large and the connections are sparse, this will be very inefficient as most requests remain unanswered. Instead of requesting to the hardware's *unique identifiers* (UID) the master node requests to highly abbreviated *calling identifiers* (CID). The master node only initiates n_f times a ranging transaction to CID = $\{0, 1, ..., n_f\text{-}1\}$. Slave nodes will respond if and only if their unique hardware identifier suffices

$$\text{mod}(\text{UID}, n_f) = \text{CID} \tag{2}$$

As multiple nodes will have an identical CID, the probability arises that multiple nodes will respond to the same request. If the ACKs of the responding slave nodes do not overlap such that the signals cannot be distinguished and decoded anymore at the receiving node, the determination of the round-trip TOF of each of them can still be performed. Parameter n_f can be chosen both offline as online to adjust for the amount of neighbouring nodes and the total signal overlap.

In order to receive all possible ACK's, the master node will wait $T_{\text{wait}} = 2T_{\text{p,max}} + T_{\text{proc}} + T_{\text{msg}}$ after transmission of a REQ before it sends a request with a next CID. Here, $T_{\text{p,max}}$ accounts for the propagation time required to reach the end of the (expected) communication range.

After all ranging transactions have been performed, the nodes will go into a low-energy sleep mode to await the start of the next sample. When $t_s > T_{\text{sample}}$, nodes will internally initiate a new sample. The sample is initiated in a sleep mode and nodes will wake up upon reception of any signal (using e.g. a threshold detection). It then starts a listening mode in which it can decode incoming signals. Before a master node starts with the first CID, it can transmit a short signal to wake up the neighbouring nodes.

3.4 Reducing the Latency

As seen in Fig. 3, the nodes just outside the communication range of the master node do not become slave node and will have to wait for itself to become master node, or will have to wait for a node within its communication range to become one.

In order to speed up this process and have the network-wide ranging cycle end sooner, an avalanche effect is induced. Nodes that receive ACK signals without having received REQ signals are likely to be just outside an already formed cluster. Their remaining delay time before they become master node is reduced by a factor $M_{\text{avalanche}}$ at the reception of any ACK signal until they become master or slave.

This reduction of the delay time T_D induces an avalanche effect throughout the network such that all nodes become either master or slave within less time after each other, therewith, reducing the ranging latency throughout the network.

3.5   Synchronization

Absolute synchronization is not required for determining distances as all distances are obtained using a round-trip TOF measurement. It is however beneficial to have nodes synchronized to a level in which samples are aligned such that the avalanche effect introduced in Sect. 3.4 allows nodes to sleep for the majority of the sample time instead of responding to nodes that are still in the previous or already in the next sample.

For this reason, in order for connected nodes to remain in the same sample, it is proposed to subdivide a sample on the node level into timeslots as illustrated in Fig. 4a. The random delay time is chosen from a uniform distribution within the range $T_D \in (T_{\text{start}}, T_{\text{end}})$ or the *active time period*. The internal sample timer t_s is reset to $t_s = T_{\text{start}}$ when becoming master; or, at reception of the first signal (any REQ or ACK) in the sample, as illustrated in Fig. 4b. This will assure that connected nodes remain synchronized to the sample level, as long as (groups of) nodes have not been disconnected from each other. In the *awaiting period* nodes do not become master and can only receive signals, in the *silent period*, nodes have already received their first signal or already became master node.

In our work, the three time periods are chosen to be of equal length, i.e. $T_{\text{start}} = \frac{1}{3} T_{\text{sample}}$ and $T_{\text{end}} = \frac{2}{3} T_{\text{sample}}$. Note that these periods do not indicate when a node is asleep or in which mode it is in.

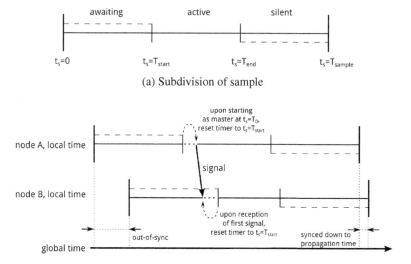

(a) Subdivision of sample

(b) Sample timer reset, either at first reception of signal or at start as master

Fig. 4. Subdivision of sample in three parts: awaiting, active and silent. Each sample, the start delay that determines when to become master is randomly chosen within active period (uniform distribution). Upon first reception of signal in the sample, or, upon becoming master node, the sample timer t_s is reset to $t_s = T_{\text{start}}$.

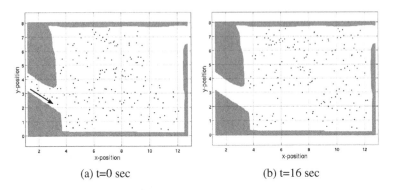

(a) t=0 sec   (b) t=16 sec

Fig. 5. Nodes positions throughout simulation in 2D tank-like environment with injection flow from left (indicated by arrow).

4 Simulations

The protocol implementation is simulated in OMNeT++ network simulator [12,13].

In order to simulate a dynamic swarm of nodes that passively flow through an enclosed environment, we use a flow simulator to generate the nodes positions over time [14]. The positions are generated based on tracer positions in a fluid flow in a 9 by 8 m 2-D tank-like environment with an inlet an outlet. The positions of the $N = 200$ nodes at the beginning and end of the simulation time are illustrated in Fig. 5.

The average node speed throughout the simulation is 0.20 ± 0.17 m/s with a maximum of 0.80 m/s. The communication range is set to a fixed 1 m and results in an average node density of 9.8 ± 3.7 neighbouring nodes within the communication range. The clock frequency offset is set to 100 ppm and is fixed throughout the simulation.

The sample time is set to $T_{sample} = 1$ s and the amount of CIDs is in this paper is swept between: $n_f = \{8, 16, 32\}$. The ultrasound transmission rate is set to 40 kb/s. The avalanche induction is studied by sweeping $M_{avalanche} = \{1, 2, 4, 8\}$ in which $M_{avalanche} = 1$ means no induced avalanche.

The output of the simulations consist of the data that are being stored on the nodes internal storage: the messages sent and received, the timestamp at transmission/reception, the timestamp when new samples start and for research purposes also the internal states the nodes are in and at what specific time.

5 Results

The protocol is analysed based on several performance metrics that assess the design goals for the ranging method. One is the ranging latency within a swarm versus the signal overlap, second is the energy usage of an individual node. And

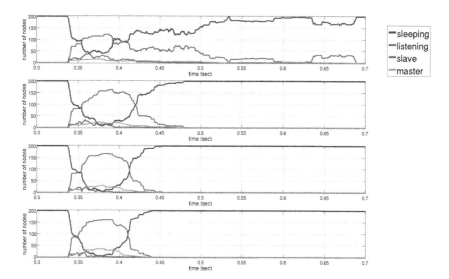

Fig. 6. Total number of nodes in specific states through first sample. From top to bottom the avalanche parameter: $M_{\text{avalanche}} = 1, 2, 4, 8$. The top graph, $M_{\text{avalanche}} = 1$ means there is no induced avalanche effect. In these simulations, the number of CIDs $n_f = 16$.

as last, the fraction of the theoretical amount of possible distances that are determined using this protocol: the coverage.

Figure 6 shows for the simulated datasets with $n_f = 16$ an overview of the amount of nodes in a specific state over a single sample. Figure 7 shows the main performance metrics of the protocol in a simulated network as described in Sect. 4 and discussed next.

5.1 Latency Versus Signal Overlap

From Fig. 6 it can be seen that the induced avalanche effect assures that the network finished a single ranging cycle sooner. In this sample, the maximum latency goes from 360 ms for $M_{\text{avalanche}} = 1$ (no avalanche) down to 150 ms, 120 ms and 110 ms for $M_{\text{avalanche}} = \{2, 4, 8\}$, respectively. As a reference; within a single cluster, the ranging latency is $n_f T_{\text{wait}} = 43$ ms.

Figure 7a shows the average latency between all ranging transactions in a ranging cycle. Increasing the avalanche effect ($M_{\text{avalanche}}$) yields a smaller latency. The latency of the full ranging cycle is approximately between 4 to 5 times larger as the average latency between the transactions.

Figure 7a also shows the average fraction of signals that are received with overlap with another signal. Increasing the avalanche effect and reducing the latency inevitably increases the signal overlap.

At lower values of n_f, the latency drops quicker, but signal overlap is higher; less CIDs are scanned but more nodes will respond to the same REQ signal. There is a clear trade-off between latency and signal overlap.

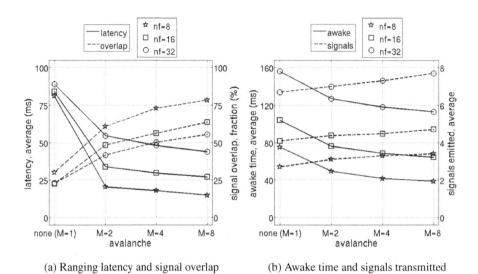

(a) Ranging latency and signal overlap   (b) Awake time and signals transmitted

Fig. 7. Average values of perfomance metrics of simulated ranging protocol using different input parameters n_f and $M_{\text{avalanche}}$. Clear trade-offs are visible.

5.2 Energy Efficiency

The energy efficiency in this paper mainly focusses on the nodes' awake time and the amount of signals transmitted (and related to that the amount of signals received and stored). The awake time is defined as the time not spend in the low-energy sleep state, but rather in an active signal transmission or receiving/decoding state.

Figure 7b shows the node's average awake time per sample and the average amount of signals transmitted per node per sample (master and slave nodes together). As the latency decreases with increasing $M_{\text{avalanche}}$, so does the time that nodes need to be awake. With increasing avalanche effect, the number of signals required for transmission increases slightly as more nodes will become master node.

Both the awake time and the number of signals transmitted increase with increasing n_f as more CIDs will have to be transmitted and decoded.

5.3 Coverage

The coverage can be defined as the fraction of connections (node pairs that are within each others communication range) for which the ranging procedure yields sufficient information to determine a distance measure. Since only distances can be calculated within a cluster, the coverage will be lower than 100% as not all connections can fall within a cluster. Throughout all simulations, the coverage was between 86%–89%.

Even though for the other 11%–14% no distances can be determined using RT-TOF, the basic connectivity information is available: the received ACK signals that did get received by the nodes outside the cluster, provides information on which nodes where within their communication range. The localization algorithm can use this information to its advantage.

The coverage does not need to be 100% to localize the entire swarm. In fact, for example in [6], studies are performed where localization is stress-tested on e.g. the loss of large amounts of connections. Also note that each sample, different clusters are formed such that this group of 11%–14% of the connections is different for each sample.

6   Discussion and Future Work

Although the current implementation of the ranging protocol has been simulated over a relative short measurement time. Simulations using extremely large clock deviations of up to 100 000 ppm have been tested and show good alignment of samples over the simulation time of 16 s (not shown here). As long as the network is sufficiently connected and not disjoint, the avalanche effect can keep the nodes' internal clock synchronized within a fraction of the sample time T_{sample}.

The simulations in this paper have been performed in a 2-D environment. Although this protocol can be directly used in 3-D, the induced avalanche effect will have quantitatively a slightly different result as the ones presented here. No qualitative differences are to be expected.

Instead of scanning all possible CIDs, all nodes can actively record which nodes it has seen in the past. Upon becoming master node, instead of scanning all available CIDs, the master node can scan the UIDs of nodes that it has seen in the previous (several) sample periods. This will reduce signal overlap and can reduce the amount of required signals for transmission.

The protocol can fairly easily be adjusted to also account for disjoint networks coming together such that they can become synchronized up to the sample level. This is part of future work.

7   Conclusion

This paper illustrates the challenges involved in performing round-trip TOF in a swarm of autonomous nodes without external contact in an unpredictable and dynamic topology with sparse connectivity. The applications require high levels of miniaturization of the nodes and introduce a specific set of constraints and challenges that has not been researched before. A novel asynchronous multi-way ranging protocol has been presented to allow round-trip TOF measurements. Control of the ranging transactions can be performed by master nodes that initiate them to their neighbouring slave nodes. Master nodes are assigned at random each time a new sample starts.

The latency between ranging measurements in the entire swarm can be reduced by inducing an avalanche effect of nodes that become master node. The avalanche effect also reduces the required time for the nodes to be actively listening for signals and allows for synchronization down to a fraction of the sample time.

The trade-off's that are involved in this ranging protocol are a direct consequence of the application: the need for resource-limited nodes, the use of ultrasound and the unpredictable and fast-changing network topology with sparse connectivity. Getting insight in these trade-off's allow for adjusting the ranging protocol based on the specific circumstances that nodes are in. In [2], this exploration method and the ability of nodes to adjust for specific circumstances, is further explored.

Acknowledgement. INCAS[3] is co-funded by the Province of Drenthe, the Municipality of Assen, the European Fund for Regional Development and the Ministry of Economic Affairs, Peaks in the Delta. This project has received funding from the European Union's Horizon 2020 research and innovation programme under grant agreement No 665347 The authors would like to thank Elena Talnishnikh, Hao Gao, Jan Bergmans, Libertario Demi and Gijs Dubbelman for their help in this research.

References

1. Talnishnikh, E., et al.: Micro Motes: a highly penetrating probe for inaccessible environments. In: Leung, H., Mukhopadhyay, S.C. (eds.) Intelligent Environmental Sensing. Smart Sensors, Measurement and Instrumentation, vol. 13, pp. 33–49. Springer, Cham (2015)
2. EU Horizon 2020 FET-Open project: PHOENIX. www.phoenix-project.eu
3. Akyildiz, I.F., Pompili, D., Melodia, T.: Underwater acoustic sensor networks: research challenges. Ad Hoc Netw. **3**, 257–279 (2005)
4. Patwari, N., et al.: Cooperative localization in wireless sensor networks. IEEE Sig. Process. Mag. **22**(4), 54–69 (2005)
5. Bachrach, J., Taylor, C.: Localization in sensor networks. In: Handbook of Sensor Networks: Algorithms and Architectures. John Wiley & Sons Inc., 23 September 2005
6. Duisterwinkel, E.H.A.: Ph.D. Dissertation, research in progress
7. Dubbelman, G., et al.: Robust sensor cloud localization from range measurements. In: IEEE International Conference on Intelligent Robots and Systems, Chicago, Illinois, USA (2014)
8. Li, H., Jung, K.W., Deng, Z.D.: Piezoelectric transducer design for a miniaturized injectable acoustic transmitter. Smart Mater. Struct. **24** (2015)
9. Kim, H.: Performance comparison of asynchronous ranging algorithms. In: IEEE Global Telecommunications Conference, Honolulu, Hawaii, December 2009
10. Green, M.P.: N-way time transfer ('nwtt') method for cooperative ranging, Contribution 802.15-05-0482-00-004a to the IEEE 802.15.4a Ranging Subcommittee, July 2005

11. Hach, R.: Symmetric double sided - two way ranging, Contribution 802.15-05-0334-00-004a to the IEEE 802.15.4a Ranging Subcomittee, June 2005
12. Puts, N.A.H.: Analysis and design of an ultrasound positioning system protocol for sensor swarms. MSc. Dissertation, Eindhoven University of Technology, The Netherlands (2016)
13. Varga, A., Homig, R.: An overview of the OMNeT++ simulation environment. In: SIMUTools, Marseille, France, 03–07 March 2008
14. Bosman, H.H.W.J.: For the adaptation and extension of Daniel V. Schroeder's Java based flow simulator (2013, 2015)

Comparison of RPL Routing Metrics on Grids

Lilia Lassouaoui[(✉)], Stephane Rovedakis, Françoise Sailhan, and Anne Wei

CEDRIC Laboratory, Conservatoire National des Arts et Métiers,
292 rue Saint Martin, 75003 Paris, France
{lilia.lassouaoui,stephane.rovedakis,francoise.sailhan,anne.wei}@cnam.fr

Abstract. The IPv6 Routing Protocol for Low power and lossy networks (RPL) is appearing as an emerging IETF standard of Wireless Sensor Networks (WSNs). RPL constructs a Direct Acyclic Graph (DAG) according to an objective function that guides the routing based on some specified metric(s) and constraint(s). In the last decade, a number of RPL simulations have been proposed for several metrics and constraints, but for the best of our knowledge there is no comparative evaluation for RPL energy-aware routing metrics. In this paper, we present the first comparative study of RPL energy-aware routing metrics on Grid topology. Our experiments show that multi-criteria metrics perform better.

Keywords: RPL · Energy-aware routing metrics · Evaluation

1 Introduction

Wireless Sensor Networks (WSNs) remain an emerging technology that has a wide range of applications including environmental monitoring, smart space and robotic exploration. WSN are characterised by constrained nodes with limited processing capabilities and memory, which are typically battery-operated and interconnected by wireless links that are operating at a low data rate. WSN are usually experiencing a high loss rate coming from the low power and lossy nature of the links. Such constraints combined with a typical large number of sensors have posed many challenges related to the configuration, management and routing. In order to tackle this issue, the IETF has standardised RPL [WTB+12], a new IPv6 routing protocol especially taylored for Low power and Lossy Networks (LLN). In compliance with the IPv6 over Low power Wireless Personal Area Networks (6LoWPAN) standard, RPL supports the idea of applying IPv6 [MKHC07] even to the smallest device by providing a mechanism whereby multipoints-to-point traffic from sensors inside the 6LoWPAN network towards a central control point (e.g., a server on the Internet) as well as point-to-multipoint traffic from the central control point to the sensors inside the 6LoPWAN are enabled. Support for point-to-point traffic is also available. For this purpose a Destination Oriented Directed Acyclic Graph (DODAG) is built. This DODAG is constructed using an objective function which defines how the routing metric is calculated. In particular, this objective function determines how routing constraints and metrics are taken into account to determine the best route. During the last decade,

© ICST Institute for Computer Sciences, Social Informatics and Telecommunications Engineering 2017
Y. Zhou and T. Kunz (Eds.): ADHOCNETS 2016, LNICST 184, pp. 64–75, 2017.
DOI: 10.1007/978-3-319-51204-4_6

several metrics and constraints have been considered, e.g., ETX, ENG-TOT, ENG-MinMax (see Sect. 2.1 for more detailed).

In this paper, we compare the performance of several RPL routing metrics proposed for saving power and maximizing lifetimes. To the best of our knowledge, this is the first comparative study of RPL routing metrics. We conduct our experiments on top of the Cooja simulator [ODE+06], using Contiki OS 3.0. Simulation results show that multi-criteria metrics perform better.

The remainder of this paper is organized as follows. We first provide in Sect. 2 an overview of RPL and its related metrics. Then, we present an evaluation of the performance of energy-aware routing metrics (Sect. 3). We conclude this article with a summary of our contribution along future work.

2   Background

The Routing Protocol for Low power and lossy network (RPL) [WTB+12] has been proposed by the IETF Routing Over Low power and Lossy networks (ROLL) working group. RPL is a distance-vector routing protocol targeting IPv6 networks. In compliance with the IPv6 architecture, it builds a Directed Acyclic Graph (DAG) so as to establish bidirectional routes between sensors. RPL is mainly designed to exchange data between each (RPL) node and a particular node, called sink node. The sink node acts as a common transit point that bridges the LLN with the IPv6 networks. It also represents a final destination node. The traffic flows supported by RPL, include sensors-to-sink, sensor-to-sensor, sink-to-sensors. A sensor network can be used for different applications and several sink nodes can coexist, i.e., we can have potentially one sink per application. A Destination Oriented DAG (DODAG) is constructed for each application according to a specific function (called Objective Function) which optimizes a specified metric for data routing, e.g., minimizes the network distance. Every DODAG is rooted at the corresponding application sink (DODAG root). Some applications can optimize objective function, which may be contradictory with another application. To this end, the concept of RPL instance has been introduced. A RPL instance brings together a subset of DODAGs in a sensor network which follow the same objective function. Several RPL instances can run concurrently, but a node belongs to at most one DODAG per RPL instance.

RPL separates packet processing and forwarding from the routing optimisation functions which may include minimising energy, latency and generally speaking satisfying constraints. In particular, RPL provisions routes towards the DODAG roots which is optimised with respects to the Objective function. In order to create and maintain a DODAG, RPL specifies a set of ICMPv6 control messages, such as DODAG Information Object (DIO) and DODAG Information Solicitation (DIS). The root starts the construction of the DODAG by broadcasting a DIO message carrying several parameters, including an affiliation with a DODAG (DODAGID), a rank which represents the position of the node with regards to the DODAG root, a routing cost and its related metrics, a Mode of Operation (MOP). The nodes that are in communication range with the root decide whether to join the DODAG or not. In particular, based on the neighbours ranks and according to the objective function, each node selects its DODAG parent. For this purpose, the node provision a routing table, for the destinations specified by the DIO message, via parent(s). Then, the node origi-nates its own DIO message. Rather than waiting for the DIO message, node may also broadcast a DIS message requesting information from the other RPL nodes. Overall this DODAG root permits to support sensors-to-root traffic, which is a dominant flow in many applications.

Sensor-to-sensor traffic flows up toward a root and then down to the final des-tination (unless the destination is on the upward route). For this purpose, RPL establishes downward routes using Destination Advertisement Object (DAO) messages. DAO message is an optional feature. RPL supports two modes of Downward traffic: Non-Storing (fully source routed) or Storing (fully stateful). In the Non-Storing case, the packet travels towards the root before traveling Down; the only device with a routing table is the root that acts as a router, hence source routing is used, i.e., the root indicates in the data packet the full route towards the destination. In the Storing case, sensors are configured as routers and maintain a routing table as well as a neighbour table that are used to look up routes to sensors. Thus, packet may be directed down towards the des-tination by a common ancestor of the source and the destination prior reaching a root.

In order to increase the network lifetime, RPL uses a dynamic dissemination algorithm, called Trickle. This algorithm adapts the rate at which DIO messages are sent by adjusting a timer. A DIO message is sent every $Imin$ ms during the DODAG construction, and when the DODAG construction has converged this interval is doubled at each time period until reaching a maximum interval corresponding to $Imax$ ms. When the DODAG reconfigurate due to e.g. the addition of new nodes or the detection of an inconsistency, RPL resets the timer to $Imin$. RPL also includes a mechanism to detect and suppress loops in the DODAG, based on the ranks in the DODAG. This loop-free property is obtained by insuring that the ranks increases in a strickly monotonically fashion, from the sink toward the leaf nodes. Therefore, every node compares the ranks of its neighbors to detect inconsistency, which is materialised by e.g., the reception of a downward data packet from a neighbor with a higher rank. When node detects a loop, it initiates a route poisoning (i.e., it broadcasts an infinite rank) so as to trigger a reconstruction of its sub-DODAG.

2.1 Objective Functions

An Objective Function (OF) specifies the objectives used to compute the (con-strained) path and to select parents in DODAG. In practice, it defines the trans-lation of metric(s) and constraint(s) into a value called Rank, which approxi-mates the node distance from a DODAG root. Regardless of the particular OF used by a node, rank always increases so that loop-free paths are always formed. The definition of the OF is separated from the core RPL protocol. It allows RPL to meet different optimization criteria for a wide variety of applications.

For a detailed survey on the OF, the interested reader may refer to [GK12]. The ROLL working group has specified two types of OFs: Objective Function zero (OF0) and Minimum Rank Hysteresis Objective Function (MRHOF). OF0 is the default objective function that uses the hop count as routing metric. The MRHOF minimizes the routing metric and uses the hysteresis mechanism to reduce the churn coming from small metric changes for a better path stability.

RPL supports constraint-based routing. A constraint may be applied to link or node, and, if a link/node does not satisfy the given constraint, it is pruned from the candidate neighbors set, hence leading to a constrained shortest path. A metric is used in association with an OF for route optimization. The ROLL working group proposes two types of metrics: the node metric and link metric. The node metric represents the node state (e.g., node energy, node hops). The Link metric reflects the route quality, e.g. latency, throughput, Expected Transmission count (ETX). These metrics can be additive or multiplicative, they can also refer to a maximum or minimum property along a path in the DODAG.

In order to construct and update the DODAG, each non-root node has to select a *preferred* parent. This selection is performed by computing the path cost for each parent (neighbor with a lower Rank). The *path cost* is a numerical value which represents a property of the path toward the sink node. It is computed by summing up the selected node/link metric to the advertised path cost. The best cost returned by the OF using the specified metric for each candidate parent is used to select the preferred parent, i.e., the parent on the path with the best cost. The path cost is computed again either if the node/link metric is updated or if a new metric is advertised. When MRHOF is used, according to the hysteresis mechanism the current preferred parent is changed if the difference between the current and the new path cost is at least equal to a specified threshold.

After selecting its preferred parent P, a non-root node q computes its rank $R(q)$ as follows: $R(q) = R(P)+\texttt{rank\_increase}$, with $R(P)$ defining the Rank advertised by P and $\texttt{rank\_increase}$ the rank increment. Note that a DODAG root advertises a Rank equal to $\texttt{rank\_increase}$. The Rank and the path cost computed by each node are disseminated in a DIO message.

2.2 Some Routing Metrics Proposed for RPL

Several routing metrics have been proposed in the litterature to increase the network lifetime, to maximize the reliability or to minimize the latency. In this paper, we focus on the energy-aware routing metrics because the energy is a key criterion of wireless sensor networks.

One of the classical and popular routing metric available in several RPL implementations is the Expected Transmission count (ETX). ETX estimates the number of transmissions that take place through a link before the reception of a correct acknowledgment. This value can be computed as: $ETX = \frac{1}{PDR_{s \to d} \times PDR_{d \to s}}$, with $PDR_{s \to d}$ defining the estimated packet delivery ratio from s to d. More particularly, this estimated packet delivery ratio is computed as the ratio between the number of transmitted packets and the number of acknowledged packets, including retransmission(s). Then, among the neighbours

N_i, using MRHOF a node i selects as preferred parent, the neighbour characterised by the minimum ETX, i.e., $\min_{j \in N_i} ETX_j$. The lower is ETX, the better
is the link quality. The Rank $R(i)$ of node i with preferred parent P is given
by: $R(i) = R(P)+\texttt{rank\_increase}$. It is disseminated by node i using a DIO
message. ETX seems to be a good candidate to reduce the end-to-end delay.
Indeed, the lower is the retransmission number, the better is transmission time
for a data packet toward the sink. In addition, since communication is the most
energy consuming activity, ETX allows to reduce the energy consumed at each
node. However, this does not permit to select a route composed of nodes with
high battery level.

In order to design an energy-aware route selection, the *residual energy*
$ResEng_i$ can be used as a RPL metric. The residual energy is computed as
the difference between the maximum battery level $MaxEng_i$ and the energy
consumed $EngCons_i$ by a node i, i.e., $ResEng_i = MaxEng_i - EngCons_i$. The
energy consumed by a sensor is due to the computation and the radio communication (i.e., transmission and listening). Demicheli [Dem14] proposed the
first RPL metric which considers the energy consumed by sensor nodes along
a path. The Rank $R(i)$ of each node i is obtained by adding an increment
(fixed to 16 by the author) to the Rank $R(P)$ of its preferred parent P, i.e.,
$R(i) = R(P)+\texttt{rank\_increase}$. Each node i sends in DIO messages its Rank as
well as the *energy consumed along the path* $PathEngCons(i)$ in the Metric Container field, with $PathEngCons(i)$ equals to the sum of $PathEngCons(P)$ sent
by its preferred parent P and $EngCons_i$. The preferred parent is the parent
with the lowest energy consumed along the path. The main drawback is that
a path toward the sink may contain a node with a very low residual energy.
To tackle this issue, Xu et al. [XL13] and Kamgueu et al. [KNDF13] consider
the residual energy as a routing metric. Xu et al. [XL13] have proposed to use
RPL with a residual energy metric: the Rank $R(i)$ of node i is equal to the
Rank of the preferred parent $R(P)$ plus the residual energy $ResEng_i$ of i, i.e.,
$R(i) = R(P) + ResEng_i$. Each node selects as preferred parent the one with
the highest Rank, and i sends a DIO message with its Rank and an idle Metric
Container field. Kamgueu et al. [KNDF13] define the cost of a path PW_i of a
node i toward the sink as the minimum among the residual energies along the
path. Therefore, each node sends in DIO messages its rank and its path cost
using the Metric Container field. Every node i computes the path cost that can
be obtained for each parent (as the minimum between its residual energy and
the path cost sent by the parent), and selects as preferred parent the one with
the maximum computed path cost, i.e., $PW_i = \min(\max_{j \in N_i}\{PW_j\}, ResEng_i)$,
where N_i refers to the neighbours of node i.

Some applications require data transmission with a low delay. Several routing
metrics have been proposed to minimize the end-to-end delay with RPL [DPZ04,
ABP+04, KB06]. Chang et al. [CLCL13] propose an energy-aware metric which
considers the number of retransmissions. For this purpose, the residual energy
is combined with the ETX. Each node i sends its Rank and its residual energy
$ResEng_i$ using the Container Metric field in DIO messages. The preferred parent

is selected among the parent j of i which gives the minimum of the weighted function: $\alpha \frac{ETX_j}{Max\_ETX} + (1 - \alpha) \times (1 - \frac{ResEng_j}{MaxEng})$, where Max_ETX and $MaxEng$ are respectively the maximum ETX value of a link and the maximum battery level of a sensor node. The Rank of each node is computed as it is done by the ETX metric (and described above). Recently, Iova et al. [ITN14] have proposed another routing metric, called Expected LifeTime (ELT), to better optimize the network lifetime. This metric takes into account the link quality, the residual energy and the traffic. First of all, each node i computes its expected lifetime ELT_i following Eq. 1:

$$ELT_i = \frac{ResEng_i}{T_i \times \frac{ETX(i,P)}{Data\_Rate} \times PowTX_i}, \tag{1}$$

where T_i is the traffic of i (in bits/s), $ETX(i, P)$ is the ETX value of the link to the preferred parent P of i, $Data\_Rate$ is the rate at which data is sent (in bits/s) and $PowTX_i$ is the power consumed by a radio transmission made by i. For each path in the DODAG, the minimum expected lifetime is propagated along the path using the Metric Container field of DIO messages. Each node i selects as preferred parent the one which gives the maximum expected lifetime, i.e., $\max_{j \in N_i} ELT_j$. The Rank associated to a node i in the DODAG is computed as for ETX.

Table 1 presents a brief summary of the RPL metrics described in this section, it also gives the topology considered by the authors to evaluate their metrics.

Table 1. Summary of the presented RPL Metrics

Paper	Energy-aware	Metrics information		Topology
		Energy	Link quality	
[GL10]	No	-	Yes	Grid & Random
[Dem14]	Yes	Energy consumption	No	Grid & Random
[XL13]	Yes	Residual energy	No	Grid & Random
[KNDF13]	Yes	Residual energy	No	Grid & Random
[CLCL13]	Yes	Residual energy	Yes	Random
[ITN14]	Yes	Residual energy	Yes	Random

3   Metric Evaluation

3.1   Simulation Setup

In order to simulate and analyze the performance of RPL, we use the Cooja simulator [ODE+06], a flexible Java-based simulator which supports C program language as the software design language by using Java Native Interface. We simulate a Wireless LLN consisting of 56 nodes which are emulated as Tmote

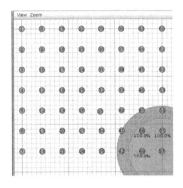

Fig. 1. Grid topology

Table 2. Node distribution on grid

Distance to the sink	Number of nodes
70 m	3
140 m	5
210 m	7
280 m	9
350 m	11
420 m	13
490 m	7

sky mote [PSC05] (a widely used sensor platform) with a 2.4 GHz CC2420 radio transceiver with IEEE 802.15.4 operating at the radio layer. These nodes are deployed over a 300×300 m grid with a sink located at the bottom right corner (Fig. 1). This sink location represents a worst case scenario (comparing to a sink located at the grid center): a higher congestion is observed around the sink because very few sensors are connected to the sink. The distance separating the nodes from the sink is given in Table 2. We use the ContikiOS 3.0 with ContikiMac [Dun11] which provides a power efficient medium access control by turning off the radio 99% of the time. We further rely on RPL as a routing protocol and we simulate a sensors-to-sink traffic wherein each node periodically sends to the sink some data packets, at a rate of 6 packets per minute, i.e., we consider Constant Bit Rate (CBR) convergecast flows. Note that each node starts sending its first data packet 65 s after the beginning of a simulation. The main parameters used for simulation are summarized in Table 3. TX rate and RX rate define respectively the success ratios in transmission and reception mode. We do not consider packet loss to better evaluate the performance of the metrics. We average the simulation results over 10 simulation runs, each one of 5 h duration.

3.2 Evaluation Criteria

In this study, we evaluate the influence of the metrics in terms of energy consumption, Packet Delivery Ratio (PDR), End-to-end delay and number of control messages exchanged.

– **Energy Consumption -** In order to compute the energy consumption, we rely on the Power-trace mechanism [DEFT11] provided by Contiki. The power-trace estimates the power consumption due to the CPU usage and the network-level activitities including packet transmission and reception. During our experiment, we focus on the period of time the radio is on. We further calculate the energy consumption $EngCons_i$ at each node i (in mJ):

$$EngCons_i = \frac{(T_{CPU} * I_{CPU} + T_{RX} * I_{RX} + T_{Tx} * I_{TX}) * Volt}{Rtimer} \quad (2)$$

Table 3. Simulation parameters

Parametres	Value
OF	MRHOF
RPL MOP	NO_DOWNWARD_ROUTE
Start Delay	65 s
Imin	2^{12} ms
Imax	2^{20} ms
Data sent interval	6 pkt/min
RX and TX ratios	100%
TX Range	45 m
Interference Range	70 m

where $Volt$ corresponds to the battery voltage (=3 V), I_{CPU} (=1.8 mAh), I_{RX} (=20 mAh) and I_{TX} (=17.7 mAh) represent the current that has been consumed respectively during the CPU run time T_{CPU}, the radio listen run time T_{RX} and the radio transmit run time T_{Tx} (all expressed in ticks). $Rtimer$ represents the number of ticks per second (=32768 ticks/s).

– **Packet Delivery Ratio (PDR)** is defined as the number of packets that are successfully received by the sink, divided by the number of packets sent by all the nodes to the sink.
– **End-to-end delay** is defined by the period of time between the packet generation by the node source in the application layer and its reception by the sink (in the application layer).
– **Control messages -** In order to reflect the cost and stability of RPL network topology, we trace the number of control messages (i.e., DIS and DIO messages) exchanged in the network.

3.3 Results

In the following, we present our results. In particular, we present the five following well known RPL metrics (that have been surveyed in Sect. 2.2) used to optimize the network lifetime:

– ETX: this is the default metric for RPL which considers the number of retransmissions for each link.
– Energy consumption: we consider the metric proposed by Demicheli [Dem14] in which the path cost represents the sum of the consumed energies, called *ENG-TOT* hereafter.
– Residual energy: we consider the metric proposed by Kamgueu et al. [KNDF13] wherein the path cost is given by the minimum residual energy on the path, called *ENG-MinMax* hereafter.

- Residual energy + ETX: we selected the metric proposed by Chang et al. [CLCL13] in which the path cost is equal to a weighted function integrating ETX and the residual energy, called R hereafter. The two parameters have the same weight in our simulation, i.e., we defined $\alpha = 0.5$.
- Expected lifetime: we choose the metric proposed by Iova et al. [ITN14] because it carefully models the network lifetime, called *ELT* hereafter. However, we do not implemented the expected traffic associated to each node, since it requires to exchange additional control messages to estimate the traffic in the sub-DODAG of each node.

The evaluation using the four criteria previously described for each of the above five metrics is presented in Fig. 2.

Energy consumption We consider first the energy consumed by each routing metric. In our simulation, all the nodes have the same characteristics and in particular have the same initial battery charge of 853 mAh. For a better use with the energy-aware routing metric, we have represented this charge in Cooja on a scale of 255 (as suggested in the RPL standard) and every step of 3.345 mAh decreases the battery level of one.

The percentage of time the radio is on reflects the energy consumed by the RPL protocol. In the first chart of Fig. 2, the energy consumption increases and then decreases as a function of the distance to the sink for the five metrics. This increase is due to the fact that the sink represents a bottleneck; packets are dropped which leads to a higher energy consumption. Then, as expected the energy consumption decreases as a function of the node distance. ETX is the routing metric which has the highest energy consumption, followed by the other energy-aware metrics ENG-MinMax and ELT, then ENG-TOT and R. The R metric achieves the lowest energy consumption, the radio is on at most 1% of the time.

Table 4 presents for each metric the percentage of energy consumed after 5 h of simulations and an extrapolation of the network lifetime expressed in days. It is noteworthy that the short network lifetime is related to the low initial battery charge (of 853 mAh instead of 2000 mAh in real Tmote sky mote platform [NF13]). We observe that the R metric outperforms all the metrics. It achieves a lifetime of 133 days, 5 to 7 times better than the other energy-aware metrics. Note that ENG-TOT has a good network lifetime after R metric, but it is a side effect of the low PDR reached (see below).

Packet Delivery Ratio. As shown in second chart of Fig. 2, the packet delivery ratio decreases as a function of the distance to the sink. ETX metric achieves good results with a PDR between 80 (for the farthest nodes) and 100% (around the sink), since it takes into account the link quality so as to choose the best parent. Comparatively, energy-aware metrics show poorer performance. The ENG-MinMax and ELT have a PDR between 25 (for the farthest nodes) and 95% (around the sink), while ENG-TOT metric has the worse PDR of 5% for the farthest nodes. The best results are given by the R metric with a PDR very close to

Table 4. Metrics lifetime

Time	Metric				
	ENG-MinMax	ENG-TOT	ETX	ELT	R
5 h	1.12%	0.38%	1.15%	0.97%	0.157%
	124698mj	42212mj	166098mj	107822mj	17359mj
Days	19	55	18	22	133

100% for any node. Better results are achieved by the R metric because it takes into account the link quality and the residual energy, contrary to ETX metric for which a certain amount of packets are lost due to the exhausted battery.

End-to-end delay. The third chart of Fig. 2 shows the results related to the end-to-end delay. The latency naturally increases along with the distance to the sink. ETX metric achieves better delays than energy-aware metrics since the link quality is not taken in account by the latter. As expected, we observe that reducing the number of retransmissions decreases the end-to-end delay. For the energy-aware metrics, ENG-TOT is close to ETX metric; ENG-MinMax metric gives the poorer results with a delay between 300 and 500 ms for most of the nodes except for the farthest nodes whose packets are transmitted with 900 ms delay. ELT metric is situated between ENG-TOT and ENG-MinMax metrics. The best end-to-end delays over all metrics are again obtained by the R metric with end-to-end delays smaller than 200 ms for the farthest nodes and 100 ms for the other nodes in the network. It is up to 5 times better than the worst delays achieved by ENG-MinMax metric.

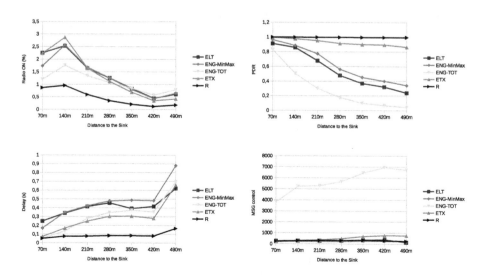

Fig. 2. Experimentation results for each implemented RPL metric.

Control messages. The overhead expressed as the amount of control messages sent by RPL increases slowly as a function of the distance to the sink. We observe a high overhead for ENG-TOT metric comparing to the other metrics; the overhead caused by ELT and ENG-MinMax metrics is relatively stable. The high amount of control messages for ENG-TOT metric is related to the low PDR achieved by this metric. In fact, the amount of control messages exchanged results from route instabilities in the DODAG. This has been analyzed by Boubekeur et al. [BBLM15] which address this problem by reducing the maximum number of children that a node can have. We note that, appart from ENG-TOT metric, the control traffic is negligible compared to the data traffic and as the DODAG stabilizes the control traffic decreases significantly.

4   Conclusion

To minimise energy consumption, guarantee a reliable communication and provide a high delivery ratio is especially challenging in WSN and necessitates to design special mechanisms at the network layer. As a result, RPL was specified by the IETF ROLL working group as a distance vector routing protocol for LLNs. A Destination Oriented Directed Acyclic Graph (DODAG) is constructed by optimizing an objective function which takes into account metrics and constraints for route selection towards the sink.

In this paper, we have presented the first comparative study of energy-aware metrics that have been proposed to enhance RPL and in particular extend the network lifetime. The default metric ETX considers the number of retransmissions and allows to reduce indirectly the end-to-end delay towards the sink. However, it reaches a poor network lifetime, despite it reduces the energy consumed for data transmission at each node. The widely used energy-aware RPL metrics achieve better network lifetime, but the end-to-end delays towards the sink may be important. Some energy-aware metrics, like ENG-TOT metric, have high end-to-end delays because of route instabilities in the DODAG. Moreover, it appears that bi-criteria metrics such as the R metric shows the best performance in terms of network lifetime and end-to-end delays. This can be explained by the fact that the parameters optimized by this metric are not orthogonal. Our results show thatere is need for devising multi-criteria metrics that consider both the lossy nature of the link and the low power of the node to improve communication guarantee in WSNs.

References

[ABP+04] Adya, A., Bahl, P., Padhye, J., Wolman, A., Zhou, L.: A multi-radio unification protocol for IEEE 802.11 wireless networks. In: First International Conference on Broadband Networks, pp. 344–354. IEEE (2004)

[BBLM15] Boubekeur, F., Blin, L., Léone, R., Medagliani, P.: Bounding degrees on RPL. In: 11th ACM Symposium on QoS and Security for Wireless and Mobile Networks, pp. 123–130 (2015)

[CLCL13] Chang, L.-H., Lee, T.-H., Chen, S.-J., Liao, C.-Y.: Energy-efficient oriented routing algorithm in wireless sensor networks. In: IEEE International Conference on Systems, Man, and Cybernetics (SMC), pp. 3813–3818 (2013)

[DEFT11] Dunkels, A., Eriksson, J., Finne, N., Tsiftes, N.: Powertrace: network-level power profiling for low-power wireless networks. Technical report T2011: 05, Swedish Institute of Computer Science (2011)

[Dem14] Demicheli, F.: Designe, Implementation and Evaluation of an Energy RPL Routing Metric. LAP Lambert Academic Publishing, Saarbrucken (2014)

[DPZ04] Draves, R., Padhye, J., Zill, B.: Routing in multi-radio, multi-hop wireless mesh networks. In: 10th Annual International Conference on Mobile Computing and Networking, pp. 114–128. ACM (2004)

[Dun11] Dunkels, A.: The contikimac radio duty cycling protocol. Technical report T2011: 13, Swedish Institute of Computer Science (2011)

[GK12] Gaddour, O., Koub, A.: RPL in a nutshell: a survey. Comput. Netw. **56**(14), 3163–3178 (2012)

[GL10] Gnawali, O., Levis, P.: The ETX objective function for RPL. Technical report, Internet Engineering Task Force (IETF) (2010)

[ITN14] Iova, O., Theoleyre, F., Noel, T.: Improving the network lifetime with energy-balancing routing: application to RPL. In: 7th IFIP Wireless and Mobile Networking Conference, pp. 1–8. IEEE (2014)

[KB06] Koksal, C.E., Balakrishnan, H.: Quality-aware routing metrics for time-varying wireless mesh networks. IEEE J. Sel. Areas Commun. **24**(11), 1984–1994 (2006)

[KNDF13] Kamgueu, P.O., Nataf, E., Djotio, T.N., Festor, O.: Energy-based metric for the routing protocol in low-power and lossy network. In: 2nd International Conference on Sensor Networks, pp. 145–148 (2013)

[MKHC07] Montenegro, G., Kushalnagar, N., Hui, J., Culler, D.: Transmission of IPv6 packets over IEEE 802.15.4 networks. Internet proposed standard RFC, 4944 (2007)

[NF13] Nataf, E., Festor, O.: Accurate online estimation of battery lifetime for wireless sensors network. In: 2nd International Conference on Sensor Networks, pp. 59–64 (2013)

[ODE+06] Osterlind, F., Dunkels, A., Eriksson, J., Finne, N., Voigt, T.: Cross-level sensor network simulation with COOJA. In: 31st IEEE Conference on Local Computer Networks, pp. 641–648 (2006)

[PSC05] Polastre, J., Szewczyk, R., Culler, D.E.: Telos: enabling ultra-low power wireless research. In: Fourth International Symposium on Information Processing in Sensor Networks (IPSN), pp. 364–369 (2005)

[WTB+12] Winter, T., Thubert, P., Brandt, A., Hui, J., Kelsey, R., Levis, P., Pister, K., Struik, R., Vasseur, J.P., Alexander, R.: Rpl: Ipv6 routing protocol for low-power and lossy networks. RFC 6550, Internet Engineering Task Force (IETF) (2012)

[XL13] Xu, G., Lu, G.: Multipath routing protocol for DAG-based WSNs with mobile sinks. In: 2nd International Conference on Computer Science and Electronics Engineering (ICCSEE 2013) (2013)

UAV and Vehicular Networks

Communication and Coordination for Drone Networks

Evşen Yanmaz[1(✉)], Markus Quaritsch[2], Saeed Yahyanejad[3],
Bernhard Rinner[2], Hermann Hellwagner[2], and Christian Bettstetter[1,2]

[1] Lakeside Labs GmbH, 9020 Klagenfurt, Austria
yanmaz@lakeside-labs.com
[2] Institute of Networked and Embedded Systems, University of Klagenfurt,
Klagenfurt, Austria
[3] Joanneum Research, Robotics, Klagenfurt, Austria
http://uav.aau.at

Abstract. Small drones are being utilized in monitoring, delivery of goods, public safety, and disaster management among other civil applications. Due to their sizes, capabilities, payload limitations, and limited flight time, it is not far-fetched to expect multiple networked and coordinated drones incorporated into the air traffic. In this paper, we describe a high-level architecture for the design of a collaborative aerial system that consists of drones with on-board sensors and embedded processing, sensing, coordination, and communication&networking capabilities. We present a multi-drone system consisting of quadrotors and demonstrate its potential in a disaster assistance scenario. Furthermore, we illustrate the challenges in the design of drone networks and present potential solutions based on the lessons we have learned so far.

Keywords: Drones · Unmanned aerial vehicle networks · Wireless sensor networks · Vehicular communications · Cooperative aerial imaging

1 Introduction

Autonomous unmanned aerial vehicles (UAVs), also called drones, are considered with increasing interest in commercial applications, such as environmental and natural disaster monitoring, border surveillance, emergency assistance, search and rescue missions, and relay communications [1–6]. Small quadrotors are of particular interest in practice due to their ease of deployment and low acquisition and maintenance costs.

Research and development of small UAVs has started with addressing control issues, such as flight stability, maneuverability, and robustness, followed by designing autonomous vehicles capable of waypoint flights with minimal user intervention. With advances in technology and commercially available vehicles, the interest is shifting toward *collaborative* UAV systems. Consideration of small vehicles for the aforementioned applications naturally leads to deployment of

© ICST Institute for Computer Sciences, Social Informatics and Telecommunications Engineering 2017
Y. Zhou and T. Kunz (Eds.): ADHOCNETS 2016, LNICST 184, pp. 79–91, 2017.
DOI: 10.1007/978-3-319-51204-4_7

multiple networked aerial vehicles. Especially, for missions that are time critical or that span a large geographical area, a single small UAV is insufficient due to its limited energy and payload. A multi-UAV system, however, is more than the sum of many single UAVs. Multiple vehicles provide diversity by observing and sensing an area of interest from different points of view, which increases the reliability of the sensed data. Moreover, the inherent redundancy increases fault tolerance.

Several projects explored the design challenges of UAV systems for different applications (see [6] and references therein). For civil applications, the design principles of a multi-UAV system, however, still need investigation and remain an open issue. In this paper, we summarize challenges for the design of a system of multiple small UAVs, which have a limited flight time, are equipped with on-board sensors and embedded processing, communicate with each other over wireless links, and have limited sensing coverage. The hardware and low-level control, on-board sensors, and interpretation of sensed data are out of the scope of this paper.

Our main goal is to provide an overview of the design blocks and gain insight toward a general system architecture. We envision that such an architecture can be exploited in the design of multi-UAV systems with different vehicles, applications of interest, and objectives. To illustrate the discussed principles, we introduce a representative network of collaborative UAVs and provide a case study in a real world disaster scenario, where we show how we can support firefighters with our aerial monitoring system. We envision that the lessons learned in our experiments will guide us toward achieving an effective multi-UAV system.

2 Multi-UAV System Overview

Important properties of a multi-UAV system to realize its full potential are its robustness, adaptivity, resource-efficiency, scalability, cooperativeness, heterogeneity, and self-configurability. To achieve these properties, the physical control of individual UAVs, their navigation, and communication capabilities need to be integrated. Design and implementation of these functionalities, by themselves, constitute well-known research topics. Algorithms and design principles proposed by wireless ad hoc and sensor networks, robotics, and swarm intelligence research communities provide valuable insights into one or more of these functionalities as well as combinations of them [7–9].

The past decade observed several projects with UAVs for civil applications (e.g., UAV-NET, COMETS, MDRONES, cDrones, OPARUS, AUGNet, RAVEN testbed, sFly, and MSUAV [6]). A classification can be made for these works, first, on the type of vehicles used, such as helicopters, blimps, fixed-wing UAVs. These vehicles have different sizes, payloads, or flight times, and these differences affect the network lifetime, distances that can be traveled, as well as the communication ranges. Second, a classification can be made on the focus of research, such as design of the vehicles (low-level control) or design of algorithms (path planning, networking, cooperation). Last but not least, the applications for which these networks are deployed also differ. Requirements from the applications add different constraints on the system design have recently been explored [6]. While these projects start from different assumptions, focus on different functionalities, and aim to address different constraints, in principle they satisfy some common design paradigms. Accordingly, one can come up with an intuitive conceptual diagram that captures the essence of multi-UAV systems, which consists of multiple vehicles (*UAVs*) that observe the environment (*sensing*) and implicitly or explicitly communicate the observations to other vehicles (*communication&networking*) to achieve a common goal via planning their paths and sharing tasks (*coordination*). Depending on the application at hand, existing multi-UAV systems focus on the design of one or more of these blocks. For instance, MDRONES focuses on the design of autonomous small-scale UAVs; COMETS consists of sensing, coordination, and communication subsystems [9], and sFly focuses on a combination of UAVs, sensing, and coordination blocks.

The optimal method of integrating these blocks, designing the necessary interaction and feedback mechanisms between them, and engineering an *ideal team* of multiple UAVs are important issues to be addressed.

3   General Collaborative Aerial System Architecture

There are several challenges in developing a system of collaborative UAVs. Especially, the interaction between the hardware, sensing, communication&networking, and coordination blocks of the high-level architecture is still an open issue. In the following, we summarize the desired functionalities in these blocks as well as the associated challenges with an emphasis on communication&networking and coordination.

A multi-UAV system can operate in a centralized or decentralized manner. In a *centralized* system, an entity on the ground collects information, makes decisions for vehicles, and updates the mission or tasks. In a *decentralized* system, the UAVs need to explicitly cooperate on different levels to achieve the system goals and exchange information to share tasks and make collective decisions. Whether centralized or decentralized, what makes a group of single UAVs into a *multi-UAV system* is the implicit or explicit *cooperation* among the vehicles. The UAVs need to

- *observe* their environment,
- evaluate their own observations as well as the information received from other UAVs, and *reason* from them, and
- *act* in the most effective way.

Reasoning can be done at the centralized control entity or on-board the UAVs with full or partial information. The possible *actions* on the other hand are determined by the capabilities of the UAVs and the goal of the multi-UAV system.

The *communication&networking* block is responsible for information dissemination. This block needs to be robust against the uncertainties in the environment and quickly adapt to changes in the network topology. Communication is

not only imperative for disseminating observations, tasks, and control information, but it needs to coordinate the vehicles more effectively toward a global goal
such as monitoring a given area or detecting events in the shortest time, which
are especially important in disaster situations. Some specific issues that need to
be addressed within this block are:

- *Maintaining connectivity*: In a disaster, it is likely that a communication
  infrastructure is lacking. Hence, use of UAVs as relays between disconnected
  ground stations might become imperative. The UAVs have limited communication ranges, are highly mobile, and have scarce energy resources (i.e., the
  UAVs can leave and enter the system based on their battery levels). This block
  has to maintain *connectivity* and the used networking and scheduling protocols
  need to adapt to the highly dynamic environment.
- *Routing and scheduling*: Beyond maintaining connectivity and meeting quality
  of service (QoS) requirements, protocols that can handle or, more desirably,
  that incorporate three-dimensional controlled mobility need to be designed.
- *Communication link models*: Small-scale quadrotors have specific layouts and
  constraints different from fixed-wing UAVs. Link models that capture the
  characteristics of such UAV-UAV and UAV-ground links are needed.

The *coordination* block is responsible for using local observations and observations from other UAVs, mission requirements, and system constraints to organize
the UAVs. In a nutshell, it needs to compute the trajectories of the vehicles and
make decisions on how to allocate tasks to achieve team behavior. The coordination can mean achieving and sustaining rigid formations or can be task distribution among vehicles in a self-organizing manner. Similarly, it can be done at
a local or global level, depending on the mission and capabilities of the vehicles.
Scalability and heterogeneity are also desired in a multi-UAV system, since a
large number of vehicles with different capabilities are expected. Therefore, the
coordination block needs to handle growing numbers of heterogeneous UAVs,
tasks, and possibly mission areas. Some specific issues that need to be addressed
within this block are:

- *Task allocation*: Reasoning and decision making protocols are necessary to
  optimally distribute tasks to individual UAVs or groups of UAVs that can handle uncertain or incomplete information and dynamic missions. Mechanisms
  to define and adapt tasks to the mission requirements or vehicle capabilities
  need to be designed.
- *Path planning*: There are several path planning strategies proposed for ground
  robots and also trajectory designs for formations of robots. More task-
  optimized, communication-aware, three-dimensional path planning methods
  are desired for multi-UAV systems that can handle scarce energy resources
  and heterogeneous vehicles.

While not in the scope of this paper, advances in UAVs and sensing blocks
are also essential. Especially, techniques for efficient data fusion from multiple
heterogeneous sensors, interpretation of the data and feedback mechanisms to the

coordination block, as well as effective obstacle and collision avoidance methods need to be developed for the small-scale vehicle networks.

This general overview can provide some guidelines in the design of multi-UAV systems with different capabilities and with different constraints imposed by different applications. We have been working on a representative system (http:// uav.aau.at/), details of which we present in the following.

4 Collaborative Drones Network

Our collaborative drone system has focused on sensing, communication, and coordination blocks of the general architecture using commercial quadrotors. Sensing capabilities and desired sensor coverage as well as resource limitations of the UAVs (e.g., flight time) are available to the coordination block [10]. The amount of sensor data to be delivered is utilized in the communication&networking block during scheduling of transmissions [11]. We also consider alternative levels of interactions between coordination and communication&networking blocks, where we have the option of centralized coordination with no interaction or decentralized coordination with communication-dependent UAV motion [10,12].

The objective of our system is to monitor a certain area in a given time period and with a given update frequency to assist rescue personnel in a disaster situation. It is designed to capture aerial images and provide an overview image of the monitored area in *real time*. Figure 1 depicts the high-level architecture. The basic operation starts with a user-defined task description, which is used to compute routes for the individual UAVs. The UAVs then fly over the area of interest and acquire images. The images are sent to the ground station and mosaicked to a large overview image. The high-level modules in this architecture are (i) the user interface; (ii) the ground station comprising mission control, mission planning, and sensor data analysis; (iii) a communication infrastructure; and (iv) the UAVs with their on-board processing and sensing capabilities.

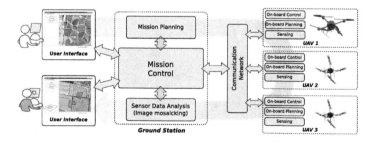

Fig. 1. System architecture: double-headed arrows indicate interactions between individual modules while the shaded arrow in the background indicates the basic operation flow.

The *User Interface* has two main purposes. First, it allows the user to define the high-level tasks to be accomplished by roughly sketching the area to be monitored on a digital map. Additionally, the user can define certain properties such as the required image resolution or update intervals (cf. Fig. 2). Second, it provides the user with the generated mosaicked image with the current position of the UAVs. During mission execution, the user can change the tasks as needed. The *Ground Station* contains three main components. *Mission Control* is the core module of our system. It takes the user's input and dispatches it to the other components. The *Mission Planning* component breaks down the high-level tasks to flight routes for individual UAVs. A flight route contains a sequence of points to visit in world coordinates (GPS coordinates) and certain actions for each waypoint (e.g., take a picture). We have developed both centralized and decentralized mission planning strategies to handle static and dynamic environments [10,12]. Finally, the *Sensor Data Analysis* component mosaics the images from the UAVs into a single large overview image, which is then presented to the user. Since mosaicking is a computationally intensive process, we exploit an incremental approach that promptly shows an overview image to the user while the UAVs are still executing their mission [13]. Our system does not impose special requirements on the communication infrastructure. As a first step, we have used standard IEEE 802.11 (a, n, ac) wireless LAN on-board our UAVs in infrastructure and mesh modes. We have tested methods to improve the wireless links for ground-UAV and UAV-UAV communication in terms of throughput and radio transmission range [14–16].

Before or during the mission, the flight routes (sequence of waypoints) are sent to the UAV's *On-board Control*. The on-board control is not only responsible for the low-level control to stabilize the UAV's altitude, but also to navigate

Fig. 2. User interface showing the observation area (green polygon) and forbidden areas (red polygons) defined by the user on a digital map. (Color figure online)

Fig. 3. Three AscTec Pelican drones taking off for a mission at the University of Klagenfurt campus.

efficiently to the computed waypoints. The *Sensing* module is responsible for capturing images and pre-processing the image data on-board before transmission to the ground station. Pre-processing includes feature extraction, annotation with meta-data, quality checks (to delete blurred images), and multi-resolution encoding.

We support heterogeneous UAVs that provide some minimum functionality, such as autonomous flight and means to specify the navigation waypoints. The computed routes are given in a platform-independent format and the UAV's on-board control translates these generic commands into the UAV-specific low-level commands. In our system, we use quadrotors from Microdrones and Ascending Technologies (Fig. 3). We consider both centralized and decentralized approaches for coordination (planning and sensor analysis) and communication modules. In the decentralized case, planning functionality is migrated from the ground station to the UAVs.

5 Disaster Management Case Study and Lessons Learned

We demonstrated our system in several real-world applications, including assistance during a disaster and documenting the progress of a large construction site. We took part in a county fire service drill with more than 300 firefighters practicing different scenarios. In total, we did five flights over a period of about three hours. The accident scenario was a leaking railroad car with hazardous goods. Our task was twofold: (i) to build an up-to-date overview image of the affected area, which allows the officers in charge to assess the situation and allocate field personnel; and (ii) to frequently update the overview image of the area during the mission to keep track of ongoing ground activities.

We have followed an approach with central control. The routes of all UAVs are pre-computed on the ground station and then sent to the UAVs' on-board

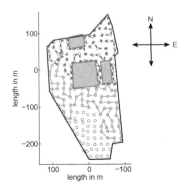

Fig. 4. Mission plan for three UAVs to cover the area of interest (Color figure online)

control for execution. The sensor data analysis, i.e., the overview image mosaicking, is done at the ground station. Figure 2 shows a screen capture of the user interface with the area of interest (green polygon) and three forbidden areas (red polygons). In this case, the forbidden areas are large buildings, which are not of interest. However, the forbidden areas can also mark obstacles or potentially dangerous areas to be avoided. Three UAVs were used to cover the whole area in a given flight time (approx. 15 min). Figure 4 depicts the computed plan using an integer linear programming strategy [10] for the three UAVs (red, blue, and green routes), the circles along the route indicate the positions where pictures are taken. In total, 187 pictures are needed to cover the area of interest (approx. 55 000 m^2) using a camera with a focal length of 28 mm and a flight altitude of 40 m. We have used an average overlap of 50% between neighboring images to create enough redundancy in case some images cannot be used because of low quality and to compute an overview image that meets the quality requirements imposed by the application. The lengths of the three routes are between 950 m and 1 350 m.

One of the challenges we have faced is transmitting the images from the UAVs to the ground station over the 802.11a wireless channel. For this aerial monitoring case study, the required throughput to transmit the images from one UAV is about 2.5 Mbps. The throughput that can be provided over various 802.11 links has been measured in field tests at the University of Klagenfurt (see Table 1). Observe that these results are encouraging the use of UAVs as communication relays between otherwise disconnected ground nodes for this disaster scenario. We use JPEG2000 multi-resolution image compression and apply a scheduled transmission scheme that transmits low-resolution image layers first and additional image layers for higher resolution images as the channel permits [11]. This enables us to immediately present low-resolution images to the user while the UAVs are still on their mission and improve the image quality over time when better quality image layers become available. Figure 5 depicts a part of the overview image computed from a set of about 40 pictures. It covers the main area of activity during this fire service drill.

Table 1. Throughput measurements of aerial Wi-Fi networks for line-of-sight links including air-air (A2A), air-ground (A2G) and ground-air (G2A)

Technology	Link	Topology	Throughput
802.11a ($P_{tx} = 20\,\text{dBm}$)	A2G, G2A,	single-hop	UDP: 14 Mbps (350 m), 29 Mbps (50 m) [14]
	A2A	single-hop	TCP: 10 Mbps (500 m), 17 Mbps (100 m) [15]
802.11n ($P_{tx} = 12\,\text{dBm}$)	A2G, G2A,	single-hop	TCP: 10 Mbps (500 m), 100 Mbps (100 m) [16]
802.11ac($P_{tx} = 10\,\text{dBm}$)	A2G, G2A,	single-hop	TCP: 5 Mbps (300 m), 220 Mbps (50 m) [16]
802.11a + 802.11s ($P_{tx} = 12\,\text{dBm}$) [15]	A2G	multi-hop	1-hop: 5 Mbps (300 m)
(Fixed PHY rate: 36 Mbps)	A2A–A2G	multi-hop	2-hop: 8 Mbps (300 m, infrastructure mode)
			2-hop: 5 Mbps (300 m, mesh mode)

Lessons Learned

In the following, we elaborate on the performance of the overall system and the individual functional blocks.

– The *User Interface* is useful and efficient in defining the tasks. The observation area and forbidden areas can be marked in less than two minutes. Capability to

Fig. 5. Part of the overview image stitched from approx. 40 pictures taken during the firefighter's practice along with the UAV's trajectory (red path). (Color figure online)

view images as they become available is also valuable to the user for assessing the situation and re-planning if necessary.

- The *Mission Planning* component generates a deterministic plan taking into account the user input, available resources, and mission requirements. This phase takes about one minute. A sequence of waypoints with corresponding GPS coordinates and a list of actions are then uploaded to the UAVs. The UAVs are ready for takeoff in about five minutes (including acquiring the current GPS position). The time needed to cover the whole area could be reduced depending on the desired image quality. This can be done by choosing less overlap between neighboring pictures and/or using a higher flight altitude.

- *Sensor Data Acquisition and Analysis.* To compute overview images of high quality, it is important to choose the appropriate equipment. High quality cameras are too heavy for small-scale UAVs. Lightweight cameras, on the other hand, are not as well-developed and require setting parameters such as focus, exposure time, and white balance. Working with dozens of high resolution images requires significant amounts of memory, computing power, and data rate. When mosaicking an overview image of large and structured areas taken from low altitude, it is important to minimize the stitching errors for every single image. State of the art mosaicking tools fail in such cases, because the optimization goal is a visually appealing panorama from single viewpoint. In our mosaicking approach, spatial accuracy is more desirable than the visual appearance.

- The multi-UAV system has to deal with *omni-present resource limitations.* Small-scale platforms impose strong resource limitations on several dimensions. The available on-board energy directly influences the total flight time but also affects the payload and possible flight behavior and flight stability, especially in windy conditions. Limited sensing, processing, and communication performance impede sophisticated on-board reasoning, such as performing real-time collision avoidance or online data analysis. Compensating a resource deficiency in one dimension often impairs another resource dimension. For example, flying at lower speed typically improves the image sensing but reduces the covered area.

- While our centralized planning approach allows for re-planning, a more adaptive coordination, where the UAVs decide their tasks on their own, would be beneficial especially in case of dynamic environments. For instance, if the goal is beyond getting an overview image, e.g., tracking changes and dynamic events, the trajectories cannot be determined beforehand. A distributed and adaptive coordination can also give further capabilities and response options in a disaster management scenario such as the fire drill. The UAVs can be used to track the boundary of the hazardous materials or guide the firefighters and the survivors to safety.

- In our case study, we used WLAN in infrastructure mode; i.e., the sensed data from each UAV is delivered to the ground control, processed there, and feedback can be given to the UAVs with new tasks if necessary. This approach is efficient, since the ground control has more computational power than the UAVs. However, it is limited by the transmission range of the ground control and the UAVs. Either the planned paths need to guarantee that the UAVs do not leave the communication coverage of the ground control or the communication&networking block needs to allow operation in ad hoc mode and maintain multi-hop routes between the UAVs and the ground control [15]. Since the wireless channel fluctuates due to motion and multi-path fading, even if the UAVs are always within the average transmission range, all-time connectivity cannot be guaranteed and this issue has to be dealt with.

6   Conclusion and Open Issues

We illustrated a high-level architecture for the design of multi-UAV systems that consist of vehicles with on-board sensors and embedded processing, and sensing, coordination, and communication&networking blocks. We presented a system consisting of quadrotors and demonstrate its potential in a disaster scenario.

From several real-world tests, we have observed that for effective design of multi-UAV networks, especially for dynamic applications, special focus should be given to better defining the interactions between the design blocks in addition to addressing the issues we summarized specific to communication&networking and coordination blocks. Our current research focus is on addressing those issues and on advanced modeling and designing a multi-UAV system. Our evaluations via simulations as well as real-world experiments so far give us the following insights into the capabilities and requirements of multi-UAV systems:

1. Strong interdependence between design blocks
   - Impact of the UAV platform and sensing block
     The flight dynamics of quadrotor platforms (e.g., tilting, sensitivity to wind and weather) as well as position and orientation of the UAVs have a great impact on the communication links. In addition, processing of the data requires a high computational power, which might not be feasible on UAVs. The routes the UAVs need to fly (regardless of being designed before or during the mission) on the other hand are affected by sensed data quality. The sensors on-board the UAVs can be imperfect or the sensor data analysis might not be able to return a conclusive finding. In such cases, a feedback from sensing needs to be given to coordination module, either to repeat the tasks or adapt the ongoing plan accordingly.
   - Impact of the communication&networking and coordination blocks
     Communications have a direct impact on the coordination of the vehicles, and hence, on the success of the mission. The sensed data need to be delivered to the ground control and new tasks or mission requirements need to be delivered to the UAVs. WLAN 802.11 is limited and can be a bottleneck, especially if large data amounts need to be transferred (e.g., in case of high quality images and real-time video streaming). Large data amounts also have impact on the mission times. Similarly, if the vehicles are coordinated such that the data needs to be collected simultaneously by many vehicles with different points of view, data exchange and processing can become a

challenge. Especially, if the on-board sensor is a camera, registering and mosaicking images from different UAVs, possibly different cameras, with different view angles and altitudes (and hence different resolution) is a great challenge.

2. Efficient evaluation methods
   It is difficult to evaluate the interdependence of the design blocks as well as the overall performance of the multi-UAV systems. Simulators are useful to a certain extent, however, real-life dynamics of the system cannot be fully grasped with only simulators, thus experimental testbeds are required. Several testbeds exist to evaluate multi-UAV control algorithms. However, there is still a lack of testbeds to evaluate the sensing, communication&networking, and coordination algorithms for the multi-UAV systems. At a minimum, the impact of flight dynamics on communication links, sensed data quality, and the impact of small-scale vehicle characteristics such as short flight times and low payload on coordination can be better modeled via input from real-world tests.

3. Autonomy and user interaction
   Finally, most applications require some autonomy in the flight operation of the UAVs. While this may be *preferable* for single-UAV applications, autonomous flight operation is *required* for multi-UAV systems. Autonomy helps to simplify and abstract the user interface. With autonomy and an efficient user interface design, the users can focus on the overall mission and do not need to deal with individual UAVs (as we have demonstrated with our map-based user interface). Methods to achieve high levels of autonomy and low levels of user interaction are required.

While there are still many open issues for achieving an ideal multi-UAV system, we are confident that the applications UAVs are deployed for will keep on increasing and multiple-UAVs will occupy our skies in the near future.

Acknowledgment. This work was funded by ERDF and KWF in the Lakeside Labs projects cDrones (grant 20214/17095/24772) and SINUS (grant 20214/24272/36084).

References

1. Ryan, A., Zennaro, M., Howell, A., Sengupta, R., Hedrick, J.: An overview of emerging results in cooperative UAV control. In: Proceedings of the IEEE Conference on Decision and Control, vol. 1, pp. 602–607, December 2004
2. Kovacina, M., Palmer, D., Yang, G., Vaidyanathan, R.: Multi-agent control algorithms for chemical cloud detection and mapping using unmanned air vehicles. In: Proceedings of the IEEE/RSJ International Conference on Intelligent Robots and Systems, vol. 3, pp. 2782–2788, October 2002
3. Palat, R.C., Annamalai, A., Reed, J.H.: Cooperative relaying for ad hoc ground networks using swarms. In: Proceedings of the IEEE Military Communications Conference, (MILCOM'05), vol. 3, pp. 1588–1594, October 2005
4. Cole, D., Goktogan, A., Thompson, P., Sukkarieh, S.: Mapping and tracking. IEEE Robot. Autom. Mag. **16**(2), 22–34 (2009)

5. Lindsey, Q., Mellinger, D., Kumar, V.: Construction of cubic structures with quadrotor teams. In: Proceedings of Robotics: Science and Systems, Los Angeles, CA, USA, June 2011
6. Hayat, S., Yanmaz, E., Muzaffar, R.: Survey on unmanned aerial vehicle networks for civil applications: a communications viewpoint. IEEE Commun. Surv. Tutorials (2016)
7. Cole, D.T., Thompson, P., Göktoğan, A.H., Sukkarieh, S.: System development and demonstration of a cooperative UAV team for mapping and tracking. Int. J. Robot. Res. **29**, 1371–1399 (2010)
8. Rathinam, S., Zennaro, M., Mak, T., Sengupta, R.: An architecture for UAV team control. In: Proceedings of the IFAC Conference on Intelligent Autonomous Vehicles, July 2004
9. Ollero, A., Lacroix, S., Merino, L., Gancet, J., Wiklund, J., Remuss, V., Perez, I., Gutierrez, L., Viegas, D., Benitez, M., Mallet, A., Alami, R., Chatila, R., Hommel, G., Lechuga, F., Arrue, B., Ferruz, J., Martinez-De Dios, J., Caballero, F.: Multiple eyes in the skies: architecture and perception issues in the COMETS unmanned air vehicles project. IEEE Robot. Autom. Mag. **12**(2), 46–57 (2005)
10. Quaritsch, M., Kruggl, K., Wischounig-Strucl, D., Bhattacharya, S., Shah, M., Rinner, B.: Networked UAVs as aerial sensor network for disaster management applications. Elektrotech. Informationstechnik (e&i) **127**(3), 56–63 (2010)
11. Wischounig-Strucl, D., Rinner, B.: Resource aware and incremental mosaics of wide areas from small-scale UAVs. Mach. Vis. Appl. **26**(7), 885–904 (2015)
12. Yanmaz, E., Kuschnig, R., Quaritsch, M., Bettstetter, C., Rinner, B.: On path planning strategies for networked unmanned aerial vehicles. In: Proceedings of the IEEE Conference on Computer Communications Workshops (INFOCOM WKSHPS), pp. 212–216, April 2011
13. Yahyanejad, S., Rinner, B.: A fast and mobile system for registration of low-altitude visual and thermal aerial images using multiple small-scale UAVs. ISPRS J. Photogrammetry Remote Sens. **104**, 189–202 (2015)
14. Yanmaz, E., Kuschnig, R., Bettstetter, C.: Achieving air-ground communications in 802.11 networks with three-dimensional aerial mobility. In: Proceedings IEEE Conference on Computer Communications (INFOCOM), Turin, Italy, April 2013
15. Yanmaz, E., Hayat, S., Scherer, J., Bettstetter, C.: Experimental performance analysis of two-hop aerial 802.11 networks. In: IEEE Wireless Communications and Networking Conference, April 2014
16. Hayat, S., Yanmaz, E., Bettstetter, C.: Experimental analysis of Multipoint-to-Point UAV communications with IEEE 802.11n and 802.11ac. In: Proceedings of the IEEE International Symposium on Personal Indoor and Mobile Radio Communications (PIMRC), August 2015

Intelligent Wireless Ad Hoc Routing Protocol and Controller for UAV Networks

Abhinandan Ramaprasath[1], Anand Srinivasan[2],
Chung-Horng Lung[1(✉)], and Marc St-Hilaire[1]

[1] Department of Systems and Computer Engineering,
Carleton University, Ottawa, Canada
{abhinandan.ramaprasath, chlung}@sce.carleton.ca,
marc_st_hilaire@carleton.ca
[2] EION Inc., Ottawa, Canada
anand@eion.com

Abstract. In this paper, we propose a novel UAV to UAV communication approach that is based on the concept of Software Defined Networking (SDN). The proposed approach uses a controller as a central source of information to assign routes that maximize throughput, distribute traffic evenly, reduce network delay and utilize all network elements. Simulation results of the proposed methodology were compared to the performance of two common ad hoc routing protocols, namely AODV and OLSR. Performance analysis shows that the proposed methodology improves throughput by over 300%. Simulation results also show a reduction in network delay for delay sensitive packets by nearly 25% and a 26 times increase in packet delivery ratio for packets with higher priority.

Keywords: UAV networks · SDN · Wireless Ad hoc networks · OLSR · AODV · Routing protocols · Performance · Simulation

1 Introduction

Networks of Unmanned Aerial Vehicles (UAVs), also known as Unmanned Aeronautical Ad hoc Networks (UAANET), can be used as an alternative when ground communication is not possible (e.g., disaster recovery, forest fire, etc.). A network of UAVs can span areas of many square kilometers and should be resilient to changes at the ground level. Since a UAV can be moved around the area on demand, a network using UAVs is flexible and scalable. Additional UAVs can be deployed on demand to expand the area of connectivity or replace dying UAVs. The density of UAVs can also be increased in areas where there is a higher demand for network resources. Moreover, this network must be able to provide Internet or network connectivity to users and be able to support transmitting priority packets that need to reach the destination before other packets.

To efficiently utilize network resources and maximize throughput, a central repository, also known as "controller", is used to store and process all the information. A controller monitors the network providing data to network administrators and

© ICST Institute for Computer Sciences, Social Informatics and Telecommunications Engineering 2017
Y. Zhou and T. Kunz (Eds.): ADHOCNETS 2016, LNICST 184, pp. 92–104, 2017.
DOI: 10.1007/978-3-319-51204-4_8

assigning routes within the UAV network for data and control packet transmissions. The controller is the main entity in Software Defined Networks (SDN) and therefore, applying SDN concepts to UAV networks can provide advantages.

In this paper, we propose a new UAV to UAV communication scheme based on the concept of SDN. In our approach, UAVs create a backbone infrastructure that is scalable, provides high network efficiency in terms of bandwidth and latency, and supports packets with different priority levels. To that end, the rest of this paper is organized as follows. Section 2 describes a brief review of relevant work. Section 3 discusses the proposed UAV to UAV communication scheme. Section 4 presents the performance evaluation of our proposed scheme and two commonly used routing protocols. Finally, Sect. 5 concludes the paper.

2   Related Work

Networks of UAVs are not new. The authors in [17] provide a good survey of flying ad hoc networks and point out that the most important design aspect is communication. An efficient communication protocol must be used to enable proper cooperation between UAVs. A typical example of UAV to UAV communication protocol was proposed in [18]. The authors proposed to combine AODV and greedy geographic forwarding (called Reactive-Greedy-Reactive (RGR)) in order to improve delay and packet delivery ratio. However, the protocol inherits from some of the drawbacks of AODV. Moreover, there is no notion of centralize controller and priority levels.

The concept of SDN has been used to improve different kind of networks. For example, the authors in [14] show the benefits of having Wireless Local Area Networks (WLANs) on top of the SDN/OpenFlow infrastructure. A SDN controller can manage the access points (APs) and the way they behave. By switching the routes and the traffic flow pattern beforehand, the authors demonstrated that a SDN-based WLAN can reduce the switching time from 2.934 s to 0.85 s. However, the problem with this approach is that the switch from one AP to the next is made by the client. This means that the network has no control over which user device is connected to which AP. In our problem, there is a need to load balance the network to ensure that users are distributed evenly and the traffic among the UAVs is also evenly distributed.

The approach proposed in [15] explains how nearby controllers can be used to create a scalable architecture using a WiFi SDN network. A similar architecture could be useful for our proposed approach as the UAV network scales. A nearby controller only controls its immediate environment. As the network scales, a hierarchical structure of controllers can be used to obtain information from these nearby controllers. The proposed approach in [15], however, does not solve the problem of finding the optimal route to the destination, providing seamless roaming of users within a network, or providing any energy management techniques. The authors in [16] also proposed a hierarchy of controllers. The proposed approach enables the deployment of UAV based WiFi networks in places where there is no existing WiFi infrastructure. It also enables the transfer the user device over to a different WiFi network when connectivity is available to reduce the load on the UAV network. The approach is proposed to solve the problem of optimizing connectivity in a dense and heterogeneous network.

Optimized Link State Routing version 2 (OLSRv2) [3] offers significant performance improvements and other benefits over its predecessor. OLSRv2 is known to show significant improvements in route discovery times, much better performance in terms of bandwidth and data transfer volume, offers support for discovery of the shortest link to a given node and lower power consumption per node [4, 5]. OLSRv2 still does not guarantee that the most optimal route in terms of both bandwidth and latency will be selected. In addition, multi-route packet transfer is still not possible and there is no provision for priority packet transfer.

Ad hoc On-Demand Distance Vector (AODV) routing is a popular routing protocol for reactive routing [1]. AODV was designed for mobile nodes when the network is constantly changing. In order to efficiently use network resources, we may need to switch routes during operation and therefore fixed routes until they expire would be disadvantageous. AODV does not support priority levels or multipath routing. Many other enhancements to AODV fail to address these concerns [6–9].

In our problem, UAVs are intended to provide Internet access over a city or for a disaster area. In most traditional routing protocols, traffic is typically routed through the same paths, causing those UAVs to drain power quickly, while some UAVs are underutilized. Since battery life is crucial for UAVs, it is important to conserve energy by distributing traffic evenly throughout the network. It is also important to prioritize important packets so that they can be delivered first.

3   Proposed UAV to UAV Communication Scheme

This section presents the proposed approach. Figure 1 shows a simplified ad hoc network. We have a source that needs to transmit information to the destination (Internet) and the packet has to go through a series of hops (as directed by the controller) to reach the sink. The sink then relays the packet to the destination completing the packet transfer. In the next sections, we will discuss different aspect of the communication scheme.

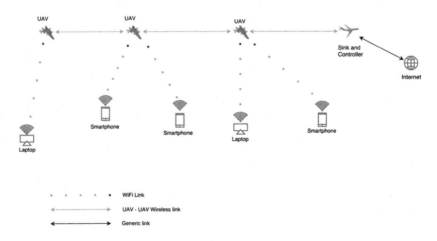

Fig. 1. Base network architecture

3.1   Priority Levels

In order to maintain the stability of the network, packets have to be prioritized. In our approach, since decisions are made by the controller, we need to make sure that these decisions are made and communicated quickly without delay. To facilitate this, packets transmitted through the network are classified into one of the four categories described below:

1. Priority control packets: These are packets with the highest priority and require immediate attention.
2. Non-priority control packets: These are control packets that are sent to the controller at regular intervals (every 30 s for example).
3. Priority data packets: Data packets are categorized into two levels: priority and non-priority. These levels are determined by the UAVs by monitoring the data sent over the network.
4. Non-priority data packets: All data packets that do not fall under the priority data packets fall under the non-priority category. For example, requests to access a web page or streaming music from the Internet.

3.2   Network Setup and Discovery

Initially, we assume that all UAVs are dispatched from a base station. During dispatch, each UAV receives an initial location from the controller. Once all UAVs have reached their destination, each UAV starts transmitting HELLO messages. UAVs will also listens to HELLO messages for a specified amount of time (e.g. 30 s) to discover neighbors.

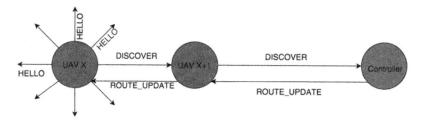

Fig. 2. Network setup and discovery

Once a UAV has discovered its neighbors, it transmits a DISCOVER packet to the controller as shown in Fig. 2. Each DISCOVER packet contains the following 6 fields: (1) UAV ID, (2) number of HELLO messages received, (3) average signal strength (dBm), (4) variance of signal strength (dBm), (5) highest received signal strength (dBm) and (6) lowest received signal strength (dBm).

Once a DISCOVER packet is constructed, all UAVs forward this packet to the controller using AODV. AODV is used because the routes have not been established by the controller yet. The algorithm for UAV network setup is summarized below.

Algorithm: Controller Update
1. Wait for random backoff time
2. Listen for HELLO message
3. Transmit HELLO message 30 times
4. If there was a collision
5.   Backoff for random time
6. End If
7. For each HELLO message received
8.   Record UAV_id and signal strength from HELLO message
9. End for each
10. If the UAV_id already has an entry in the table
11.   Increase readings column and replace signal strength with average value
12. End If
13. Transmit DISCOVER message to the controller with all neighboring UAV info
14. Wait for ROUTE_UPDATE packet from controller
15. Update the routing tables

The controller waits for DISCOVER packets from all the UAVs in the network until a timeout has elapsed. Any UAVs that failed to communicate are marked as lost so the network administrators can take appropriate actions. Using the information within the DISCOVER packets, the controller constructs an adjacency table. Using the average signal strength values, the average modulation scheme for the communication is calculated (explained below). Since the frequency of communication is fixed, the maximum throughput for the given measurements is calculated using the Shannon-Hartley theorem [13]. This way, the controller knows how much information can be transmitted with each link.

Dijkstra's algorithm [10] and the Ford-Fulkerson algorithm [12] are used to calculate next hops for control and data priority levels, respectively. Since control packets are required to reach the destination at its earliest, we need a route with the smallest number of hops to the controller. An alternate route is also calculated by removing all the links from the main route and re-running Dijkstra. This is based on the assumption that there are multiple connections available between the UAVs. If a UAV has no other connections, then the main and alternative hops point to the same UAV in the network. Similarly, the Ford-Fulkerson algorithm is used to calculate the main and alternate routes for data packet priority levels. After the calculations, SYNC_TIMEOUT is set by the network administrator. SYNC_TIMEOUT specifies how often a UAV sends reports to the controller under stable conditions. If the network is not stable, reports are generated immediately. All of the above information is bundled into a ROUTE_UPDATE packet that contains the following information: (1) UAV ID, (2) SYNC_TIMEOUT (in seconds or ms), (3) Optimal route for priority control packets (UAV ID), (4) Alternate route for priority control packets (UAV ID), (5) Optimal route for non-priority control packets (UAV ID), (6) Alternate route for non-priority control packets (UAV ID), (7) Optimal route for priority data packets (UAV ID), (8) Alternate route for priority data packets (UAV ID), (9) Optimal route for non-priority data packets (UAV ID) and (10) Alternate route for non-priority data packets (UAV ID). This process is summarized in the algorithm below.

Algorithm: Network Setup - Controller
1. Wait for DISCOVER packet from all UAVs within a TIMEOUT
2. Update adjacency tables with information from DISCOVER packets
3. Calculate average throughput and update tables
4. Compute the routes for each UAV
5. Specify SYNC_TIMEOUT for each of UAVs in packet
6. Send ROUTE_UPDATE packet to UAVs

3.3 Reports and Route Updates

Once a route is established, communications can take place in the network. Data can flow from the source to the sink using the routes defined by the controller during the setup. In order to keep the network functional and for routes to be periodically updated, it is necessary that each UAV sends regular reports to the controller which processes them as shown by the algorithm below.

Algorithm: Controller Update
1. Wait for update packet
2. Parse update packet information
3. Update adjacency tables according to packet information
4. Recalculate average bandwidth and update table, ignore all UAVs with POWER_LOW flag turned on
5. Re-compute all routes
6. For each route change
7. Generate a UPDATE_REPLY packet to send to that UAV
8. Insert Primary and Alternate routes for each priority level
9. Send packet to corresponding UAV
10. End for each

Other components of the UAV (such as flight control or altimeter) need to be synchronized with the controller. We also developed an algorithm for UAVs to generate an UPDATE packet that is sent to the controller. The algorithm is not presented here due to the page limit.

3.4 UAV-UAV Communication

UAV to UAV communications take place using all the components mentioned above. Once a packet is queued at a UAV, an RTS packet is generated with an Allowance flag according to the priority level of the packet. It waits for a CTS packet with a certain timeout. If the timeout is exceeded, then the sender assumes that the receiving UAV is lost and restarts the RTS/CTS communication using the alternate route. A UAV marked as lost triggers the SYNC_REQUIRED flag to be set forcing the UAV to notify the controller of this change described in the UAV communication algorithm.

Algorithm: UAV Communication
1. Wait for packets in queue
2. Look for packets with highest priority
3. Look up next UAV for that priority level
4. Prepare and send RTS packet
5. If CTS is received within timeout
6.   Transfer all packets in that priority level according to Allowance flag on CTS
7. Else
8.   Set Flag SYNC_REQUIRED to true
9.   Mark destination UAV as unreachable
10.   Lookup table for alternate route for that priority level
11.   Go to Step 7
12. End if

4 Performance Evaluation

This section begins with the simulation of a single source with a single priority level to compare base performance against AODV and OLSR. Then, the complexity of the network is increased by adding more priority levels and more sources. The simulations were performed in a Linux environment using NS-3 version 3.24 (NS-3.24).

4.1 Network Characteristics and Parameters

The controller was implemented as a class called the "UAVController". The frequency used for the UAV-UAV wireless link was 1 GHz and a bandwidth of 100 MHz. Each UAV was positioned before the simulation began and UAVs were made to form connections as soon as the simulation began. The wireless channels follow properties defined by the NS-3 framework, which are listed in Table 1.

To visualize the network during simulations, we have used a tool called NetAnim. NetAnim is a Qt based visualizer tool and is part of the "ns3-allinone" package. This paper uses NetAnim version 3.106.

Table 2 describes the network discovery parameters used in the proposed approach. In this paper, we will generate 30 HELLO messages with intervals of one second between each HELLO message. If there is a collision of HELLO packets, the proposed approach backoff for a random time (any value from 5 ms to 500 ms). Since packets are transmitted at regular intervals if there are no collisions during the transmission of the first HELLO message, it is not likely that collisions will happen during subsequent transmissions.

Table 1. Network characteristics

Parameter	Type	Value(s)
No of UAVs	Int	1, 2, 5, 10, 20, 100, 500, 1000
Frequency of WiFi communication	GHz	2.4, 5
Frequency of UAV-UAV communication	MHz	1000
Bandwidth of UAV-UAV communication	MHz	100 (950 MHz - 1050 MHz)
Full/Half Duplex	-	Full Duplex
Radio Technology	-	OFDM
Modulation Method	-	Adaptive Modulation
Supported Modulation schemes	-	BPSK, QPSK, 16 - QAM, 64 - QAM, 256 – QAM
No of Radio modules in each UAV	Int	2
Link latency	Milliseconds	2
Weight of priority control packet	-	4
Weight of non-priority control packet	-	3
Weight of priority data packet	-	2
Weight of non-priority data packet	-	1
Max communication range for UAV-UAV radio	Meters	50
Max communication range for WiFi radio	Meters	30
Threshold for 256-QAM modulation scheme	Meters	5
Threshold for 64-QAM modulation scheme	Meters	10
Threshold for 16-QAM modulation scheme	Meters	20
Threshold for 8-QAM modulation scheme	Meters	30
Threshold for QPSK modulation scheme	Meters	40

Table 2. Proposed approach - discovery and setup

Parameter	Type	Value(s)
INIT TIMEOUT	Seconds	90
No of HELLO Messages	Int	30
HELLO INTERVAL	Milliseconds	1000
HELLO COLLISION INTERVAL	Milliseconds	Rand (5, 500)
NETWORK REDISCOVERY	Seconds	30

4.2 Single Priority Test

In this setup, there is a single source that produces packets and all packets have the same priority level. None of the intermediary nodes generate traffic; they are present only to relay the packets generated by the source to the sink. The source is set to

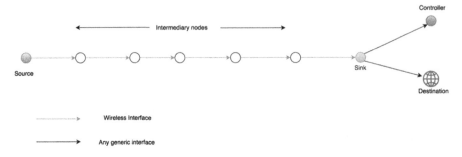

Fig. 3. Linear topology for simulation

produce packets at the rate of 100 Mbps with 5 intermediate hops (see Fig. 3). We set the priority level to the lowest to compare the proposed approach against AODV and OLSR.

Each wireless link is capped at 16 Mbps maximum throughput. The UAVs are placed sufficiently close to each other so that the modulation scheme is no longer a factor. Throughput was measured in the sink in intervals of one second. In order to average out any errors and randomness, the experiment was run 10 times and the average was calculated and plotted in Fig. 4. As can be seen, the throughput performance of the proposed approach is comparable to AODV and OLSR. The average throughput for AODV and OLSR were 15.065 Mbps and 14.892 respectively, while the average throughput of the proposed approach was 14.452 Mbps. This is in line with the expected results for the proposed approach, i.e., the proposed approach has a slightly lower throughput in this case due to the overhead of transmitting update packets.

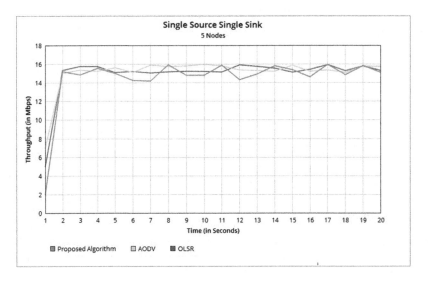

Fig. 4. Results for a single source single sink linear topology

4.3   Multiple Priority Levels

We conducted the same experiment as conducted above with packets of different priority levels to test the packet drop among different priority levels (single source network with linear topology). The program was designed to accommodate 4 UDP packet generators in this scenario producing traffic at 25 Mbps each.

For AODV and OLSR, the results were almost identical to each other as shown in Fig. 5. For AODV and OLSR, the percentage of packets dropped was almost equally distributed amongst the priority levels. The variation is due to the Random Early Detection (RED) queueing mechanism [11]. For the proposed approach, the percentage of priority packets dropped as a percentage of total packets is significantly lower than AODV and OLSR, which is demonstrated in Fig. 6. Instead of dropping higher priority packets, the algorithm dropped the lowest priority packets more often. As shown in Fig. 6, a packet marked as a priority control packet is 20 times less likely to be dropped than a packet marked as a non-priority data packet when the proposed approach is used. This is due to priority differentiation built into the network.

In conclusion, for the single source linear topology, the proposed approach has a comparable throughput but drops less important traffic. This behavior was expected since, as mentioned, there is only a single path and the proposed approach was designed to perform well when multiple paths exist.

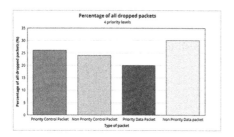

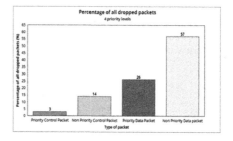

Fig. 5. Dropped packets for each priority level-AODV and OLSR

Fig. 6. Dropped packets for each priority level-proposed approach

4.4   Dual Source, 4-Tier Network

Now, let us consider a 4-tier network with two-source UAVs that produce non-priority data packets at the rate of 40 Mbps, control packets at the rate of 10 Mbps and non-priority control packets at the rate of 10 Mbps. An illustration is shown in Fig. 7. A total of 35 UAVs (including the controller) were generated and placed in a grid.

Figure 8 displays the average throughput per second at the sink for both sources combined. On average, the proposed approach provides approximately 15% greater throughput over AODV and OLSR. This is because different packets have different routes. AODV and OLSR have a fixed route for sending all types of packets. The throughput of priority packets however is interesting. We define higher priority packets

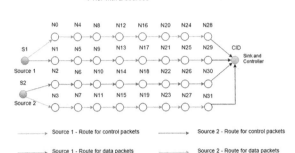

Fig. 7. Experiment with 4 tier network consisting of 2 sources and 1 sink

as any packet with a priority level greater than non-priority data packets. The proposed approach transmitted ∼95% of all higher priority packets when compared to at most 55% in AODV and OLSR, as shown in Fig. 9.

Figure 10 shows the average delay of priority packets. Evident fluctuations in delay are due to the Random Early Detection (RED) queuing mechanism in a single queue for all packets when we use AODV and OLSR. However, for the proposed approach, the delay is consistently low with minimal variations. This consistency is due to the refined MAC protocol that prioritizes transmission of priority packets before other packets. A priority control packet is 4 times more likely to be transmitted than a non-priority data packet. For priority control packets, the routes are calculated using Dijkstra's algorithm which guarantees the shortest path to the controller.

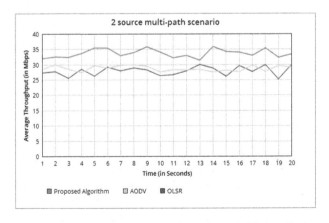

Fig. 8. Combined throughput of sources 1 and 2 (in Mbps)

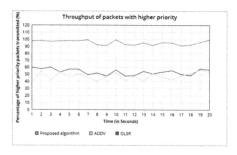
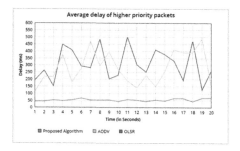

Fig. 9. Priority packets successfully transmitted

Fig. 10. Average delay of higher priority packets

5 Conclusions

In this paper, we proposed a novel UAV-UAV communication scheme based on the concept of SDN. The UAV backbone is scalable, provides high network efficiency in terms of bandwidth and latency, and supports packets with different priority levels.

The proposed methodology relies on the use of a controller acting as the central hub that monitors all the control information. This hub is used for calculation of routes and to monitor information with regards to the network, which is hard to do in a typical ad hoc network. The routes communicated by the controller provide a means to distribute traffic throughout the network evenly, hence increasing the efficiency of the global network. The important contributions of the proposed approach are as follows:

- Design a more scalable approach for UAV-UAV communication with support for packet prioritization.
- Increase overall throughput of network by evenly distributing traffic throughout the network.
- Find and transmit via faster routes for packets with low delay tolerance, i.e., priority packets. Reduce latency by prioritizing transmission of packets with higher priority.

The simulation results showed that the proposed method provided up to four times as much throughput and reduce latency to less than 1/4 for critical packets compared to AODV and OLSR. High throughput is essential for delivering a jitter free experience for the user and low latency for high priority packets is crucial for maintaining the robustness and stability of the network.

References

1. Perkins, C.E., Belding-Royer, E., Das, S.: Ad hoc On-demand Distance Vector (AODV) routing. RFC 3561 (2003)
2. Clausen, T., Jacquet, P., Adijh, C., Laouiti, A., Minet, P., Muhlethaler, P., Qayyum, A., Viennot, L.: Optimized Link State Routing Protocol (OLSR). RFC Standard 3626 (2003)

3. Barz, C., et al.: NHDP and OLSRv2 for community networks. In: Proceedings of International Conference on Wireless and Mobile Computing, Networking and Communications, pp. 96–102 (2013)

4. Clausen, T., Hansen, G., Christensen, L., Behrmann, G.: The optimized link state routing protocol, evaluation through experiments and simulation. In: Proceedings of IEEE Symposium on Wireless Personal Mobile Communications, pp. 34–39 (2001)

5. Vara, M., Campo, C.: Cross-layer service discovery mechanism for OLSRv2 mobile ad hoc networks. Sensors 15(7), 17621–17648 (2015)

6. Royer, E.M., Perkins, C.E.: Multicast operation of the Ad hoc On-demand Distance Vector routing protocol. In: Proceedings of International Conference on Mobile Computer and Network (1999)

7. Zapata, M.G.: Secure Ad hoc On-demand Distance Vector routing. ACM SIGMOBILE Mob. Comput. Commun. Rev. 6(3), 106–107 (2002)

8. Marina, M.K., Das, S.R.: On-demand multipath distance vector routing in ad hoc networks. In: Proceedings of IEEE 9th International Conference on Network Protocols, pp. 14–23 (2001)

9. Narasimhan, B., Balakrishnan, R.: Energy Efficient Ad hoc On-demand Distance Vector (EE-AODV) routing protocol for mobile ad hoc networks. Int. J. Adv. Res. Comput. Sci. 4 (9) (2013)

10. Cormen, T.H., Leiserson, C.E., Rivest, R.L., Stein, C.: Section 24.3: Dijkstra's Algorithm. Introduction to Algorithms, 2nd edn., pp. 595–601. MIT Press, Cambridge (2001)

11. Floyd, S., Jacobson, V.: Random early detection gateways for congestion avoidance. IEEE/ACM Trans. Netw. 1(4), 397–413 (1993)

12. Ford, L.R., Fulkerson, D.R.: Maximal flow through a network. Can. J. Math. 8, 399 (1956)

13. Hartley, R.V.L.: Transmission of information. Bell Syst. Tech. J. 7, 535–563 (1928)

14. Monin, S., Shalimov, A., Smeliansky, R.: Chandelle: smooth and fast WiFi roaming with SDN/OpenFlow. A Poster Presented at the US Ignite (2014)

15. Schulz-Zander, J., Sarrar, N., Schmid, S.: Towards a scalable and near-sighted control plane architecture for WiFi SDNs. In: Proceedings of the 3rd Workshop on HotSDN (2014)

16. Ali-Ahmad, H., et al.: CROWD: an SDN approach for DenseNets. In: Proceedings of the 2nd European Workshop on Software Defined Networks, pp. 25–31 (2013)

17. Bekmezei, I., Sahingoz, O.K., Temel, S.: Flying ad-hoc networks: a survey. Ad Hoc Netw. 11, 1254–1270 (2013)

18. Shirani, R., St-Hilaire, M., Kunz, T., Zhou, Y., Li, J., Lamont, L.: Combined reactive-geographic routing for unmanned aeronautical ad-hoc networks. In: IWCMC (2012)

Theoretical Analysis of Obstruction's Influence on Data Dissemination in Vehicular Networks

Miao Hu[✉], Zhangdui Zhong, Minming Ni, and Zhe Wang

State Key Laboratory of Rail Traffic Control and Safety,
Beijing Jiaoting University, Beijing, China
humiao@outlook.com

Abstract. In vehicular networks (VNs), the radio propagation characteristics between two vehicles is greatly affected by the intermediate vehicles as obstruction, which has also been verified in many measurements. This property will definitely influence the routing protocol design in VNs, where the estimation of the one-hop transmission distance is of utmost importance on the relay selection. However, to the authors' best knowledge, the obstruction's influence has not been taken into consideration theoretically. In this paper, we propose an analytical model on the obstructed light-of-sight (OLOS) transmission distance. Based on a probabilistic method, the probability density function (PDF) of the one-hop OLOS transmission distance is obtained. Monte Carlo simulations are conducted to verify our proposed analytical model.

Keywords: Light-of-sight · Obstruction · Transmission range · Vehicular networks

1 Introduction

In vehicular networks (VNs), the received signal strength is easily affected by obstructions, like the buildings, trees or the vehicles between the transceivers. With obstructions, the line-of-sight (LOS) transmission will degrade into obstructed line-of-sight (OLOS) transmission. Many experiments had shown that obstructions cause significant impact on the channel quality, where an additional 10 to 20 dB attenuation can be found on the received signal strength [1,2]. Therefore, it is of great importance to study the influence of possible obstructions on the system design and performance evaluation.

The radio range differs between LOS and OLOS scenarios because of different attenuation degrees, where the OLOS radio range is much shorter. The shorter the radio range is, the less the one hop transmission distance is, which is important for routing protocol design in data dissemination. For example, the end-to-end delay might increase in obstructed scenario because of more transmission hop count requirement. However, not much researches considered the influence of obstructions on the protocol design or performance evaluation, especially in the theoretical aspect, which will be the focus of this paper.

© ICST Institute for Computer Sciences, Social Informatics and Telecommunications Engineering 2017
Y. Zhou and T. Kunz (Eds.): ADHOCNETS 2016, LNICST 184, pp. 105–116, 2017.
DOI: 10.1007/978-3-319-51204-4_9

Extensive experiment works had shown that the vehicles as obstructions between transceivers can cause obvious decrease of signal power [2–6]. Based on this observation, many researchers considered the obstructed light-of-sight (OLOS) transmission in the simulations for routing protocol verification [7]. However, to the authors' best knowledge, although some works had conducted analysis for the routing performance in VNs [8], the analytical model for the OLOS circumstance is still an open issue.

In this paper, we analyze and model the influence of vehicle as obstruction on the one hop link transmission range in a two-lanes highway scenario given the traffic density information. We use the widely accept condition than the OLOS radio range is shorter than the LOS radio range, which is taken into consideration. With a dedicated routing protocol, the one hop transmission range changes with the vehicular density, which is modeled using a probabilistic method. Our proposed theoretical framework can give the probability density function of the one hop link transmission range.

The remainder of this paper is organized as follows. The related work of the discussed issue is introduced in Sect. 2. Section 3 presents the model hypotheses and definitions. The analysis architecture is proposed in Sect. 4. In Sect. 5, simulations are carried out to verify the accuracy of the proposed anaytical model. Last but not the least, Sect. 6 concludes the paper and proposes some possible direction in future.

2   Related Work

Many researchers have conducted experiment on the influence of obstructions on the radio propagation characteristics for VNs. The influence of the buildings, especially in the intersection scenario, on the signal attenuation is target in [3,4], where an obvious decrease of signal power can be found. In the straight road scenario, signal obstructed by vehicles between the transceivers is the main target. Meireles *et al.* [5] found that a single obstacle can cause a drop of over 20 dB on received signal strength when two cars communicate at a distance of 10 m. Measurements were also conducted by placing a bus between two cars acting as an obstruction, and found that this obstruction can create an additional 15- to 20-dB attenuation [2,6]. In [9], the propagation path losses are presented based on the uniform theory of diffraction in the OLOS cases, with several intermediate vehicles, for the inter-vehicle communications in the 60-GHz band. Many other literatures also found such obvious signal strength attenuation from different measurement campaigns [10–14].

In recent years, some literatures also considered the influence on system performance evaluation and routing protocol design. Some researchers focused on developing simulation framework for a more realistic fading environment description [7]. However, although the results from these simulation frameworks can be more accurate, the time consumption problem cannot be neglected. On the other hand, it is an accurate and effective analytical model that can provide more clear understanding for the fundamental trade-off between obstruction features (e.g. vehicles' position) and performance expectation (e.g. transmission distance, hop count, throughput, and delay etc.). The influence of radio range on the system performance has been modeled analytically in some work [8]. However, the influence of some obstructions on the signal attenuation is not taken into consideration. In recent years, some researchers conducted analysis with obstructed radio range. Chen *et al.* [15] modeled the joint effects of radio environment and traffic flow on link connectivity to investigate the relation between the obstruction probability and inter-vehicle connectivity probability. However, they did not give way to calculate the obstruction probability and not derive the influence of obstruction on the transmission distance. As far as we know, no literatures derived the transmission distance distribution with obstructions in the theoretical aspects.

3   System Model

All the vehicles are assumed moving on a highway with two lanes. The vehicle's location can be obtained by the Global Position System (GPS) unit, which is assumed to be installed in each vehicle. A vehicle can know all its neighbour's position information from the continuous exchanged beacon information or triggering information. A transmitter or relay will choose the furthest vehicle as the next hop relay according to the aforementioned assumption. Vehicles are distributed along the road in accordance with a spatial one-dimensional Poisson point process (1-PPP), which has been deemed to be appropriate under free flow conditions. The width of road is ignored and the traffic flows are independent of each other. All drivers tend to maintain a constant spacing with their leader based on the car-following model, where all vehicles in the same lane have the same velocity.

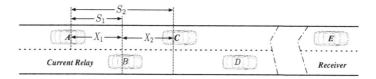

Fig. 1. An example for the adopted relay selection policy

Suppose that no static infrastructure exists or incomplete covered dedicated base stations are built, therefore, many transmission, especially when the transceiver distance is long, should be finished through multi-hop transmission. The baseline routing protocol chosen for this paper is the Greedy Perimeter Stateless Routing (GPSR) [16], based on which many work proposed some revised versions. The principle for this kind of protocols is that the furthest vehicle in current relay's radio range will be selected as the next hop relay. For example, in Fig. 1, suppose that vehicles, B, C and D are neighbours of the current relay vehicle A. Therefore, the vehicle D will be selected as the next hop relay, which is also named as the furthest vehicle in A's radio range. The distance between two relay vehicles is defined as *one-hop transmission distance*.

For the analysis tractability, the LOS radio range R_{LOS} and the obstructed radio range R_{OLOS} is assumed to be a constant.

4   Theoretical Analysis

Since the analysis is conducted on a two-lanes scenario, the derivation of one-hop transmission distance distribution is also divided into two cases: the intra-lane one-hop transmission distance and the inter-lane one-hop transmission distance. After obtaining both of the distribution of the single lane's one-hop transmission distance distribution, the two-lanes one-hop transmission distance distribution will be derived at the end of this section.

4.1   One-Hop LOS Transmission Distance

Let X_i denote the inter-vehicle distance between the $(i-1)$-th and the i-th nearest vehicle in the neighbouring vehicle set in the intra-lane scenario, which can be illustrated in Fig. 1. Since the inter-vehicle distance X_i are positive, independent, identically distributed, random variables, the n vehicles cumulative distance S_n is defined as:

$$S_n = \sum_{i=1}^{n} X_i, \ n \geq 1. \tag{1}$$

The cumulative density function (CDF) of the one-hop LOS transmission distance X_{L} will be derived as following. First, the probability of $F_{X_{\mathrm{L}}}(0)$ can be represented as:

$$F_{X_{\mathrm{L}}}(0) = \Pr\{X_{\mathrm{L}} = 0\} = \Pr\{X_1 > R_{\mathrm{L}}\}. \tag{2}$$

We have the density function of inter-vehicle distance X_1 as:

$$F_{X_1}(x) = 1 - e^{-\lambda x}. \tag{3}$$

and

$$f_{X_1}(x) = \lambda_1 e^{-\lambda x}. \tag{4}$$

Therefore, $F_{X_{\mathrm{L}}}(0)$ can be obtained as:

$$F_{X_{\mathrm{L}}}(0) = e^{-\lambda R_{\mathrm{L}}}. \tag{5}$$

Otherwise, when the one-hop LOS transmission distance is greater than zero, $F_{X_{\mathrm{L}}}(x)$ can be obtained as:

$$
\begin{aligned}
F_{X_{\mathrm{L}}}(x) &= F_{X_{\mathrm{L}}}(0) + \Pr\{x > 0, X_{\mathrm{L}} \leq x\} \\
&= e^{-\lambda R_{\mathrm{L}}} + \Pr\{X_1 \leq R_{\mathrm{L}}\}\Pr\{S_N \leq x, S_{N+1} > R_{\mathrm{L}}\} \\
&= e^{-\lambda R_{\mathrm{L}}} + (1 - e^{-\lambda R_{\mathrm{L}})} \sum_{n=0}^{\infty} \Pr\{N_{[0,x]} = n\} \cdot \Pr\{N_{[x,R_{\mathrm{L}}]} = 0\} \\
&= e^{-\lambda R_{\mathrm{L}}} + (1 - e^{-\lambda R_{\mathrm{L}}}) \sum_{n=0}^{\infty} \frac{[\lambda x]^n}{n!} e^{-\lambda(x)} \cdot e^{-\lambda(R_{\mathrm{L}}-x)} \\
&= e^{-\lambda R_{\mathrm{L}}} + (1 - e^{-\lambda R_{\mathrm{L}}})e^{-\lambda(R_{\mathrm{L}}-x)}.
\end{aligned}
\tag{6}
$$

Consequently, we have the probability density function (PDF) of X_{L} as:

$$
f_{X_{\mathrm{L}}}(x) = \lambda(1 - e^{-\lambda R_{\mathrm{L}}})e^{-\lambda(R_{\mathrm{L}}-x)}.
\tag{7}
$$

In summarization, the CDF of the one-hop LOS transmission distance can be obtained as:

$$
F_{X_{\mathrm{L}}}(x) = \begin{cases} e^{-\lambda R_{\mathrm{L}}}, & x = 0 \\ e^{-\lambda R_{\mathrm{L}}} + (1 - e^{-\lambda R_{\mathrm{L}}})e^{-\lambda(R_{\mathrm{L}}-x)}, & \text{otherwise} \end{cases}.
\tag{8}
$$

4.2 One-Hop OLOS Transmission Distance

As for the CDF of the one-hop OLOS transmission distance $F_{X_{\mathrm{O}}}(x)$, the derivation should be split into three cases, that is $x = 0$ (case I), $0 < x \leq R_{\mathrm{O}}$ (case II), and $R_{\mathrm{O}} < x \leq R_{\mathrm{L}}$ (case III), respectively.

Case I: The CDF value $F_{X_{\mathrm{O}}}(0)$ can be obtained similarly as that for the LOS circumstance, we have

$$
F_{X_{\mathrm{O}}}(0) = e^{-\lambda R_{\mathrm{L}}}.
\tag{9}
$$

Case II: When $0 < x \leq R_{\mathrm{O}}$, it means that at least one vehicle existing in R_{O} distance. Therefore, we have

$$
\begin{aligned}
F_{X_{\mathrm{O}}}(x) &= F_{X_{\mathrm{O}}}(0) + \Pr\{X_1 \leq R_{\mathrm{O}}, X_{\mathrm{O}} \leq x\} \\
&= e^{-\lambda R_{\mathrm{L}}} + \Pr\{X_1 \leq R_{\mathrm{O}}\}\Pr\{S_{N-1} \leq x - X_1, S_N > R_{\mathrm{O}} - X_1 | x \leq R_{\mathrm{O}}\} \\
&= e^{-\lambda R_{\mathrm{L}}} + (1 - e^{-\lambda R_{\mathrm{O}}}) \sum_{n=0}^{\infty} \Pr\{N_{[0,x]} = n\} \cdot \Pr\{N_{[x,R_{\mathrm{O}}]} = 0\} \\
&= e^{-\lambda R_{\mathrm{L}}} + (1 - e^{-\lambda R_{\mathrm{O}}}) \sum_{n=0}^{\infty} \frac{(\lambda x)^n}{n!} e^{-\lambda x} \cdot e^{-\lambda(R_{\mathrm{O}}-x)} \\
&= e^{-\lambda R_{\mathrm{L}}} + (1 - e^{-\lambda R_{\mathrm{O}}})e^{-\lambda(R_{\mathrm{O}}-x)}.
\end{aligned}
\tag{10}
$$

Take the derivation at x, we can obtain the corresponding PDF as:

$$f_{X_O}(x) = \lambda(1 - e^{-\lambda R_O})e^{-\lambda(R_O - x)}. \tag{11}$$

Case III: When $R_O < x \leq R_L$, it means that the first vehicle's position X_1 is greater than R_O. In this case, the first vehicle will be selected as the next hop relay, and we have

$$\begin{aligned} F_{X_O}(x) &= F_{X_O}(R_O) + \Pr\{X_1 > R_O, X_O \leq x\} \\ &= F_{X_O}(R_O) + \Pr\{R_O < X_1 \leq x\} \\ &= 1 + e^{-\lambda R_L} - e^{-\lambda x}. \end{aligned} \tag{12}$$

Take the derivation at x, we can obtain the corresponding PDF as:

$$f_{X_O}(x) = \lambda e^{-\lambda x}. \tag{13}$$

In summarization, the CDF of the one-hop OLOS transmission distance can be obtained as:

$$F_{X_O}(x) = \begin{cases} e^{-\lambda R_L}, & x = 0 \\ e^{-\lambda R_L} + (1 - e^{-\lambda R_O})e^{-\lambda(R_O - x)}, & 0 < x \leq R_O \\ 1 + e^{-\lambda R_L} - e^{-\lambda x}, & \text{otherwise} \end{cases} \tag{14}$$

Meanwhile, the PDF of the one-hop OLOS transmission distance can be obtained as:

$$f_{X_O}(x) = \begin{cases} \lambda(1 - e^{-\lambda R_O})e^{-\lambda(R_O - x)}, & 0 \leq x \leq R_O \\ \lambda e^{-\lambda x}, & \text{otherwise} \end{cases} \tag{15}$$

5   Performance Evaluation

In this section, Monte Carlo simulations are conducted to verify our proposed analytical model.

5.1   Simulation Setup

In our simulations, vehicles move within a fixed region of a two-way highway road segment with the length of L. The vehicular density is assumed to be a constant value for a relative short time period, which is denoted as λ vehicles per second (vehs/s). To have a fixed number of vehicles in the target road segment, we assume that the exit vehicle will enter the highway immediately and start to move toward the opposite direction [17]. The default value of major parameters for this simulation is shown in Table 1.

Simulations were run using different parameters and system settings. The performance analysis is designed to compare the effects of different parameters, such as the LOS radio range, the OLOS radio range, and the vehicular density etc. For each simulation parameter set, the values of the one-hop transmission distance distribution are obtained by collecting a large number of samples such that the confidence interval is reasonably small. In most cases, the 95% confidence interval for the measured data is less than 10% of the sample mean.

Table 1. Default Value of the Simulation Parameters

Parameter	Description	Value
R_L	The LOS radio range	250 m
R_O	The OLOS radio range	150 m
λ	The vehicular density	0.01 vehs/s
N	The number of Monte Carlo simulations	10^5
N_h	The number of histogram	10

5.2   LOS Scenario

Figure 2 depicts the PMF of the one-hop LOS transmission distance, where the results are compared between the simulations and our analytical model. Since the statical results are from the extensive Monte Carlo simulations, only an estimation of probability mass function (PMF) can be obtained. For the tractability of the comparison, the PMF value is estimated from the proposed analytical model with a integral function. As can be seen from Fig. 2, results from our proposed analytical model matches with well with that from the simulations, which is verified using the chi-square goodness fit test. In general, the chi-square test statistic is of the form

$$\chi^2 = \sum_{i=1}^{N_h} \frac{\rho_i^{ana} - \rho_i^{simu}}{\rho_i^{simu}} \tag{16}$$

where N_h denotes the number of histogram, ρ_i^{simu} and ρ_i^{ana} represent the values from the Monte Carlo simulations and the proposed analytical model, respectively. Based on the chi-square test statistic theory, $\chi^2 = 2.0311$, which is less than 55.758, the threshold value corresponding to the 0.05 significance level.

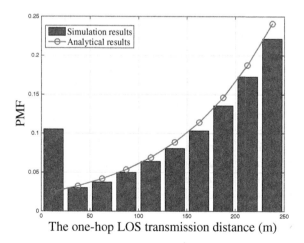

Fig. 2. The PMF of one-hop LOS transmission distance

That is, we can accept the hypothesis at the 0.05 significance level that the PDF from our proposed one-hop LOS transmission model fits with that from the statistical results with Monte Carlo simulations. Moreover, it's a significant tendency that the PMF value increases with the distance since the adopted routing protocol tries to select the furthest vehicle in its radio range as the next-hop relay vehicle.

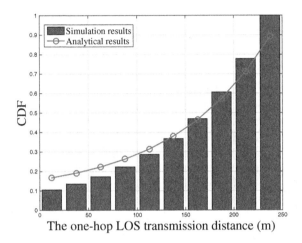

Fig. 3. The CDF of one-hop LOS transmission distance

Figure 3 conducts a comparison on the CDF of one-hop LOS transmission distance. Based on the chi-square test statistic theory, $\chi^2 = 1.4$, which is less than 55.758, the threshold value corresponding to the 0.05 significance level. We can accept the hypothesis at the 0.05 significance level that the CDF from our proposed one-hop LOS transmission model fits with that from the statistical results with Monte Carlo simulations.

5.3   OLOS Scenario

Figure 4 compares the PMF of the one-hop OLOS transmission distance between Monte Carlo simulations and our proposed analytical model. First, compared to the simulation results from that in Fig. 2, we can see that curve shape is quite different. By considering the intermediate vehicle's obstruction, the PMF of the one-hop OLOS transmission distance shows a significant fluctuation at the OLOS radio range. Although with one singular point, our analytical model can better describe the actual circumstance. Based on the chi-square test statistic theory, $\chi^2 = 1.5714$, which is less than 55.758, the threshold value corresponding to the 0.05 significance level. That is, we can accept the hypothesis at the 0.05 significance level that the PDF from our proposed one-hop OLOS transmission model fits with that from the statistical results with Monte Carlo simulations.

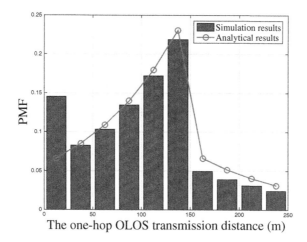

Fig. 4. The PMF of one-hop OLOS transmission distance

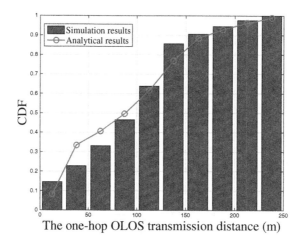

Fig. 5. The CDF of one-hop OLOS transmission distance

Figure 5 conducts a comparison on the CDF of one-hop OLOS transmission distance. Based on the chi-square test statistic theory, $\chi^2 = 1.8686$, which is less than 55.758, the threshold value corresponding to the 0.05 significance level. We can accept the hypothesis at the 0.05 significance level that the CDF from our proposed one-hop OLOS transmission model fits with that from the statistical results with Monte Carlo simulations.

5.4 Comparison Between LOS and OLOS Scenarios

Figure 6 presents the average one-hop transmission distance verse OLOS radio range R_O. Since the LOS circumstance is assumed no affected by the obstructions,

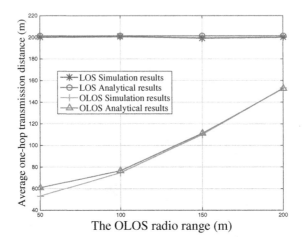

Fig. 6. The average one-hop transmission distance verse OLOS radio range

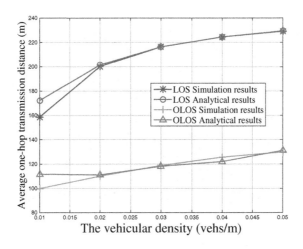

Fig. 7. The average one-hop transmission distance verse vehicular density

the average one-hop transmission distance keep stable with different OLOS radio range. In contrast, in the OLOS circumstance, the one-hop OLOS transmission distance increase with the OLOS radio range. Overall, in both LOS and OLOS circumstances, the simulation results of average one-hop transmission range match well with that from our proposed analytical model.

Figure 7 is plotted to show the influence of the vehicular density on the one-hop transmission distance. Both LOS and OLOS circumstances show a increasing tendency with the vehicular density. Again, in both LOS and OLOS circumstances, the simulation results of average one-hop transmission range match well with that from our proposed analytical model.

6   Discussion

This paper proposed an analytical model for the obstructed light-of-sight (OLOS) scenario in vehicular networks (VNs). The influence of the OLOS/LOS radio range and the vehicular density is carefully derived for the probability density function (PDF) of the one-hop transmission distance. With the PDF of the one-hop link distance, the traditional routing protocol can be modified to adapt to the real scenario, which will be one of our future works. Moreover, the performance evaluation is conducted with Monte Carlo simulations in this paper. An experiment-based model verification work will our another future work.

References

1. Meireles, R., Boban, M., Steenkiste, P., Tonguz, O., Barros, J.: Experimental study on the impact of vehicular obstructions in VANETs. In: Proceedings of IEEE Vehicular Networking Conference (VNC), pp. 338–345, December 2010
2. He, R., Molisch, A., Tufvesson, F., Zhong, Z., Ai, B., Zhang, T.: Vehicle-to-vehicle channel models with large vehicle obstructions. In: Proceedings of IEEE International Conference on Communications (ICC), pp. 5647–5652, June 2014
3. Mangel, T., Klemp, O., Hartenstein, H.: A validated 5.9 GHz non-line-of-sight path-loss and fading model for inter-vehicle communication. In: Proceedings of 11th International Conference on ITS Telecommunications (ITST), pp. 75–80, August 2011
4. Mangel, T., Schweizer, F., Kosch, T., Hartenstein, H.: Vehicular safety communication at intersections: Buildings, non-line-of-sight and representative scenarios. In: Proceedings of Eighth International Conference on Wireless On-Demand Network Systems and Services (WONS), pp. 35–41, January 2011
5. Boban, M., Vinhoza, T., Ferreira, M., Barros, J., Tonguz, O.: Impact of vehicles as obstacles in vehicular ad hoc networks. IEEE J. Sel. Areas Commun. **29**(1), 15–28 (2011)
6. He, R., Molisch, A., Tufvesson, F., Zhong, Z., Ai, B., Zhang, T.: Vehicle-to-vehicle propagation models with large vehicle obstructions. IEEE Trans. Intell. Trans. Syst. **15**(5), 2237–2248 (2014)
7. Sommer, C., Joerer, S., Dressler, F.: On the applicability of two-ray path loss models for vehicular network simulation. In: Proceedings of IEEE Vehicular Networking Conference (VNC), pp. 64–69, November 2012
8. Wang, B., Yin, K., Zhang, Y.: An exact Markov process for multihop connectivity via intervehicle communication on parallel roads. IEEE Trans. Wirel. Commun. **11**(3), 865–868 (2012)
9. Yamamoto, A., Ogawa, K., Horimatsu, T., Kato, A., Fujise, M.: Path-loss prediction models for intervehicle communication at 60 GHz. IEEE Trans. Veh. Technol. **57**(1), 65–78 (2008)
10. Paier, A., Bernado, L., Karedal, J., Klemp, O., Kwoczek, A.: Overview of vehicle-to-vehicle radio channel measurements for collision avoidance applications. In: Proceedings of IEEE 71st Vehicular Technology Conference (VTC 2010-Spring), pp. 1–5, May 2010
11. Boban, M., Meireles, R., Barros, J., Tonguz, O., Steenkiste, P.: Exploiting the height of vehicles in vehicular communication. In: Proceedings of IEEE Vehicular Networking Conference (VNC), pp. 163–170, November 2011

12. Abbas, T., Tufvesson, F., Karedal, J.: Measurement based shadow fading model for vehicle-to-vehicle network simulations. Int. J. Antennas Propag. (2012)

13. Wang, C.-X., Cheng, X., Laurenson, D.: Vehicle-to-vehicle channel modeling and measurements: recent advances and future challenges. IEEE Commun. Mag. **47**(11), 96–103 (2009)

14. Boban, M., Meireles, R., Barros, J., Steenkiste, P., Tonguz, O.: TVR: tall vehicle relaying in vehicular networks. IEEE Trans. Mob. Comput. **13**(5), 1118–1131 (2014)

15. Chen, R., Zhong, Z., Leung, V.C.M., Michelson, D.G.: Performance analysis of connectivity for vehicular ad hoc networks with moving obstructions. In: Proceedings of IEEE 80th Vehicular Technology Conference (VTC2014-Fall), pp. 1–5, September 2014

16. Karp, B., Kung, H.-T.: GPSR: Greedy perimeter stateless routing for wireless networks. In: Proceedings of the 6th Annual International Conference on Mobile Computing and Networking. ACM, pp. 243–254 (2000)

17. Zhang, Y., Cao, G.: V-PADA: vehicle-platoon-aware data access in VANETs. IEEE Trans. Veh. Technol. **60**(5), 2326–2339 (2011)

Performance Analysis for Traffic-Aware Utility in Vehicular Ad Hoc Networks

Zhe Wang[✉], Zhangdui Zhong, Minming Ni, and Miao Hu

State Key Laboratory of Rail Traffic Control and Safety,
Beijing Jiaotong University, Beijing, China
zhe.wang@outlook.com

Abstract. As one of the key research topics in Vehicular Ad Hoc Networks (VANETs), analysis of traffic-aware utility is always difficult to be solved properly. In order to make the utility function be more suitable for the realistic network environment, a performance analysis for the utility function of different traffic is conducted in this paper. We consider two types of traffic, best effort traffic and real time traffic, and develop the form of utility functions for various traffic in VANETs. To model the dynamic features in VANETs more generally, the Poisson process and the traffic flow theory are used to describe the vehicle's mobility. According to the theoretical analysis proposed in this paper, the conditional probability density function (pdf) of utility function for different traffic in VANETs can be deduced, which is much easier to be applied to the design of resource allocation algorithm for VANETs. Performance evaluation is conducted to verify the accuracy of our analysis.

Keywords: VANETs · Utility function · Best effort traffic · Real time traffic · pdf

1 Introduction

The tremendous advances in vehicular technologies has led to large amount of multimedia and data traffics in addition to the traditional traffic in Vehicular Ad Hoc Networks (VANETs). In order to improve the network efficiency and user experience, it is necessary to transmit different kinds of traffic simultaneously in VANETs. However, due to the performance requirements for different traffic is various in VANETs, resource allocation for multi-traffic is a challenging topic which has drawn lots of attention.

When the utility theory is introduced to describe the degree of user satisfaction, the complexity of resource allocation algorithm can be reduced. The utility function is measured on the basis of allocated resource (e.g., bandwidth, transmission rate), which is a non-decreasing function with respect to the given amount of resource [1]. In general, the more resource is allocated to user, the more satisfaction will be achieved.

© ICST Institute for Computer Sciences, Social Informatics and Telecommunications Engineering 2017
Y. Zhou and T. Kunz (Eds.): ADHOCNETS 2016, LNICST 184, pp. 117–127, 2017.
DOI: 10.1007/978-3-319-51204-4_10

With the development of VANETs, the traffic evolves toward mixed one. Different performance requirement for various traffics can be represented as different utility function. The various traffic could be divided into two categories in VANETs. The first one is the best effort traffic which does not have strict quality of service (QoS) requirement, such as multimedia content downloading without delay restrictions. The other one is the real time traffic with strong QoS requirement, like on-line game, or interactive media.

Most of the research (e.g., [1–6]) on utility uses a simple proportion value to represent the channel quality. While the mobility model for VANETs is almost not been considered. They are not applied to the traffic in practical system for VANETs. Meanwhile, due to the high mobility of vehicles and the rapid change of network topology, it is a very challenging task to accurately describe the utility function for the traffic in VANETs, which is an important metric to evaluate the degree of user satisfaction.

To address these issues, we focus on the characteristics of utility function for different traffic in VANETs. The contributions of this paper are two twofold. First, we use Poisson process model to model the mobility characteristics of the vehicles in VANETs, which is more realistic and suitable for VANETs scenario. We also applied the path-loss model to describe the channel's condition. Second, according to the theoretical analysis proposed in this paper, the conditional probability density function (pdf) of utility function for different traffic in VANETs can be deduced, which is much easier to applied to the design of resource allocation algorithm for VANETs.

The rest of this paper is organized as follow. In Sect. 2, we summarize the related work in utility analysis. Section 3 proposes the considered network scenario and channel model. The details of theoretical analysis for traffic-aware utility function in VANETs is presented in Sect. 4. Section 5 investigates the numerical and simulation results to verify the analysis. The conclusion of this paper is drawn in Sect. 6.

2   Related Work

The research on utility in wireless networks has obtained a few achievements, including utility-based resource allocation and utility-based data dissemination. The utility-based resource allocation in wireless networks was studied in [1,2], the authors considered two types of traffic in the wireless networks and three resource allocation schemes were proposed. [6] adopted a resource allocation which was based on the unified utility function, and the channel was indicated by Signal to Interference plus Noise Ratio (SINR).

The utility function for different traffic was discussed in [4]. [5] proposed a data dissemination approaches, which is aiming to improve the system efficiency by introducing utility. The utility based relay vehicle selection algorithm was studied in [6]. We can see that either the resource allocation or the data dissemination can be used as the objective to explore the utility. To the authors' best knowledge, the practical channel model and realistic mobility model for VANETs almost not been considered in the previous work.

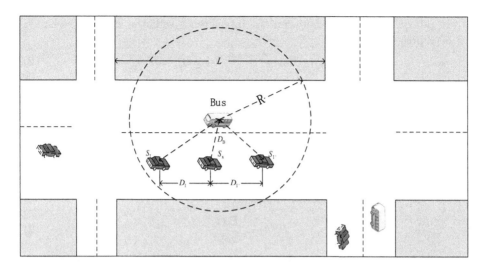

Fig. 1. Network scenario of traffic-aware utility based content downloading

3   System Model

3.1   Network Scenario

In this paper, we consider an application scenario for VANETs which is consisted of buses and sedans. We consider a bus B_P driving along an unidirectional road segment L by a constant speed to provide downloadable contents to its nearby sedans, which is shown in Fig. 1. Each vehicle is assumed to be equipped with a wireless device by which it can communicate with other vehicles within its communication range R. There is a link between a sedan S_X and the bus B_P when they are within the communication range of each other. However, one sedan's downloading procedure might be affected by other sedans. Due to the space limit, only the interferences from the closest sedans are considered in this paper. As shown in Fig. 1, the sedan S_I and $S_{I'}$ are interference sedans for sedan S_X. As denoted in the figure, the two closest interference signals' transmission distance are marked as D_I and $D_{I'}$, respectively. The distance between bus B_P and sedan S_X is denoted as D_B.

To model the sedans' mobility, it is assumed that the sedans enter the highway according to the Poisson process with intensities λ. Based on traffic flow theory [7], the velocity v of a vehicle can be expressed as $v = \lambda/\rho$, where ρ is the traffic density of the target highway scenario, the average number of vehicles within per unit length of the highway (vehicles per metre), and λ is the observed Poisson process density. According to the characteristic of Poisson process, the sedans' arrival time interval T should follow an exponential distribution with parameter λ. Hence, the pdf of T can be given as

$$f_T(t) = \lambda e^{-\lambda t}, (t \geq 0). \tag{1}$$

So the pdf of the distance between the interference sedan s_I and the sedan s_x can be represented as

$$f_{D_I}(d) = \frac{\lambda}{v_I} e^{-\lambda \frac{d}{v_I}}, (d \geq 0).$$ (2)

where v_I is the average moving velocity of the sedan S_I. The distance $D_{I'}$ follows the same distribution.

3.2   Channel Model

In this paper, we assume all vehicles send out the contents with identical transmission power P_t and the commonly used path-loss model is applied here to describe the signal power's attenuation:

$$P_r(d) = \frac{P_t G}{d^\beta}.$$ (3)

where $P_r(d)$ is the average signal power at distance d from the base station, G is a constant which depend on the characteristics of radio transceivers, and β is the path loss exponent. Since fading gain of small scale fading changes over much smaller timescale, and in a frequency selective channel (such as one using OFDM) can be averaged or mitigated in the frequency domain [8], we assume that it does not impact transmission performance.

4   Theoretical Analysis

4.1   The Analysis for Transmission Rate

Theorem 1. *Let X be a continuous random variable having probability density function f_X. Suppose that g(x) is a strictly monotonic (increasing or decreasing), differentiable (and thus continuous) function of x. Then the random variable Y defined by Y = g(X) has a probability density function given by*

$$f_Y(y) = \begin{cases} f_X[g^{-1}(y)] \left| \frac{d}{dy} g^{-1}(y) \right| & if \ y = g(x) \ for \ some \ x \\ 0 & if \ y \neq g(x) \ for \ all \ x \end{cases}$$ (4)

where g^{-1} is defined to equal that value of x such that g(x) = y.

According to the system model and the previous Lemma, the pdf of S_I's interference power could be presented as

$$f_{Z_I}(z) = f_{D_I}\left(\sqrt[\beta]{\frac{P_t G}{z}}\right) \cdot \left| \frac{d}{dz} \sqrt[\beta]{\frac{P_t G}{z}} \right|,$$ (5)

which could be easily obtained the pdf of $f_{Z_{I'}}$. Hence, the total interference power accumulated at sedan S_X is the sum of two independent random variables, as

$$f_{Z_{I+I'}}(z) = \int_0^\infty f_{Z_I}(z-y) \cdot f_{Z_{I'}}(y) dy.$$ (6)

The SIR at sedan S_X is the ratio of two random variables when the distance D_B is given. Then the conditional pdf could be presented as

$$f_{\text{SIR}}(s) = f_{Z_{I+I'}}\left(\frac{P_B}{s}\big|d_B\right) \cdot \left|\frac{d}{ds}\left(\frac{P_B}{s}\right)\right|, \tag{7}$$

where $P_B = P_t G/d^\beta$.

Based on above analysis and Shannon theorem, the transmission rate in bps of the sedan when the bandwidth is B Hz

$$R = B\log_2(1+s). \tag{8}$$

Therefore, the conditional pdf of transmission rate R_X for sedan S_X is given as

$$f_{R_X}(r|d_B) = f_{\text{SIR}}(2^{\frac{r}{B}} - 1|d_B) \cdot \left|\frac{d}{dr}(2^{\frac{r}{B}} - 1)\right|. \tag{9}$$

where B is the bandwidth allocated to the traffic.

4.2   Utility Function Modeling

In VANETs, more and more studies focus on "user satisfaction" for resource allocation to avoid such a "throughput-fairnes" dilemma. We use the utility function $U(r)$ to describe the degree of user satisfaction, which is a non-decreasing function with respect to the amount of transmission rate R. However, as VANETs evolve, the traffic evolves toward a mixed one. As a result, classification of user data in terms of traffic type is required to effectively achieve the differentiated QoS performance [4]. In general, the types of traffic in VANETs could be roughly classified into two categories [6]. The utility function has various characteristics according to different traffic.

Best Effort Traffic: For best effort traffic, such as data traffics without hard delay requirement, the utility function should be steadily increasing with the growing transmission rate. When the transmission rate is small, the utility increases significantly with transmission rate. While the transmission rate is large enough, the degree of increment will keep decreasing. In summary, the utility function for best effort traffic should be a convex function according to the transmission rate.

Therefore, the utility function for best effort traffic could be obtained as

$$U_{\text{BE}}(r) = 1 - e^{\frac{k_1 \cdot r}{A}}. \tag{10}$$

where A, k_1 can be used to adjust the slope of the utility curve. The utility function for best effort traffic is shown in Fig. 2. In this paper, the form of utility function is refer to [1,6].

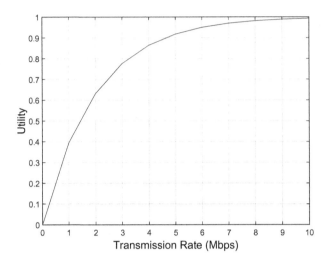

Fig. 2. Utility function of best effort traffic

Based on the analysis in subsection A, the conditional pdf of best effort traffic's utility function can be presented as

$$f_{U_{BE}}(u|d_B) = f_{R_X}(\frac{A}{k_1}\ln(1-u)|d_B) \cdot \left| \frac{d}{dr}(\frac{A}{k_1}\ln(1-u)) \right|. \tag{11}$$

Then the cumulative distribution function (CDF) of utility function for a best effort traffic is

$$F_{U_{BE}}(u|d_B) = \Pr\{U_{BE}(r) < u\} = \int_0^u f_{U_{BE}}(t|d_x)dt. \tag{12}$$

Real Time Traffic: For the real time traffic, such as streaming media traffic with guaranteed QoS requirements, the utility function should be a monotonically increasing function with the growing transmission rate. Due to the characteristics of real time traffic, the QoS requirement need certain resource to satisfy this requirement. The higher transmission rate allocated to the traffic plays a greater role in improving utilities when the transmission rate obtained by the traffic is less than the critical value. Whereas the transmission rate obtained by the traffic exceeds the critical value, the utility will be gradually decreased in increments. In general, the utility function for real time traffic should be a sigmoid function respect to the transmission rate.

Therefore the utility function for real time traffic could be presented as

$$U_{RT}(r) = \frac{1}{1 + e^{-k_2(r-r_0)}}. \tag{13}$$

where the parameter k_1 is used to adjust the slope of the utility curve around r_0. It reflects the characteristics of different real time traffic. The larger r_0 is,

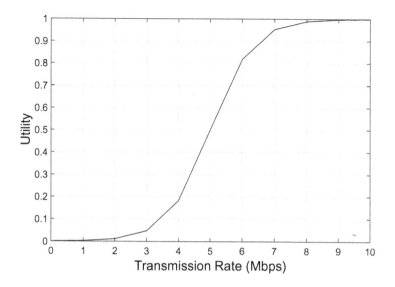

Fig. 3. Utility function of real time traffic

the more transmission rate should be allocated to the traffic to guarantee the QoS requirements. The utility function for real time traffic is shown in Fig. 3, where $r_0 = 5$ Mbps.

Given the $f_{R_x}(r|d_B)$, the conditional pdf of real time traffic's utility function can be obtained as

$$f_{U_{RT}}(u|d_B) = f_{R_x}(-\frac{1}{k_2} \cdot \ln(\frac{1}{u} - 1) + r_0|d_B) \cdot \left| \frac{d}{dr}(-\frac{1}{k_2} \cdot \ln(\frac{1}{u} - 1) + r_0) \right|. \quad (14)$$

Then the CDF of utility function for a real time traffic is

$$F_{U_{RT}}(u|d_B) = \Pr\{U_{RT}(r) < u\} = \int_0^u f_{U_{RT}}(t|d_B)dt. \quad (15)$$

5 Numerical and Simulation Results

To verify the analysis results, a series of simulations has been conducted with Matlab. The Nakagami fading model is utilized with different fading factors, and the value of pathloss exponent β is selected by referring to a vehicular communication-based filed test result [9] as 3.18. The other major parameters for simulation are shown in Table 1. In the following part of this section, the simulation results are demonstrated in groups to show the effect of different parameters on the conditional pdf for utility function.

5.1 Impact of d_X on Conditional pdf for Utility Function

Intuitively, decreasing the distance between the sedan S_X and the resource pool B_P could support higher transmission rate with the same given bandwidth.

Table 1. Reference value for main parameters

Parameter	Description	Value
β	Pathloss exponent of the fading model	3.18
k_1	The slope parameter of the utility function for the best effort traffic	0.08
k_2	The slope parameter of the utility function for the real time traffic	0.05
P_t	The vehicular transmission power	400 mw
λ	The intensity of Poisson process	0.8
ρ	The traffic density of the target highway	0.15

In Fig. 4, the conditional pdf of utility function for best effort traffic is illustrated with different d_B, while the allocated bandwidth for the traffic is 1 MB. Generally, when the signal propagation distance is increased, the probability for sedan to obtain a high utility is decreased, which is mainly due to the obvious decrease of the received signal power. Moreover, it is hard to get a high utility when the bandwidth is limited. This is because that, for the best effort traffic, the utility function is steadily increasing with the growing transmission rate which is mainly decided by the allocated bandwidth and the SIR.

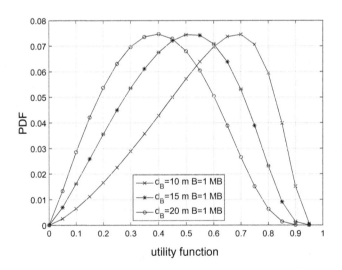

Fig. 4. Conditional pdf of utility function for best effort traffic with different d_x

Figure 5 compares the impacts of the different d_B on the performance of utility function for real time traffic. Generally, the probability is decreased with the increased utility value. Moreover, when the d_B is increased, the probability

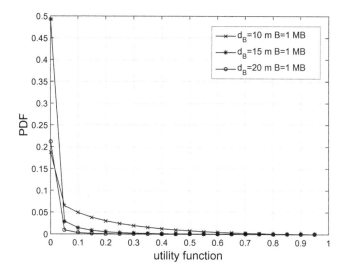

Fig. 5. Conditional pdf of utility function for real time traffic with different d_x

for the real time traffic to obtain a high utility value is decreased. This is mainly due to the fact that, the increased d_B decreases the probability for a higher SIR. As described earlier, the utility function for real time traffic is a sigmoid function respect to the transmission rate. Therefore, when the utility value is ranging from 0.1 to 0.9, the fluctuation of the pdf is not large, which is because of the limited bandwidth.

Therefore, once the bandwidth is given, we could make use of the analysis results fore utility function to design the resource allocation algorithm. For example, when d_B is 20 m, the probability for effort best traffic and real time traffic to obtain a high utility is extremely low. Therefore, in order to achieve the maximum network throughput, the network should allocate the resource to other sedans which have a shorter distance between the sedan and the bus.

5.2   Impact of B on Conditional pdf for Utility Function

For demonstrating the impact of B on conditional pdf for utility function, the SIR's conditional pdf is depicted with different given bandwidth in Figs. 6 and 7. The conditional pdf of utility function for best effort traffic is illustrated in Fig. 6. For the best effort traffic with larger allocated bandwidth B, the probability for the sedan S_X to have a high utility value is increased. This could be explained as that the increased B will increase the transmission rate, which is represented as the rise of utility value.

Finally, the conditional pdf of utility function for real time traffic is illustrated in Fig. 7 with different B. In general, when B is increased, the probability for the best effort traffic to obtain a high utility value is decreased. According to the figure, when utility value is increased from 0 to 0.1, the probability is

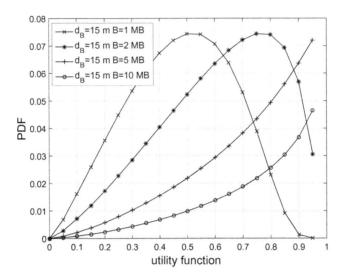

Fig. 6. Conditional pdf of utility function for best effort traffic with different B

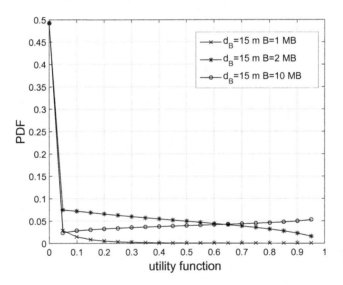

Fig. 7. Conditional pdf of utility function for real time traffic with different B

decreased slightly, which is mainly due to the characteristics of the real time traffic. The utility function for real time traffic is a sigmoid function respect to the transmission rate, which has a significantly increasing round a critical value of transmission rate.

Therefore, we could develop the resource allocation algorithm based on the above simulation results of utility function when the d_B is known.

6   Conclusion

In this paper, we analyzed the conditional probability density function (pdf) of utility function for different traffic in VANETs, which is based on Poisson process and the traffic flow theory. The stochastic characteristic of the utility value observed at a sedan was derived under the realistic channel model and mobility model. We believe this work will provide useful sights for the design and optimization of the source allocation with the help of utility function. Using the similar method, we plan to study the performance of utility function with the more realistic mobility model for VANETs, which could describe the spacial constrain revealed in the actual vehicular movement. These will be the follow-on work for this paper in the near future.

Acknowledgment. This work is partly supported by the NSFC (Grants No. 61401016, U1334202), and the State Key Laboratory of Rail Traffic Control and Safety (RCS2016ZT011).

References

1. Kuo, W., Liao, W.: Utility-based resource allocation in wireless networks. IEEE Trans. Wirel. Commun. **6**(10), 3600–3606 (2007)
2. Kuo, W., Liao, W.: Utility-based radio resource allocation for QoS traffic in wireless networks. IEEE Trans. Wirel. Commun. **7**(7), 2714–2722 (2008)
3. Chen, L., Wang, B., Chen, X., Zhang, X., Yang, D.: Utility-based resource allocation for mixed traffic in wireless networks. In: IEEE Conference on Computer Communications Workshops, pp. 91–96 (2011)
4. Balakrishnan, R., Canberk, B., Akyildiz, F.: Traffic-aware utility based QoS provisioning in OFDMA hybrid smallcells. In: IEEE International Conference on Communications, pp. 6464–6468 (2013)
5. Schwartz, S., Ohazulike, E., van Dijk, W., Scholten, H.: Analysis of utility-based data dissemination approaches in VANETs. In: Vehicular Technology Conference, pp. 1–5 (2011)
6. Chen, Q., Li, H., Yang, B., Chai, R.: A utility based relay vehicle selection algorithm for VANET. In: International Conference on Wireless Communications and Signal Processing, pp. 1–6 (2012)
7. Bellomo, N., Coscia, V., Delitala, M.: On the mathematical theory of vehicular traffic flow-Part I: fluid dynamic and kinetic modeling. Math. Models Methods Appl. Sci. **12**(12), 1801–1843 (2002)
8. Dhillon, S., Andrews, G.: Downlink rate distribution in heterogeneous cellular networks under generalized cell selecetion. IEEE Wirel. Commun. Lett. **3**(1), 42–45 (2014)
9. Altinatas, O., et al.: Field tests and indoor emulation of distributed autonomous multi-hop vehicle-to-vehicle communications over TV white space. In: The 18th Annual International Conference on Mobile Computing and Networking (2012)

Investigation and Adaptation of Signal Propagation Models for a Mixed Outdoor-Indoor Scenario Using a Flying GSM Base Station

Alina Rubina, Oleksandr Andryeyev, Mehdi Harounabadi, Ammar Al-Khani, Oleksandr Artemenko$^{(\boxtimes)}$, and Andreas Mitschele-Thiel

Integrated Communications System Group, Ilmenau University of Technology, 98693 Ilmenau, Germany
{alina.rubina,oleksandr.andryeyev,mehdi.harounabadi,
ammar.al-khani,mitsch}@tu-ilmenau.de,lestoa@gmail.com

Abstract. In this paper, we consider a disaster scenario where a Micro Aerial Vehicle (MAV) is flying around the urban area and tries to localize wireless devices such as mobile phones. There is a high chance of those devices being in the vicinity of their human owners. Fast and simple approach to map the received signal strength to distance is the Received Signal Strength Indicator (RSSI). The more accurate mapping ensures higher localization accuracy. As a consequence, an accurate signal propagation model is required.

The Free Space model, ITU indoor and outdoor model, SUI model, Hata model, COST-231 Hata model and Log-distance model have been chosen to be investigated in this work. The goal was to determine whether analytically chosen models fit to our scenario, as well as develop a suitable model for outdoor-indoor scenario. A real-world experiment was carried out to collect RSS measurements. An MAV was placed outside of a building while mobile phones were located inside a building. A measure for the evaluation was a root mean square error (RMSE).

The main contribution of this paper is an adapted log-distance model for GSM which is suited for outdoor-indoor scenario with the RMSE value of 6.05 m. The ITU indoor model represents the second best fit to our measured data with the RMSE value of 6.3 m.

Keywords: MAV · Signal propagation models · Quadrocopter · GSM · Log-distance model

1   Introduction

In order to predict radio propagation behavior, different signal propagation models have been developed as the low-cost, convenient and suitable alternative to site measurements, since the later approach is expensive and complex [1]. Today several models have emerged for indoor and outdoor environments in urban, suburban and rural areas. The nature of those areas plays a significant role

© ICST Institute for Computer Sciences, Social Informatics and Telecommunications Engineering 2017
Y. Zhou and T. Kunz (Eds.): ADHOCNETS 2016, LNICST 184, pp. 128–139, 2017.
DOI: 10.1007/978-3-319-51204-4_11

in the development of signal propagation models. Predicting the behavior of a wireless signal can be limited due to the following reasons: (1) distance between a transmitter and a receiver ranges from a couple of meters to few kilometers; (2) thickness of walls in a building can significantly affect the signal propagation; (3) the environment in which the signal propagates is usually not known [2].

A signal propagation model can be used to map received signal strength (RSS) to the distance. This mapping is crucial for localization purposes, when a technique called Received Signal Strength Indication (RSSI) is used. The more accurate the model is, the higher localization accuracy can be achieved. Other techniques, such as the Time of Arrival (TOA), Time Difference of Arrival (TDoA) and Angle of Arrival (AoA) can also be applied. However, those methods usually require additional hardware and precise time synchronization [3].

In this work, we consider a disaster scenario where the communication infrastructure has been ruined. As a consequence, victims of the disaster will not be able to make a call or send a message. A Micro Aerial Vehicle (MAV) flies around this area and locates mobile devices which could be in the vicinity of their human owners. We call this scenario – mixed outdoor-indoor, as mobile phones are inside a building and the MAV is flying outside.

For the development of a signal propagation model, we have chosen the GSM standard, as it is more reliable than the Wi-Fi network. Authors in [4] have observed that a GSM signal is more stable over time, than a Wi-Fi signal. Also, the bandwidth of the GSM channel is 200 KHz [5]. In contrast, the bandwidth of the Wi-Fi channel is 22 MHz and the channels are overlapping [6]. As a result, the interference in Wi-Fi channels can be significant.

Moreover, as previously stated, we are considering a disaster scenario in which it cannot be guaranteed that all mobile devices are running Wi-Fi access points. Whereas in the GSM network, mobile nodes will automatically perform the International Mobile Subscriber Identity (IMSI) attach procedure when they detect a network they can connect to [7]. There exist several propagation models in the literature [8–11] for calculating distance using RSS values, that were measured in a GSM network. However, none of them suit our purposes for the following reasons: (1) our scenario is unique and considers mixed communication between an indoor and an outdoor environment, while the most of models consider either an indoor or an outdoor scenario; (2) physical parameters of previous approaches do not fit our work, e.g., different receiver and transmitter heights, low transmitted power (6 dBm in our case); (3) necessity for the accurate RSS to distance mapping, as this directly affects localization accuracy. Therefore, a measurement campaign was performed according to our scenario to modify log-distance signal propagation model. The rest of the paper is organized as follows. In Sect. 2, an overview of existing models is given. Section 3 describes the conducted real-world experiment to collect RSS values. In Sect. 4, we present a developed model, as well as a comparison to the existing ones. Finally, Sect. 5 concludes the paper.

2   State of the Art

Path loss or path attenuation is a reduction in the power density of an electromagnetic wave as it propagates through space [8]. The signal propagation models
describe how the path loss is dependent on path attenuation factor, transmitter
antenna height, receiver antenna height, distance, operating frequency, etc. All
models are designed using different assumptions and experimental data, obtained
in the field. Therefore, a model should be carefully chosen in order to fulfill needs
of a specific scenario.

Fingerprinting represent another very popular solution for the localization.
However, the fingerprint technique represents an unstable solution for indoor
scenarios and requires a priori knowledge about the site. Every small change in
the environment causes drastic changes in the database of fingerprints. Therefore,
it is essential to update the database frequently [12]. As a result, we will not
follow this method in our work. In contrast, log-distance path-loss models are
much less susceptible to changes in the environment and produce more stable
results.

The most well-known signal propagation models that can be applied for a
mixed outdoor-indoor scenario, where the transmitter is a flying GSM base station, are summarized in Table 1. As follows a list of used variables and constants
is given:

- λ is a wavelength in meters,
- f is a frequency in MHz,
- γ is a path loss exponent,
- d is a distance between a transmitter and a receiver in meters,
- h_t is a transmitter antenna height above ground level in meters,
- h_r is a receiver antenna height above ground level in meters,
- X_σ represents a Gaussian random variable with zero mean and standard deviation of σ dBm and denotes shadow fading [13],
- P_{r_0} is a signal strength at 1 m from the transmitter,
- X_h is a correction factor for receiving antenna height,
- S is a correction factor for shadowing in the range between 8.2 and
  10.6 dBm [14],
- $d_0 = 100$ m in the case of SUI model,
- L_f is a floor penetration loss factor in dB,
- n is a number of floors between the transmitter and the receiver,
- L_{out} is an outdoor path loss,
- L_{tw} is through-wall penetration loss,
- α is an attenuation coefficient for indoors (0.5),
- d_{in} is an indoor distance from wall to a receiver in meters,
- d_{out} is a distance from a transmitter to the wall next to the receiver in meters,
- L_0 is a loss in the free space,
- Q represents a field amplitude factor,
- L_{rts} is a roof to street diffraction loss,
- L_{msd} is a multiscreen diffraction loss.

Free space model is one of the most basic and well-known models for predicting path loss. The main limitation of this model is consideration of a line-of-sight path through free space without any reflection or diffraction effects which are present in our scenario [13].

Another well-known model is the log-distance model [15]. As can be seen in the respective equation from Table 1, it is a general model and thus is suitable for a variety of scenarios. This implies the main disadvantage of the log-distance model - tuning is required for each scenario in order to provide accurate results.

IEEE 802.16 Broadband Wireless Access working group proposed the standards for the frequency band below 11 GHz containing the channel model developed by Stanford University, namely the Stanford University Interim (SUI) model [16]. The SUI model has limitations, namely minimal antenna heights and transmission distance, which can lead to a significant accuracy reduction in the considered scenario due to the small altitude and transmission power of a flying GSM base station.

Table 1. The most well-known signal propagation models.

Title	Signal model	Frequency range [MHz]	Environment
Free space model [13]	$L = 32.44 + 20\log_{10}d + 20\log_{10}f$	NA	Outdoor
Log-distance path loss model [13]	$L = P_{r_0} - 10\gamma\log_{10}d + X_\sigma$	NA	Outdoor/Indoor
SUI model [14]	$L = 20\log_{10}(\frac{4\pi d_0}{\lambda}) + 10\gamma\log_{10}(\frac{d}{d_0}) + X_f + X_h + S$	2500–2700	Outdoor/Indoor
Hata model [8]	$L_{50}(urban) = 69.55 + 26.16\log_{10}f_c - 13.82\log_{10}h_t - a(h_r) + (44.9 - 6.55\log_{10}h_t)\log_{10}d$	150–1500	Outdoor/Indoor
COST-231 Hata model [9]	$L_{50} = 46.3 + 33.9\log_{10}f - 13.82\log_{10}h_t - ((1.1\log_{10}f - 0.7)h_r - (1.56\log_{10}f - 0.8)) + (44.9 - 6.55\log_{10}h_t)\log_{10}d + c_m$	1500–2000	Outdoor/Indoor
Walfisch and Bertoni model [10]	$S = L_0 Q^2 L_{rts}$	800–2000	Outdoor/Indoor
Walfisch and Ikegami model [11]	$L_b = L_0 + L_{rts} + L_{msd}$	800–2000	Outdoor/Indoor
ITU indoor short-range model [17]	$L = 20\log_{10}f + \gamma\log_{10}d + L_f(n) - 28$	900–100000	Indoor
ITU outdoor short-range model [18]	$L = L_{out}(d_{out} + d_{in}) + L_{tw} + \alpha d_{in} + X_\sigma$	900–100000	Outdoor

One of the most widely used models for predicting path loss in mobile wireless systems is the Hata model (also known as Okumura-Hata model) [19]. The Hata model is not designed for frequencies beyond 1500 MHz. Moreover, it is assumed that the transmitter antenna is located at least 30 m above the ground level, which is not always the case for a flying GSM base station.

In order to overcome the main limitation of this model, namely support for frequencies beyond 1500 MHz, a modified model, called COST-231 Hata, was proposed in [9]. Nevertheless, COST-231 Hata model still assumes that the transmitter antenna height is 30 m or more.

COST-231 project proposed models to consider buildings in the vertical plane between a transmitter and a receiver - Walfisch and Bertoni, Walfisch and Ikegami models [10,11]. The application of these two models is limited to the case when the transmitter is mounted above the rooftop levels of tall buildings. Our scenario has different assumptions and due to this fact we will not investigate these models in details.

The International Telecommunication Union (ITU) proposed an indoor propagation model [17]. According to [17], this model can be used in case of a coexistence in both indoor and outdoor environments. However, to apply this model, the floor penetration loss factor and the number of floors between the flying base station and the mobile phone should be known, which is not the case in a disaster scenario. ITU indoor short-range propagation model is designed to be used mainly for predicting signal propagation in indoor environments, so it should be evaluated in our mixed outdoor-indoor scenario.

For the scenarios where both indoor and outdoor conditions exist, ITU proposed the outdoor short-range propagation model [18]. It is assumed that the receiver is most likely to be held by a pedestrian, who can travel inside and outside of the building. The coefficients for different environments can be found in [17]. This model consists of many special cases for different scenarios. In a disaster scenario, where precise characteristics of the environment are not known in advance, applying this model can be difficult or impossible.

It can be seen that all models are designed to be used in specific scenarios. Our goal is to find radio propagation models which are accurate in path loss prediction, easily tunable and applicable in a disaster scenario, where it is impossible to know the site information in advance.

Thus, we have chosen the most appropriate models to be evaluated according to our mixed outdoor-indoor scenario - SUI model, Hata model, COST-231 Hata model, free space model, log-distance model, ITU indoor and outdoor models. The purpose of our experiment was to collect RSS data and determine whether the chosen signal propagation models fit experimental data or not. For that, the path loss exponent and the intercept (the sum of the transmitted power and wall attenuation factor) was determined. Next, the evaluation and comparison between models is presented.

3   Evaluation Scenario

For the evaluation, we have chosen the following scenario. The experiment was performed at Leonardo da Vinci building in the campus of TU Ilmenau. We have used seven different mobile phones as receivers placed inside of the building and one MAV equipped with a GSM base station as a transmitter placed outside the building. The MAV with the mounted GSM base station is shown in Fig. 1. In such a way, we achieve a mixed outdoor-indoor scenario. The building plan and the distribution of the nodes is seen in Fig. 2. Weather and experiment setup are given in Table 2. The GSM base station was implemented using Software Defined Radio (SDR) with omni-directional antennas and the open-source software – OpenBTS [20]. The goal of our experiment was to collect RSSI measurements in

Fig. 1. The quadcopter with GSM BS on board.

order to develop an empirical GSM model for outdoor-indoor scenario. We have used collected RSSI from all seven phones to develop our model. We did not analyze data separately for every phone, it was averaged among all used mobile phones. In such a way we wanted to avoid dependencies on the specific model, antenna or transceiver unit of the mobile phone. Outdoor measurements were taken both in front and rear of the building, by placing MAV at the distances of $5, 10, 15, 20, 25, 30, 35$ m in the front and $5, 10, 15, 20, 25, 30, 35, 40, 45, 50, 55$ m in the rear. For every new placement of the MAV, 120 measurements have been taken for each mobile phone. We have also performed a series of indoor measurements, as some models require a reference measurement at the distance $d = 1$ m.

Table 2. Weather and experiment setup.

Parameter	Name/Value
Air temperature	-2°
Humidity	83%
Speed of wind	3 km/h
Air pressure	1027 mbar
Building size	20×30 m^2
Uplink frequency	1710.2 MHz
Downlink frequency	1805.2 MHz
Number of nodes	7
Measured parameter	RSSI

As the log-distance model is a very general model which can be applied indoors and outdoors, as well as extended with a wall attenuation factor, it was decided to adapt this model to our scenario. For the empirical model development, we had to determine a path loss exponent γ and an intercept I.

Our goal was to develop an easily tunable model, for that we avoid having too many parameters, like wall attenuation factor, floor penetration loss factor, etc.

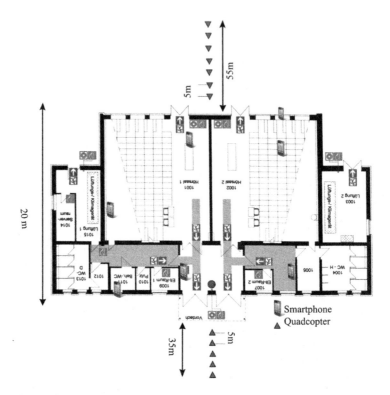

Fig. 2. Floor plan of Leonardo da Vinci building at technical university of Ilmenau. Mobile phones were located inside the building. MAV was placed outside the building in front and rear. Positions are marked accordingly.

Table 3. Propagation parameters for the evaluation.

Parameters	Values
Frequency	1805.2 MHz
Distance d_0	1 m
Receiving antenna height (smartphone)	0.5 m
Transmitting antenna height (MAV)	1.5 m
Path loss exponent γ	[1, 5]
Intercept	[0, 100] dBm
Floor penetration loss factor $L_f(n)$	0

As opposite to the existing models, we combine all these factors to one parameter and call it intercept. We will call it adapted log-distance model and express it as:

$$L_{adapted} = 10\gamma \log_{10} d + I \tag{1}$$

Algorithm 1. Adaptation of signal propagation models using RMSE

1: $D \leftarrow$ array of distances
2: $P_r \leftarrow$ array of experimentally obtained received signal strengths
3: $P_{rmodel} \leftarrow$ array of calculated received signal strengths
4: $error_{min} \leftarrow \infty$
5: **for** $\gamma \leftarrow$ 1.0, 5.0 step 0.001 **do**
6:     **for all** D **do**
7:         $error, intercept \leftarrow calculateRMSE(\gamma, D, P_r, P_{rmodel})$
8:         **if** $error <= error_{min}$ **then**
9:             $error_{min} = error$
10:             $\gamma_{best} = \gamma$
11:             $intercept_{best} = intercept$
12:         **end if**
13:     **end for**
14: **end for**
15: **return** $error_{min}, \gamma_{best}, intercept_{best}$

Propagation parameters used for the evaluation of the results are given in Table 3. We have used these parameters to find the signal propagation model which has the closest fit to our experimental data. The brute force method was used to iterate over all possible values for the path loss exponent (from 1 to 5) and the intercept (from 0 to 100 dBm), as it can be seen in Algorithm 3. For every combination of γ and I, RMSE has been calculated as a measure of similarity to our empirical data and can be expressed as:

$$RMSE = \sqrt{\frac{\sum_{j=1}^{N}(L_{model_j} - L_{measured_j})^2}{N}} \qquad (2)$$

where L_{model_j} is the value of chosen model in dBm, $L_{measured_j}$ is the measured value in dBm and N is the amount of measurements.

The next chapter will present our evaluation results for the chosen models.

4   Evaluation Results

The path loss in dBm with respect to the distance between the MAV and mobile phones was measured. In Fig. 3 the average signal strength values using both indoor and outdoor measurements are plotted. As predicted, the measured data follows logarithmic distribution.

Furthermore in Fig. 3 five curves are plotted: the adapted log-distance model, COST-231 Hata model, ITU indoor and outdoor models, as well as the log-distance model. These models represent the best fit according to the RMSE. As can be seen from figure, all five models are located quite close to developed empirical model presented by the adapted log-distance model. To identify the best fit, RMSE values should be analyzed.

As depicted in Fig. 4, the adapted log-distance model has the lowest RMSE and represents the best fit to our measured data. It outperforms its opponents

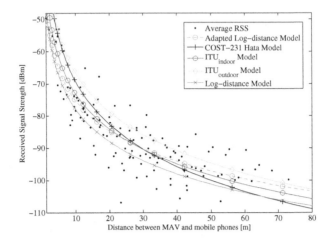

Fig. 3. Path loss in dBm vs. distance in a mixed outdoor-indoor scenario.

with the RMSE value of 6.05 m. Taking into account all measurements, the adapted log-distance model presented $\gamma = 3.1$ and intercept equal to 44.14 dBm. Values of γ and intercept for all five models are given in Table 4. Moreover, ITU indoor model has the second best fit to the measured data with the RMSE value of 6.3 m. Also, the third best fit was presented by the log-distance model. It is worth noticing that COST-231 Hata and ITU outdoor models have γ values of 4.4 and 4.0 accordingly, which is not realistic for our scenario. The chosen building for the experiment was distanced from the other buildings, there was low traffic and as a consequence expected path loss value should be in the range of 3–3.5. Furthermore, it was stated in [16] that values of γ from 4 to 6 are only typical for indoor areas.

As a result, we have developed an empirical GSM model which is suited for a scenario where a transmitter is located outside the building and mobile phones are placed inside. Using Eq. 1 resulting adapted log-distance model can be written as:

$$L_{adapted} = 10 \cdot 3.1 \log_{10} d + 44.1 \tag{3}$$

Table 4. Path loss exponent and intercept values for the chosen models.

Model	γ	Intercept, dBm
Adapted log-distance	3.1	44.1
COST-231 Hata	4.4	25.6
ITU$_{indoor}$	3.6	37.1
ITU$_{outdoor}$	4.0	26.7
Log-distance	3.3	44.8

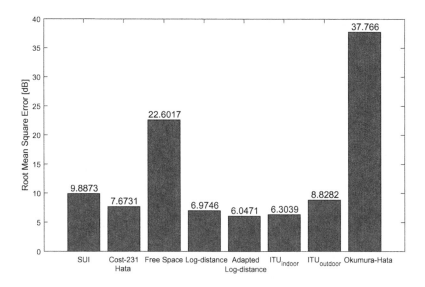

Fig. 4. Evaluation of chosen models in terms of root mean square error.

The transmitter and the receiver can be separated by up to four walls. Nevertheless we have collected our measurements only in one site, this model could also be used in other similar outdoor-indoor scenarios. This should be validated and will be a part of our future work.

5   Conclusion

In this paper a comprehensive study and analysis of wireless propagation models were made. Our goal was to find a model which can accurately map distance to the RSS as this will signficantly increase a localization accuracy. Analytical analysis has shown that SUI model, Hata model, COST-231 Hata model, free space model, log-distance model, ITU indoor and outdoor models could be the best candidates for prediction of path loss in a mixed outdoor-indoor scenario. In order to validate chosen models a measurement campaign was performed.

Evaluation results were the following:

- Adapted log-distance model presents the best fit to our data with RMSE value of 6.05 m with path loss exponent being $\gamma = 3.1$ and intercept $= 44.1$ dBm.
- The second best fit was presented by ITU indoor model with RMSE value being 6.3 m. Altered parameters for this model were $\gamma = 3.6$ and intercept $= 37.1$ dBm.
- Worst fit was presented by COST-231 Hata and ITU outdoor models which had γ values of 4.4 and 4.0 accordingly, which is not realistic for our scenario.

As a result we can state that developed log-distance model can be used in any scenario where mobile phones are located inside the building and flying GSM base station is placed outside. However, this should be validated and will be a part of our future work.

References

1. Ezzine, R., Braham, R., Al-Fuqaha, A., Belghith, A.: A new generic model for signal propagation in Wi-Fi and WiMAX environments. IEEE (2008)
2. Artemenko, O., Rubina, A., Simon, T., Mitschele-Thiel, A.: Evaluation of different static trajectories for the localization of users in a mixed indoor-outdoor scenario using a real unmanned aerial vehicle. In: 7th International Conference on Ad Hoc Networks (ADHOCNETS 2015), San Remo, Italy, September 2015
3. Artemenko, O., Rubina, A., Golokolenko, O., Simon, T., Rämisch, J., Mitschele-Thiel, A.: Validation and evaluation of the chosen path planning algorithm for localization of nodes using an unmanned aerial vehicle in disaster scenarios. In: 6th International Conference on Ad Hoc Networks (ADHOCNETS 2014), Rhodes, Greece, Best Paper Award, August 2014
4. Otsason, V., Varshavsky, A., LaMarca, A., de Lara, E.: Accurate GSM indoor localization. In: 7th International Conference, UbiComp 2005: Ubiquitous Computing 2005, Proceedings, Tokyo, Japan, 11–14 September 2005, pp. 141–158 (2005)
5. Goldsmith, A.: Wireless Communications. Stanford University, California (2004)
6. IEEE: Part 11: Wireless LAN Medium Access Control (MAC) and Physical Layer (PHY) specifications. IEEE (2012)
7. Rao, G.S.: Cellular Mobile Communication. Pearson, London (2013)
8. Hata, M.: Empirical formula for propagation loss in land mobile radio services. IEEE Transactions on Vehicular Technology $29(3)$, 317–325 (1980)
9. Damosso, E., Correia, L.M.: COST action 231: digital mobile radio towards future generation systems: final report, European commission (1999)
10. Walfisch, J., Bertoni, H.L.: A theoretical model of UHF propagation in urban environments. IEEE Transactions on Antennas and Propagation $36(12)$, 1788–1796 (1988)
11. Löw, K.: Comparison of urban propagation models with CW-measurements. In: IEEE 42nd Vehicular Technology Conference, pp. 936–942. IEEE (1992)
12. Bruning, S., Zapotoczky, J., Ibach, P., Stantchev, V.: Cooperative positioning with magicmap. In: 4th Workshop on Positioning, Navigation and Communication, 2007. WPNC 2007, 17–22 March 2007
13. Rappaport, T.: Wireless Communications: Principles and Practice. Prentice Hall PTR, Upper Saddle River (2001)
14. Erceg, V., Hari, K., Smith, M., Baum, D.S., Sheikh, K., Tappenden, C., Costa, J., Bushue, C., Sarajedini, A., Schwartz, R., et al.: Channel models for fixed wireless applications (2001)
15. Faria, D.B.: Modeling signal attenuation in IEEE 802.11 wireless LANs, vol. 1. Technical report TR-Kpp, 06-0118, Kiwi Project, Stanford University, January 2006
16. Abhayawardhana, V., Wassell, I., Crosby, D., Sellars, M., Brown, M.: Comparison of empirical propagation path loss models for fixed wireless access systems. In: 2005 IEEE 61st Vehicular Technology Conference, VTC 2005-Spring, vol. 1, pp. 73–77. IEEE (2005)

17. Union, I.T.: P.1238: Propagation data and prediction methods for the planning of indoor radiocommunication systems and radio local area networks in the frequency range 900 MHz to 100 GHz (2015)
18. Union, I.T.: P.1411: propagation data and prediction methods for the planning of short-range outdoor radiocommunication systems and radio local area networks in the frequency range 300 MHz to 100 GHz (2015)
19. Leppänen, R., Lähteenmäki, J., Tallqvist, S.: Radiowave propagation at 900 and 1800 MHz bands in wooded environments. IEEE Transactions on Antennas and Propagation **92**, 112 (1992)
20. Loula, A.: OpenBTS installation and configuration guide (2009). http://openbts. org/documentation/

Modelling and Analysis

Optimizing Power Allocation in Wireless Networks: Are the Implicit Constraints Really Redundant?

Xiuhua Li$^{(\boxtimes)}$ and Victor C.M. Leung

Department of Electrical and Computer Engineering,
The University of British Columbia, Vancouver V6T 1Z4, Canada
{lixiuhua,vleung}@ece.ubc.ca

Abstract. The widely considered power constraints on optimizing power allocation in wireless networks, e.g., $p_n \geq 0, \forall n$, and $\sum_{n=1}^{N} p_n \leq P_{\max}$ where N and $P_{\max}$ are given constants, imply the constraints, i.e., $p_n \leq P_{\max}, \forall n$. However, the related implicit constraints are regarded as redundant in the most current studies. In this paper, we explore the question "Are the implicit constraints really redundant?" in the optimization of power allocation especially when using iterative methods that have slow convergence speeds. Using the water-filling problem as an illustration, we derive the structural properties of the optimal solutions based on Karush-Kuhn-Tucker conditions, propose a non-iterative closed-form optimal method, and use subgradient methods to solve the problem. Our theoretical analysis shows that the implicit constraints are not redundant, and their consideration can effectively speed up convergence of the used iterative methods and reduce the sensitivity to the chosen step sizes. Numerical results for the water-filling problem and another existing power allocation problem confirm the effectiveness of considering the implicit constraints.

Keywords: Power allocation · Water-filling · Subgradient method

1 Introduction

Future wireless communication networks are required to support a large number of users with various requirements, especially the large bandwidth demands of multimedia services. To fulfill the requirements, radio resource management (RRM) plays an essential role as the system level control of co-channel interference and other radio transmission characteristics in wireless communication systems [1]. RRM involves strategies and algorithms for controlling parameters such as transmit power, user allocation, beamforming, data rate, handover criteria, modulation scheme and error coding scheme, etc., aiming at maximizing the utilization of the limited radio-frequency spectrum and radio network infrastructure [2]. Among these RRM techniques, optimization of power allocation is an important aspect of wireless communication system design that is well-studied in the past decades [1, 2].

© ICST Institute for Computer Sciences, Social Informatics and Telecommunications Engineering 2017
Y. Zhou and T. Kunz (Eds.): ADHOCNETS 2016, LNICST 184, pp. 143–154, 2017.
DOI: 10.1007/978-3-319-51204-4_12

On one hand, as an iterative first-order method, the subgradient method is widely used in many studies [3–13] to solve various power allocation problems or other optimization problems on RRM in wireless systems. In [3,4], the subgradient method was used to solve the problem of maximizing the throughput under the constraints of interference power and individual transmit power in cognitive radio networks. In [5], subgradient methods were utilized based on dual decomposition to solve the simultaneous routing and resource allocation problem. In [6], a subgradient solution was achieved to compute the maximum rate and the optimal routing strategy to solve the maximum multicast rate problem in the general undirected network model. In [7], a distributed subgradient method was used to solve the problems of how to choose opportunistic route for users to optimize the total utility or profit of multiple simultaneous users in wireless mesh networks. In [8], distributed subgradient methods were applied to optimize global performance in delay tolerant networks with limited information. In [9], a subgradient solution was proposed to solve the problem of jointly optimizing channel pairing, channel-user assignment, and power allocation in a single-relay multiple-access system. In [10], an α-approximation dual subgradient algorithm was proposed to optimize the total utility of multiple users in a load-constrained multihop wireless network. Based on the subgradient method, the study in [11] proposed a distributed optimal data gathering cost minimization framework with concurrent data uploading in wireless sensor networks. With the dual subgradient method, the study in [12] focused on convergence analysis of decentralized min-cost subgraph algorithms for multicast in coded networks. In [13], the subgradient method was used for joint power and bandwidth allocation in an improved amplify and forward cooperative communication scheme. Though subgradient methods can be operated in a distributed manner, they usually have **slow convergence speeds** and are very **sensitive to the chosen iteration step sizes** [14,15], which need to be improved to reduce the computation costs and even signaling overhead in wireless networks and to reduce the sensitivity to the chosen step sizes since *(1) the subgradient method may not converge under an improper step size,* and *(2) it is not easy to choose the proper step size, especially when the formulated optimization problem is very complex.*

On the other hand, mathematically, the formulated optimization problems of power allocation in wireless systems are generally subject to at least two inequality constraints [1–13] on p_n, the transmit power allocated at a base station (BS) for the n-th user, e.g., *(1) nonnegative:* $p_n \geq 0, \forall n$, and *(2) limited sum:* $\sum_{n=1}^{N} p_n \leq P_{\max}$, where N and $P_{\max}$ respectively denote the total number of users served by the BS and the BS's maximum transmit power. These two power constraints imply another set of (implicit) constraints, i.e., $p_n \leq P_{\max}, \forall n$. However, in most currently studied power allocation optimization problems or other similar optimization problems with the above two inequality constraints, the implicit constraints are regarded as redundant and useless in the design of strategies and algorithms for solving the problems. From the perspective of mathematics, the implicit constraints obviously hold, but are they really redundant in optimization algorithms? To the best of our knowledge, this question is unexplored.

The above motivates us to answer the question "Are the implicit constraints really redundant?" in power allocation optimization especially when using subgradient methods in the solution algorithms. Specifically, we study the water-filling problem as a typical example of power allocation optimization. Based on Karush-Kuhn-Tucker (KKT) conditions, we derive the structural properties of the optimal solutions to the water-filling problem and evaluate the performance of the proposed methods with and without considering the implicit constraints. Our contributions are summarized below:

- To explore the first studied question, using the water-filling problem as an illustration, our theoretical analysis shows that considering the implicit constraints can effectively speed up the convergence of the subgradient method, reduce the sensitivity to the chosen step size and lead to convergence even when an improper step size is used, while the opposite is true if the implicit constraints are not considered. This finding can be extended to other optimization problems and applied to other iterative methods. Besides, we propose a non-iterative closed-form optimal method.
- Numerical results on the water-filling problem and the power allocation problem for multiuser systems in [16] show that considering the implicit constraints in the algorithm design can effectively improve the performance of the used subgradient methods.

The rest of this paper is organized as follows. Section 2 introduces the water-filling problem as an illustration of power allocation. Section 3 derives the structural properties of the optimal solutions. Section 4 proposes and analyzes the algorithms to solve the optimization problem. Section 5 evaluates the performance of the proposed algorithms. Finally, Sect. 6 concludes this paper.

2 The Water-Filling Problem Typical in Resource Allocation

In this section, we provide a general form of the resource allocation problem and its formulation as the widely studied water-filling problem, to explore whether the implicit constraints are really redundant for optimization.

2.1 General Resource Allocation Problem

Many existing optimization problems for allocation of power or other resource can be formulated or transformed into a general form as

$$\max_{\mathbf{p},\mathbf{y}} f(\mathbf{p},\mathbf{y}) \tag{1a}$$

$$s.t. \sum_{n=1}^{N} p_n \leq P_{\max}; p_n \geq 0, \forall n \in \mathcal{N}, \tag{1b}$$

$$\mathbf{y} \in \mathcal{S}_Y; g_i(\mathbf{p},\mathbf{y}) \leq 0, \forall i \in \mathcal{I} \tag{1c}$$

where N is a given number (e.g., number of users), $\mathcal{N} = \{1, 2, \ldots, N\}$, $\mathcal{I}$ and $\mathcal{S}_Y$ are two given sets about resource constraints; $\mathbf{p} = [p_1, p_2, \ldots, p_N]^T$ and $\mathbf{y}$, respectively, are variable vectors of power and other resource allocations; $f(\mathbf{p}, \mathbf{y})$ and $g_i(\mathbf{p}, \mathbf{y})$ are, respectively, the given objective function (e.g., sum data rate) and constraint functions w.r.t. $\mathbf{p}$ and $\mathbf{y}$; $P_{\max}$ is a positive constant scalar (e.g., maximum sum power). From (1b), we can get the implicit constraints as

$$p_n \leq P_{\max}, \forall n \in \mathcal{N}. \tag{2}$$

In existing studies, the same or similar implicit constraints in (2) are usually overlooked and are regarded as redundant. Besides, whether problem (1) is convex or nonconvex, it can be solved with a family of iterative methods (e.g., subgradient method) to get the optimal or suboptimal solutions.

2.2   Water-Filling Problem

The water-filling problem given below is a typical formulation of the general resource allocation optimization problem described above, in which the sum capacity of users is maximized under transmit power constraints [17].

$$\max_{\mathbf{p}} \sum_{n=1}^{N} \log_2(1 + \alpha_n p_n), \ s.t. \ \sum_{n=1}^{N} p_n \leq P_{\max}; \ p_n \geq 0, \forall n \in \mathcal{N} \tag{3}$$

where $\boldsymbol{\alpha} = [\alpha_1, \alpha_2, \ldots, \alpha_N]^T$ is a strictly positive constant vector. Clearly, (3) is a simple case of (1) without loss of generality.

We incorporate the implicit constraints into problem (3) as

$$\min_{\mathbf{p}} z = -\sum_{n=1}^{N} \log_2(1 + \alpha_n p_n) \tag{4a}$$

$$s.t. \ -p_n \leq 0, \ \forall n \in \mathcal{N}, \tag{4b}$$

$$p_n - P_{\max} \leq 0, \ \forall n \in \mathcal{N}, \tag{4c}$$

$$\sum_{n=1}^{N} p_n - P_{\max} \leq 0, \tag{4d}$$

which is a strictly convex optimization problem. Thus, a local optimal solution is also globally optimal and the optimal solution is unique. Moreover, we can get the Lagrangian for problem (4) as $L(\mathbf{p}, \boldsymbol{\lambda}, \boldsymbol{s}, \nu) = -\sum_{n=1}^{N} \log_2(1 + \alpha_n p_n) + \sum_{n=1}^{N} (\nu - \lambda_n + s_n) p_n - (\nu + \sum_{n=1}^{N} s_n) P_{\max}$, where $\boldsymbol{\lambda} \in \mathbb{R}^N$, $\boldsymbol{s} \in \mathbb{R}^N$ and $\nu \in \mathbb{R}$ are the nonnegative Lagrange multiplier vectors and scalar for constraints (4b), (4c) and (4d), respectively. Thus, we can get the dual objective as $g(\boldsymbol{\lambda}, \boldsymbol{s}, \nu) = \inf_{\mathbf{p}} L(\mathbf{p}, \boldsymbol{\lambda}, \boldsymbol{s}, \nu)$, and then the dual problem as $\max_{\boldsymbol{\lambda}, \boldsymbol{s}, \nu} g(\boldsymbol{\lambda}, \boldsymbol{s}, \nu)$. Since problem (4) is convex, the corresponding duality gap reduces to zero at the optimum.

3   Structural Properties of the Optimal Solutions

According to the KKT conditions [14,15], if a feasible solution $\mathbf{p}^* \in \mathcal{S}_P$ is a local (and global) minimizer of the convex optimization problem (4), then there exist multipliers $(\boldsymbol{\lambda}^*, \boldsymbol{s}^*, \nu^*)$, not all zero, $(\boldsymbol{\lambda}^* \succeq 0, \boldsymbol{s}^* \succeq 0, \nu^* \geq 0)$, such that

$$\frac{\partial L}{\partial p_n} = -\frac{\alpha_n}{(1+\alpha_n p_n^*)\ln 2} - \lambda_n^* + s_n^* + \nu^* = 0, \ \forall n \in \mathcal{N}, \tag{5}$$

$$\lambda_n^* p_n^* = 0, \ \lambda_n^* \geq 0, \ p_n^* \geq 0, \ \forall n \in \mathcal{N}, \tag{6}$$

$$s_n^*(p_n^* - P_{\max}) = 0, \ s_n^* \geq 0, \ p_n^* \leq P_{\max}, \ \forall n \in \mathcal{N}, \tag{7}$$

$$\nu^*\left(\sum_{n=1}^{N} p_n^* - P_{\max}\right) = 0, \ \nu^* \geq 0, \ \sum_{n=1}^{N} p_n^* \leq P_{\max}. \tag{8}$$

Define $\mathcal{N}_1 \triangleq \{n | s_n^* > 0, n \in \mathcal{N}\}$, $\mathcal{N}_2 \triangleq \{n | s_n^* = 0, n \in \mathcal{N}\}$, and $\boldsymbol{\omega}^* \triangleq \nu^* \cdot \mathbf{1} + \boldsymbol{s}^* \in \mathbb{R}^N$, where $\mathbf{1} = [1, 1, \ldots, 1]^T \in \mathbb{R}^N$. Thus, we have $\mathcal{N}_1 \cup \mathcal{N}_2 = \mathcal{N}$ and $\mathcal{N}_1 \cap \mathcal{N}_2 = \emptyset$. Specifically, if there exists $n \in \mathcal{N}$ such that $p_n^* = P_{\max}$, then denote the index as $k^\#$, i.e., $p_{k^\#}^* = P_{\max}$; otherwise, $k^\#$ does not exist. We give some remarks for the above KKT conditions and derive some structural properties of the optimal solutions via some theorems below.

Remark 1. Specifically, if $\mathcal{N}_1 \neq \emptyset$, then for $\forall n \in \mathcal{N}_1$, we have $p_n^* = P_{\max}$ according to (7), and thus $p_k^* = 0$ and $s_k^* = 0$ for $\forall k \in \mathcal{N}, k \neq n$ according to (7) and (8). Thus, there exists at most one positive element in $\boldsymbol{s}^*$, i.e., $|\mathcal{N}_1| \leq 1$. Besides, if $|\mathcal{N}_1| = 1$, the only element in $\mathcal{N}_1$ is equal to $k^\#$ and we have $\mathcal{N}_2 = \mathcal{N} \setminus \{k^\#\}$. Note that even if $\mathcal{N}_1 = \emptyset$, i.e., $\mathcal{N}_2 = \mathcal{N}$, it is possible that $k^\#$ also exists.

Remark 2. For $\forall n \in \mathcal{N}$, $\lambda_n^* s_n^* \equiv 0$. If there exists any $k \in \mathcal{N}$, such that $\lambda_k^* > 0$ and $s_k^* > 0$, then according to (6) and (7), we can get $p_k^* = 0$ and $p_k^* = P_{\max}$ simultaneously, which is clearly contradictory.

Theorem 1. *With $\boldsymbol{\alpha}$ fixed, if P_{max} is not fixed and can be adjusted, then the optimal objective z^* is strictly decreasing with the increase of P_{max}.*

Theorem 2. *The optimal ν^* satisfies that: if $\mathcal{N}_1 = \emptyset$, $\nu^* = \max\limits_{n \in \mathcal{N}}\{\dfrac{\alpha_n}{(1 + \alpha_n p_n^*)\ln 2}\}$; Otherwise, $\nu^* = \dfrac{\alpha_n}{(1+\alpha_n P_{max})\ln 2} - s_n^*, n \in \mathcal{N}_1$, and $\nu^* \geq \max\limits_{n \in \mathcal{N}_2}\{\dfrac{\alpha_n}{\ln 2}\}$.*

Remark 3. Based on *Theorem 2*, we have $\nu^* > 0$, and thus the optimal solution $\mathbf{p}^*$ satisfies $\sum_{n=1}^{N} p_n^* = P_{\max}$ according to (8).

Remark 4. In terms of $\boldsymbol{\omega}^*$, if $\mathcal{N}_1 = \emptyset$, then we have $\boldsymbol{\omega}^* = \nu^* \cdot \mathbf{1} \succ 0$, which indicates that (5) is reduced to the form in existing studies (i.e., $\frac{\partial L_1}{\partial p_n}$) in this case. Otherwise, based on *Remarks* 1 and 3, we have $\omega_{k^\#}^* = \nu^* + s_{k^\#}^* > \nu^* > 0$ for $k^\# \in \mathcal{N}_1$, and $\omega_n^* = \nu^* > 0$ for $\forall n \in \mathcal{N} \setminus \{k^\#\}$, which indicates that $\boldsymbol{\omega}^*$ is divided into two positive parts.

Theorem 3. *The optimal solution p^* and the corresponding multiplier scalar ν^* satisfy $p_n^* = min\{\left[\frac{1}{\nu^* \ln 2} - \frac{1}{\alpha_n}\right]^+, P_{max}\}, \forall n \in \mathcal{N}$ where $[x]^+ \triangleq max\{x, 0\}$.*

Remark 5. From *Theorem 3*, in the optimal solution's closed-form expression, the multiplier vectors $(\boldsymbol{\lambda}^*, \boldsymbol{s}^*)$ can be eliminated, while the the multiplier scalar ν^* is dominating. However, $\boldsymbol{\lambda}^*$ is to operate on $[\mathbf{x}]^+$ such that the optimal solution $\mathbf{p}^* \succeq 0$, while $\boldsymbol{s}^*$ is to operate on $min\{\mathbf{x}, P_{max}\}$ such that the optimal solution $\mathbf{p}^* \preceq P_{max}$. In most works, their corresponding solutions are in the form of $[\mathbf{x}]^+$ and have no operation of $min\{\mathbf{x}, P_{max}\}$.

Theorem 4. *If $\alpha_{n_1} \geq \alpha_{n_2}, n_1 \in \mathcal{N}, n_2 \in \mathcal{N}$, then $p_{n_1}^* \geq p_{n_2}^*$ holds in the optimal solution $\mathbf{p}^*$.*

Remark 6. Let $n_1 \in \mathcal{N}$ and $n_2 \in \mathcal{N}$ be two indices such that $\alpha_{n_1} \geq \alpha_{n_2}$. Specifically, for the optimal solution $\mathbf{P}^*$, if $p_{n_1}^* = 0$, then $p_{n_2}^*$ must also be zero based on *Theorem 4*.

Theorem 5. *There exists the only $k^\# \in \mathcal{N}$ such that $p_{k^\#}^* = P_{max}$, if and only if both $k^\# = arg \underset{n \in \mathcal{N}}{max}\{\alpha_n\}$ and $\underset{n \in \mathcal{N} \setminus \{k^\#\}}{max}\{\alpha_n\} \leq \dfrac{\alpha_{k^\#}}{1 + \alpha_{k^\#} P_{max}}$ hold.*

Theorem 6. *Let $\boldsymbol{\pi}$ be the vector obtained by sorting $\boldsymbol{\alpha}$ in a descending order. Then the number of strictly positive elements in the optimal solution $\mathbf{p}^*$ is*

$$\chi = max\{n \in \mathcal{N} \mid \frac{1}{\pi_n} - \frac{1}{n}(\sum_{r=1}^{n} \frac{1}{\pi_r} + P_{max}) < 0\}, \tag{9}$$

and then the corresponding optimal multiplier ν^* can be expressed as

$$\nu^* = \begin{cases} any\ value\ in\ \left[\frac{\pi_2}{\ln 2}, \frac{1}{(\frac{1}{\pi_1} + P_{max}) \ln 2}\right], \chi = 1, \\ \dfrac{\chi}{\left(\sum_{r=1}^{\chi} \frac{1}{\pi_r} + P_{max}\right) \ln 2}, \chi \in \mathcal{N}, \chi \geq 2. \end{cases} \tag{10}$$

Remark 7. From (10) in *Theorem 6*, ν^* can take the value of $\frac{\chi}{\left(\sum_{r=1}^{\chi} \frac{1}{\pi_r} + P_{max}\right) \ln 2}$ for all the possible values of χ, which holds in most works where the implicit constraints in (2) are not considered. However, if the implicit constraints in (2) is considered, ν^* may take multiple values as shown in *Theorem 6*. Most importantly, *Theorem 6* provides a simple **non-iterative closed-form method** to get the optimal solution $\mathbf{p}^*$, denoted as Direct Search Method (DSM).

From the above analysis, not considering the implicit constraints in (2) can be regarded as a special case of considering them in this paper, which can be extended to other optimization problems. To get the optimal solution $\mathbf{p}^*$ to problem (3), whether the implicit constraints are considered in this paper or not in most works, it is very important to get the optimal multiplier ν^* by using either non-iterative methods (i.e., the proposed DSM) or iterative methods (e.g., subgradient method). We will show that considering the implicit constraints can greatly improve the convergence speed in iterative methods. In this paper, we only discuss the widely used subgradient method.

4 Algorithms to Solve the Optimization Problem

Based on the above analysis, the subgradient method without considering the implicit constraints and the proposed subgradient method that considers the implicit constraints are described in Algorithm 1 (Alg. 1) and Algorithm 2 (Alg. 2), respectively. Once the convergence condition is given, the only difference between Alg. 1 and Alg. 2 is the updating of $\mathbf{p}^{(t)}$ at each iteration, i.e., the operation of $\min\{\mathbf{x}, P_{\max}\}$ in the proposed Alg. 2 is not found in Alg. 1. Note that these algorithms share a common form of those applying the subgradient method in most works and that $\mathbf{p}^*$ is unique while ν^* may not be unique.

Algorithm 1. Existing Subgradient Method without Implicit Constraints.

1: **Input:** α, $P_{\max}$.
2: Initialize $t = 0$, $\mathbf{p}^{(0)} = \mathbf{0}_{N\times 1}$, $\nu^{(0)} = 0.1$, accuracy $\eta = 10^{-5}$.
3: **while** not converge **do**
4:    Update $\mathbf{p}^{(t)}$ as $p_n^{(t)} = \left[\frac{1}{\nu^{(t)}\ln 2} - \frac{1}{\alpha_n}\right]^+, \forall n \in \mathcal{N}$.
5:    Check convergence condition: $\left|\nu^{(t)}\left(\sum\limits_{n=1}^{N} p_n^{(t)} - P_{\max}\right)\right| < \eta$.
6:    Set $t \leftarrow t + 1$.
7:    Update $\nu^{(t)} = \left[\nu^{(t-1)} + \theta^{(t)}\left(\sum\limits_{n=1}^{N} p_n^{(t-1)} - P_{\max}\right)\right]^+$.
8: **end while**
9: **Output: p**, ν.

Algorithm 2. Proposed Subgradient Method with Implicit Constraints.

1: **Input:** α, $P_{\max}$.
2: Initialize $t = 0$, $\mathbf{p}^{(0)} = \mathbf{0}_{N\times 1}$, $\nu^{(0)} = 0.1$, accuracy $\eta = 10^{-5}$.
3: **while** not converge **do**
4:    Update $\mathbf{p}^{(t)}$ as $p_n^{(t)} = \min\left\{\left[\frac{1}{\nu^{(t)}\ln 2} - \frac{1}{\alpha_n}\right]^+, P_{\max}\right\}, \forall n \in \mathcal{N}$.
5:    Check convergence condition: $\left|\nu^{(t)}\left(\sum\limits_{n=1}^{N} p_n^{(t)} - P_{\max}\right)\right| < \eta$.
6:    Set $t \leftarrow t + 1$.
7:    Update $\nu^{(t)} = \left[\nu^{(t-1)} + \theta^{(t)}\left(\sum\limits_{n=1}^{N} p_n^{(t-1)} - P_{\max}\right)\right]^+$.
8: **end while**
9: **Output: p**, ν.

Besides, note that the achieved nonnegative solution $\mathbf{p}$ may be infeasible in the iteration process with subgradient method. Then we provide the sketch proof that considering the implicit constraints can improve the convergence speed of subgradient method as follow.

   (1) If there exists $k < +\infty$ such that $\nu^{(k)} = 0$ in the iteration process, we have

- Without the implicit constraints considered, we have $p_n^{(k)} = \left[\frac{1}{\nu^{(k)} \ln 2} - \frac{1}{\alpha_n}\right]^+ = +\infty, \forall n \in \mathcal{N}$, and $\nu^{(k+1)} = \left[\nu^{(k)} + \theta^{(k+1)}\left(\sum_{n=1}^N p_n^{(k)} - P_{\max}\right)\right]^+ = +\infty$. Then we have $p_n^{(k+i)} = 0, \forall n \in \mathcal{N}$ and $\nu^{(k+i)} = +\infty$ for all $1 \leq i < +\infty$, which means the iteration process will not converge in a limited number of iterations.

- With the implicit constraints considered, we have $p_n^{(k)} = P_{\max}, \forall n \in \mathcal{N}$, and $\nu^{(k+1)} = \theta^{(k+1)}(N-1)P_{\max}$, which can avoid the above bad case and thus guarantee the convergence in a limited number of iterations.

(2) Otherwise, we have $\nu^{(t)} > 0, \forall t$. Thus, we have $\nu^{(t)} = \nu^{(t-1)} + \theta^{(t)}\left(\sum_{n=1}^N p_n^{(t-1)} - P_{\max}\right) > 0$ for $\forall t \geq 1$, and then $|\nu^{(t)} - \nu^{(t-1)}| = \theta^{(t)}\left|\sum_{n=1}^N p_n^{(t-1)} - P_{\max}\right|$, which indicates that the convergence speed of $\nu^{(t)}$ depends on the steps $\theta^{(t)}$ and the value of $\left|\sum_{n=1}^N p_n^{(t-1)} - P_{\max}\right|$. Refer to [14,15] on how to choose a proper specific series of steps $\theta^{(t)}$. Since $\left|\sum_{n=1}^N \min\left\{\left[\frac{1}{\nu^{(t)} \ln 2} - \frac{1}{\alpha_n}\right]^+, P_{\max}\right\} - P_{\max}\right| \leq \left|\sum_{n=1}^N \left[\frac{1}{\nu^{(t)} \ln 2} - \frac{1}{\alpha_n}\right]^+ - P_{\max}\right|$ for $\forall t$, with the same steps $\theta^{(t)}$, the subgradient method considering the implicit constraints can achieve a higher convergence speed and needs fewer iterations than the subgradient method that does not consider the implicit constraints.

In terms of the step size $\theta^{(t)}$, we will use three common categories as: (1) C1: constant step size, e.g., $\theta^{(t)} = 0.1$ for $\forall t$; (2) C2: nonsummable diminishing step size, e.g., $\theta^{(t)} = \frac{1}{\sqrt{t}}$ for $\forall t$; (3) C3: square summable but not summable step size, e.g., $\theta^{(t)} = \frac{100}{t+100N}$ for $\forall t$.

As shown in [14,15], for C1, the subgradient method converges to the optimal value within a small range, i.e., $\lim_{t \to \infty} |\nu^{(t)} - \nu^*| < \varepsilon$. This indicates that the subgradient method finds an $\varepsilon-$suboptimal point within a finite number of iterations. The value ε is a decreasing function of the step size. Moreover, for C2 and C3, the subgradient method is guaranteed to converge to the optimal value if the chosen steps are small enough. With **proper** initialization, both the proposed and existing subgradient methods always converge but need different numbers of iterations. Most importantly, the solution found in each iteration may not be feasible. Considering the implicit constraints in each iteration can accelerate the iterative search of the optimal solution by excluding infeasible solutions from the search subspace, and thus achieve a higher convergence speed.

In terms of the sensitivity of the subgradient method to the chosen step sizes, the existing method may not converge if the step sizes are **improper**, but the proposed method will converge for these step sizes. The detailed theoretical analysis is omitted due to the page limit, but we will provide some numerical examples in Sect. 5 to show that considering the implicit constraints leads to convergence even using step sizes that are **improper** to the existing method.

5   Numerical Results

In this section, we evaluate by simulations the effectiveness of the subgradient methods considering the implicit constraints in solving the water-filling problem and the power allocation problem in [16], in terms of the convergence speed and sensitivity to the chosen step sizes. Since the only difference between the subgradient methods with and without considering the implicit constraints is the operation of $\min\{x, P_{max}\}$, which computation complexity can be neglected compared to that of the whole algorithm, we use the number of iterations required to satisfy the convergence condition as the convergence speed of the respective algorithms [14,15]. Note that in the following, convergence accuracy refers to the gap between the achieved value (e.g., optimization objective value and multiplier value) and its optimal value, and that we use the ratio of the required numbers of iterations of the subgradient methods without/with considering the implicit constraints for the following comparison of convergence speed x.

5.1   Numerical Examples in Water-Filling Problem

In the water-filling problem in (3), by using the proposed DSM to get the optimal values (i.e., the optimal objective value z^*, the optimal dominating multiplier ν^* and the optimal solution $\mathbf{p}^*$) as the baseline, we mainly evaluate the performance of the subgradient methods with/without considering the implicit constraints and then explore whether the implicit constraints are really redundant in the above three scenarios. We provide two numerical examples as follow.

– **Example 1:** $N = 3$, $P_{max} = 1$, $\alpha = [0.75, 2, 3]^T$. DSM gives the optimal solution $\mathbf{p}^* = [0, 0.416667, 0.583333]^T$, the optimal objective $z^* = -\sum_{n=1}^{N} \log_2(1 + \alpha_n p_n^*) = -2.333901$, and the optimal multiplier $\nu^* = 1.573849$.
– **Example 2:** $N = 20$, $P_{max} = 2$, $\alpha = [2, 1.5, \alpha_3, \ldots, \alpha_{20}]^T$, where α_n is a random value in $(0, 2]$ for $\forall n \in \mathcal{N}, n \geq 3$. DSM can also yield the optimal solution, the optimal objective and the optimal multiplier directly.

Figure 1 compares the values of $|z^{(t)} - z^*|$ and $|\nu^{(t)} - \nu^*|$ versus the iteration number for the subgradient methods in **Example 1** and **Example 2**. Here, the step sizes are set as $\theta^{(t)} = 0.1$, $\theta^{(t)} = \frac{1}{\sqrt{t}}$ and $\theta^{(t)} = \frac{100}{t+100N}$ for C1, C2 and C3, respectively. For **Example 1**, Fig. 1(a) and (b) show that with each of C1, C2 and C3, Alg. 2 has similar convergence accuracy and a higher convergence speed compared with Alg. 1. Specifically, **Alg2-C1** is about 1.3 times as fast as **Alg1-C1** while **Alg2-C2** is about 22.3 times faster than **Alg1-C2**. Besides, **Alg2-C3** is about 2.7 times as fast as **Alg1-C3**. For **Example 2**, Fig. 1(c) and (d) show that with each of C1, C2 and C3, Alg. 2 has higher or since convergence accuracy and higher convergence speed than Alg. 1. Specifically, **Alg2-C1** is about 4.9 times as fast as **Alg1-C1** while **Alg2-C2** is about 35.9 times faster than **Alg1-C2**. Besides, **Alg2-C3** is about 3.9 times as fast as **Alg1-C3**.

Moreover, by setting different step sizes, we show that considering the implicit constraints can reduce the sensitivity to the step sizes and lead to convergence while the existing method fails to converge. Here, if the iteration process does not stop within 10^7 iterations, the method is regarded as not convergent from the perspective of practical engineering implementation. For instance, we set $\theta^{(t)} = \frac{N}{t+N}$ for $\forall t$ in C3 and use the above same convergence conditions. In this

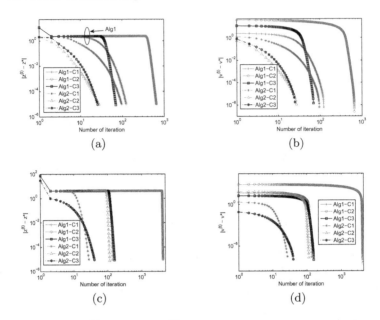

Fig. 1. The values of $|z^{(t)} - z^*|$ and $|\nu^{(t)} - \nu^*|$ versus iteration number, (a) and (b) for **Example 1**, (c) and (d) for **Example 2**.

setting, **Alg1-C3** does not converge in **Example 1** while **Alg2-C3** converges with 29 iterations.

5.2 Numerical Results of Optimizing Power Allocation in [16]

To further evaluate the effectiveness of considering the implicit constraints, we consider the power allocation optimization problem in [16], which maximizes the total system throughput in a multiuser orthogonal frequency-division multiplexing system under the constraints of total power and minimum data rate required by each user. To solve the problem, we use the subgradient method without considering the related implicit constraints (Alg. 3) and with this consideration (Alg. 4). Note that Alg. 3 is the subgradient method used in [16]. In our simulations, we set the subcarrier number $N = 10$, user number $K = 4$, total power $P_{tot} = 5$ Watt and all the users' required minimum data rate is 5 bps/Hz. We also use the same channel model as in [16], the same convergence condition as in Alg. 1 and Alg. 2, and the same step sizes in C1, C2 and C3 as in the previous examples. Note that the convergence condition used in this paper is much stricter than that in [16] and thus the algorithms require more iterations to converge but they achieve more accurate solutions. We use the gap between the achieved objective and the optimal objective as the performance metric. All the results are averaged over 100 channel realizations.

Figure 2 compares the performance of the subgradient methods w.r.t. the values of objective gap versus iteration number. Figure 2 shows that Alg. 4 always

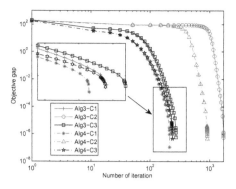

Fig. 2. The values of objective gap versus iteration number.

has a higher convergence accuracy and higher speed than Alg. 3. Specifically, in C1, C2 and C3, Alg. 4 is about 1.13, 1.87 and 1.15 times as fast as Alg. 3. Thus, considering the implicit constraints can effectively speed up the convergence of the subgradient method for optimizing the power allocation in [16].

6   Conclusions

In this paper, we have explored the question "Are the implicit constraints really redundant?" in power allocation optimization especially when using subgradient methods. Specifically, by illustrating the water-filling problem to answer the question, we have derived the structural properties of the optimal solutions based on KKT conditions, proposed a non-iterative closed-form optimal method and applied subgradient methods to solve the water-filling problem. Besides, our theoretical analysis has shown that the implicit constraints are not redundant, and their consideration can effectively improve the subgradient methods' convergence speed and reduce the sensitivity to the chosen step sizes. Numerical results have shown that considering the implicit constraints in the water-filling problem can greatly speed up the methods' convergence speed by up to about 36 times and reduce the sensitivity to the chosen step sizes, and that it can also effectively accelerate the convergence speed by up to 87% in the power allocation problem in [16]. Thus, the implicit constraints are not redundant in the algorithm design. Most importantly, the corresponding theoretical analysis and conclusions about the implicit constraints can be extended to many other resource allocation problems and to other iterative methods.

Acknowledgment. This work is support in part by a China Scholarship Council Four Year Doctoral Fellowship, the Canadian Natural Sciences and Engineering Research Council through grants RGPIN-2014-06119 and RGPAS-462031-2014 and the National Natural Science Foundation of China through Grant No. 61271182.

References

1. Zander, J., Kim, S.L., Almgren, M.: Radio Resource Management. Artech House Publishers, Norwood (2001)
2. Peng, M., Wang, C., Li, J., Xiang, H., Lau, V.: Recent advances in underlay heterogeneous networks: interference control, resource allocation, and self-organization. IEEE Commun. Surv. Tutorials 17(2), 700–729 (2015)
3. Zhang, L., Xin, Y., Liang, Y.C., Poor, H.V.: Cognitive multiple access channels: optimal power allocation for weighted sum rate maximization. IEEE Trans. Commun. 57(9), 2754–2762 (2009)
4. Nguyen, D.N., Krunz, M.: Spectrum management and power allocation in MIMO cognitive networks. In: Proceedings of IEEE INFOCOM, pp. 2023–2031, April 2012
5. Xiao, L., Johansson, M., Boyd, S.: Simultaneous routing and resource allocation via dual decomposition. IEEE Trans. Commun. 52(7), 136–1134 (2004)
6. Li, Z., Li, B.: Efficient and distributed computation of maximum multicast rates. In: Proceedings of IEEE INFOCOM, pp. 1618–1628, March 2005
7. Fang, X., Yang, D., Xue, G.: Consort: node-constrained opportunistic routing in wireless mesh networks. In: Proceedings of IEEE INFOCOM, pp. 1907–1915, April 2011
8. Masiero, R., Neglia, G.: Distributed subgradient methods for delay tolerant networks. In: Proceedings of IEEE INFOCOM, pp. 261–265, April 2011
9. Hajiaghayi, M., Dong, M., Liang, B.: Optimal channel assignment and power allocation for dual-hop multi-channel multi-user relaying. In: Proceedings of IEEE INFOCOM, pp. 76–80, April 2011
10. Fang, X., Yang, D., Xue, G.: Resource allocation in load-constrained multihop wireless networks. In: Proceedings of IEEE INFOCOM, pp. 280–288, April 2012
11. Guo, S., Yang, Y.: A distributed optimal framework for mobile data gathering with concurrent data uploading in wireless sensor networks. In: Proceedings of IEEE INFOCOM, pp. 1305–1313, April 2012
12. Zhao, F., Médard, M., Ozdaglar, A., Lun, D.: Convergence study of decentralized min-cost subgraph algorithms for multicast in coded networks. IEEE Trans. Inf. Theor. 60(1), 410–421 (2014)
13. Tous, H.A., Barhumi, I.: Resource allocation for multiuser improved AF cooperative communication scheme. IEEE Trans. Wirel. Commun. 14(7), 3655–3672 (2015)
14. Boyd, S., Vandenbergh, L.: Convex Optimization. Cambridge University Press, New York (2004)
15. Gäler, O.: Foundations of Optimization. Springer, New York (2010)
16. Chung, Y.J., Paik, C.H., Kim, H.G.: Subgradient approach for resource management in multiuser OFDM systems. In: Proceedings of International Conference on Communications and Electronics, pp. 203–207, October 2006
17. He, P., Zhao, L., Zhou, S., Niu, Z.: Water-filling: a geometric approach and its application to solve generalized radio resource allocation problems. IEEE Trans. Wirel. Commun. 12(7), 3637–3667 (2013)

Towards Dynamic Wireless Capacity Management for the Masses

Aikaterini Vlachaki, Ioanis Nikolaidis$^{(\boxtimes)}$, and Janelle J. Harms

University of Alberta, Edmonton, AB, Canada
{vlachaki,nikolaidis,janelleh}@ualberta.ca

Abstract. In this paper we speculate that, with the technological elements already in place, an automated dynamic management of the RF spectrum in urban residential settings will soon be possible. Dense urban environments are increasingly facing RF spectrum congestion, in particular in the ISM bands. The Internet of Things is only expected to add to the pressures. In this work we outline an architecture that will analyze and resolve spectrum congestion. We are motivated by the adaptive and modifiable nature of existing protocols, inspired by existing capacity planning and channel allocation schemes from cellular networks, and emboldened by the synergies possible via software–defined radios. The cloud computing infrastructure can be leveraged to perform most compute-intensive tasks required towards this goal. We are encouraged that the approach is viable by the relatively static, in the local sense, topology that most residential networks exhibit. To be able to support a wide range of device capabilities we consider the possibility of using a mix of techniques, ranging from advanced physical layer, to special MAC coordination, to higher-layer protocol operations to indirectly influence the operation of legacy equipment.

Keywords: RF spectrum congestion · Urban environments · Wireless capacity planning · Channel allocation · Software-defined radios · Cloud computing · Medium access control

1 Introduction

Today's landscape of wireless networks in residential and, in many cases, in enterprise settings, is characterized by diversity and elevated expectations of efficiency and flexibility. It is now commonplace for wireless networks of various technologies to co-exist in the same space. Conceived and deployed separately from each other, the deployed networks result in a challenging operating environment. The cross-technology interference (CTI) gets further compounded by interference originating from non-communicating devices that exist in abundance in today's environments (microwave ovens, lighting equipment, electrical motors, etc.). The combined impact is poor performance which is only expected to get worse as two technological trends continue: (a) ever-increasing density of deployments, such as the ones heralded for supporting the Internet of Things (IoT),

© ICST Institute for Computer Sciences, Social Informatics and Telecommunications Engineering 2017
Y. Zhou and T. Kunz (Eds.): ADHOCNETS 2016, LNICST 184, pp. 155–166, 2017.
DOI: 10.1007/978-3-319-51204-4_13

and (b) ever-decreasing operating power, meant to extend autonomy for battery-powered devices but at the same time jeopardizing the possibility for error-free reception in dense environments, because of poor Signal-to-Noise+Interference (SINR) at the receivers.

Yet, all of today's wireless networks interface, in one way or another, to the Internet, i.e., to a wired infrastructure. We speculate that this common denominator, i.e., the access to a common wired backbone network, and hence to services residing there, might be sufficient to improve the performance of co-existing wireless networks via a dynamic capacity management. Residential network deployments lack any capacity planning, which is a characteristic available only to a limited degree in some high-end enterprise networking products [9]. The spectrum utilization of the ISM bands is continuously changing with new technologies added every day making a dynamic wireless capacity management service among heterogeneous devices a necessity. This dynamic spectrum management application aims to improve the capacity of the wireless network, to efficiently and fairly share the spectrum and to support high Quality of Service (QoS).

At a very high level of abstraction, spectrum sensing and decision making algorithms can run on the cloud, which can then inform the local user equipment of actions to take. An example is the control of multiple access points (APs) in a neighborhood, subject to the channels and power used by other APs, as well as the existence of interference in certain geographical areas. We are motivated by the fact that a modicum of adaptivity and modifiability already exists in legacy protocols, such as 802.11, and an increasing number of access points exhibit "smart" software-controlled behavior. For example, the cloud-based control could inform the APs which channels, and what power, each should use.

We are inspired by the traditional cellular capacity and resource allocation schemes [16] but we speculate how, instead of being an exceptional one-time design, it is a continuous control and refinement process, in particular given the dynamics of the residential wireless environment. The overall proposed architecture is shown in Fig. 1. We separate the role of the wireless network elements as being access points (APs), clients (C) if they depend on APs, and peers (P) if they are autonomously operating. We note that AP refers to any element that has to function in a coordinating role for a wireless network to operate, i.e., it is not restricted to 802.11 APs. A role of interferer (I) is also noted but it is an abstract representation of a non-communicating source of interference. APs (and some Ps) may be connected to the wired infrastructure. A second dimension is whether they are legacy devices (superscript l), configurable by control at higher-layer protocol (superscript h), and configurable based on control of the physical layer (superscript p). A p device is assumed to be at least as capable as an h, but additionally it sports high-end programmable physical interfaces, as is the case with software-defined-radio (SDR). Clearly, "smart" devices (h and p) have to work around the fact that l devices and I are inflexible to control actions. The most challenging aspect of the architecture, not captured by Fig. 1 is that the sets of devices are dynamically added/removed and their traffic demands can change over time. The task of h and p devices is both to collect

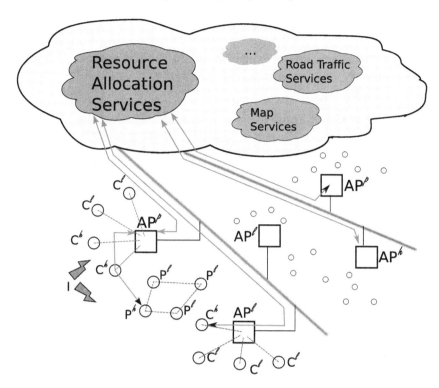

Fig. 1. Example layout: bidirectional brown arrows capture the measurement and control data flows. All the network elements in the lower left (under the left line representing the wired network) can be assumed to be within range of each other. Dashed colored lines indicate associations (AP with clients, or peers among them). (Color figure online)

measurements and to respond to control decisions. Early approaches with similar intent e.g., monitoring and diagnosis, have already been proposed [30]. Current efforts in the direction of IEEE 1900 and 802.11af integrate aspects of spectrum sensing as well. One notable challenge is how such systems will handle legacy and external (non-communication) interference sources.

This paper is structured as a survey of fields that come together under the same ambitious plan. Subsequent sections indicate the general system characteristics (Sect. 2), and new opportunities available because of recent technological advances (Sect. 3). We then provide a more thorough review of related work (Sects. 4 and 5), honing down eventually to a few works that open the possibilities we described in the previous paragraphs. The final, concluding, section is a compilation of open research opportunities that can be immediately pursued.

2   System Characteristics

A number of characteristics are evident in urban wireless networks; some aris-
ing from regulatory and device capability considerations and others from their
typical use scenarios.

1. **Prevalence of ISM bands:** Despite the considerable attention paid to cognitive wireless networks [1], the immediately pressing needs for wireless co-
   existence of multiple devices is in ISM bands. In ISM bands, where exists no
   distinction between primary and secondary users, thus creating more options
   for network control. A test case for ISM band co-existence is the co-existence
   in the 2.4 GHz ISM band, where one of the main concerns, given their prevalence, is the performance attained by 802.11 devices. Nevertheless, several
   sub-GHz ISM devices [6] exist and are extensively used in IoT applications,
   especially exploring the ability to reach longer distance yet operating at very
   low power. Additionally, the 5 GHz ISM band, while not yet crowded, will
   eventually become crowded as well. We therefore argue that *RF co-existence
   in the ISM bands is of immediate concern.*
2. **Cross-technology interference:** Secondly, wireless communication standards are typically conceived, designed, and implemented, independently of
   other wireless protocols. For example, what constitutes a "channel" for one
   protocol can be completely different from what is a channel in another one, or
   for what the power and timing relation between transmissions might be. The
   exception is the purposeful backward compatibility when a standard evolves,
   albeit still the entire family of protocols for one standard tends to ignore
   other protocol families. Interestingly, the ISM bands have given the opportunity for completely proprietary protocols to emerge as well. We use the term
   cross-technology interference (CTI), as adopted by Hithnawi et al. [12] to
   express the impact the operation of the protocol has, seen as "interference",
   on another protocol operating in the same location.
3. **Manifestations of interference:** Generally, the RF front-end and the associated signal processing chain, is highly specialized to (and in this sense efficient for) the needs of the particular protocol. We will call such RF designs
   "monolithic" as opposed to flexible Software Defined Radio (SDR) designs,
   even though some degree of agility is still possible in monolithic designs (channel selection, transmit power control, etc.). Correspondingly, the impact of
   another protocol's transmissions manifest themselves as interference leading
   to either a poor Signal to Noise-plus-Interference (SINR) figure, or simply
   as indication that the medium is busy, i.e., when Clear Channel Assessment
   (CCA) is being attempted. From the viewpoint of a legacy, monolithic, RF
   design, there exists no real difference between a source of interference due to
   another protocol's operation or from non-communication entities (microwave
   ovens, fluorescent light ballasts, internal combustion engine sparking, etc.).
4. **Predictability of user behavior:** While the user population is diverse,
   the per-individual, and the per-household behavior is remarkably consistent
   [26]. Today's residential users' traffic is dominated, in terms of data volume,

by video streaming of various compression qualities. Moreover, the network topology of residential wireless networks rarely changes drastically, and typically, only the location of a subset of the devices changes within the premises (e.g., smartphones and laptops). A residential network includes also a subset of IoT nodes and other low-power devices, such as Bluetooth Low Energy (BLE). While we cannot make a claim that the exact distance between devices is constant, the compartmentalization of devices within the premises of their corresponding owners suggests that the distances between nodes of the same residential unit remain bounded by a certain maximum distance.

Of the above characteristics, we note that items 1 to 3 present challenges, while item 4 provides also a glimmer of hope, in that making capacity management decisions can be helpful for an extended period of time, hopefully in the order of hours. Additionally, previous control actions can be repeated due to the predictable (diurnal, seasonal, etc.) user demands.

3   Technological Opportunities

We note the emergence of three technologies whose synergy could accelerate the deployment of dynamic wireless capacity management schemes.

1. **Spectrum sensing:** The emergence of SDRs has allowed the development of distributed spectrum analyzers [25], providing a comprehensive view of the spectrum use (including non-communication interference sources). We speculate that the features of SDRs will be integrated in APs in the near future. As such, the AP^p access points are assumed to be endowed with such capabilities, allowing them to switch between serving traffic, to, during idle periods, sample the background noise. A minimal form of spectrum sensing is also possible by legacy devices if access to Received Signal Strength Indicator (RSSI) values is supported. For example, almost all inexpensive sub-GHz RF transceivers allow RSSI values to be sampled, and this is a feature found in many P^l devices. Yet another, coarser and indirect indicator of spectrum condition, is the frame/packet error statistics collected by even the least flexible legacy devices (as interface statistics).

2. **Cloud processing:** The connectivity to the wired Internet by AP (as well as, indirectly, by C, and P) devices allows significant amounts of data collected by spectrum sensing to be shipped over for processing in the cloud. Trends and location-specific behavior of the interference and load demands can then be analyzed by computationally-intensive algorithms. In other words, the limited in-situ processing is overcome by off-site cloud-based processing. The idea of using the same cloud-based infrastructure to control the network has been an open research direction in 5G networks [27]. We augment this by (a) including sensing of the spectrum to ascertain the presence of multiple protocol device and sources of interference and (b) assume that legacy, l–type, devices cannot form part of the set of controllable devices, and hence it is up to the h and p devices to infer the behavior of co-located legacy devices.

3. **Web services:** While usually deployed as cloud-based application themselves, additional web-based services can assist in augmenting the decision making process. For example, WiFi Service Set Identifiers (SSIDs) mapped to geographic locations (such as wigle.net) or live traffic updates, provide, correspondingly, approximate information about the spatial separation of APs and a basis for anticipating residential data traffic load fluctuations. Moreover, persistent non-communication interference sources can be localized and their locations related to map coordinates [14]. While this does not imply a mitigation strategy, it can enable actions outside the automated network control. Another related example is a database service for area-specific white spaces demonstrated in SenseLess [24], geared towards non-ISM cognitive networks.

4   Related Work: Resource Allocation

The literature on wireless resource management is vast. We note, for example, the channel assignment literature for cellular networks, e.g., [16] as relevant, but one that follows the cellular network model of operation, i.e., single (or few) providers, no outside interferers, known locations (hence one-time, or infrequent, computation of allocation plans), etc. We need to abandon these assumptions in order to capture the essence of residential ISM, and in particular dense urban, environments. A starting point can thus be seen in the channel allocation schemes specific to WiFi, such as those surveyed by Chieochan et al. [8], which take into account the characteristic irregularities of AP coverage and the variety of traffic and QoS demands in different areas. They study schemes aimed for centrally managed environments where interference constitutes the metric of interest (which translates to a capacity measure) as well as schemes for uncoordinated environments.

The control of APs suggests a need for an Inter-AP Protocol (IAPP), a protocol essential for the cooperation and communication between APs of the l and h variety. Mahonen et al. have described a scheme using the IAPP protocol [21]. The channel allocation is expressed as a classical vertex coloring approach, DSATUR, where APs are modeled as vertices of an interference graph. The ingredient of their scheme which we consider essential in a cloud-supported allocation system is the fact that their proposal for channel allocation is dynamic and cooperative. Every new station that arrives within range automatically becomes part of the interference graph. APs detect other APs by means of hearing their beacons and construct vectors of information consisting of AP MAC addresses, signal-to-noise ratios, and received signal strength [8,21]. After the identification of their neighbours, the APs can share their knowledge with other APs in the network and the procedure can be repeated whenever the topology changes.

The lessons learned from 802.11 can be transformed to a broader class of wireless networks. The underlying coloring problems, and any other heavily computational optimization problems, can be relegated to the cloud. Additionally, there are a number of extensions, that could allow the easier transformation of 802.11 schemes to more general ones. First, the AP-centric view has been exchanged for a client-/peer-specific view. For example, Mishra et al. [22], introduce load balancing into the channel assignment scheme where interference is examined from the clients' point of view and it is claimed that even hidden (from the APs view) interference is captured and accounted for. A client is considered to suffer from channel conflict in two cases. In the first, the interference seen by a client comes from APs located within a communication range of the client of interest and in the second from APs or wireless clients (in neighboring BSSs), located within a one-hop distance of the AP-client link of interest [8,22]. Their client-centric algorithm called CFAssign-RaC is based on conflict set coloring. In this algorithm, the goal is to distribute the clients to APs in a way that the conflict is minimized while the load is balanced.

Following on similar logic, Chen et al. [7] present, among other algorithms, Local-Coord. Local-Coord coordinates the APs in a network aiming to minimize the interference as seen by both APs and wireless clients. This is work that uses in-situ, real time interference power measurements at clients and/or APs, on all the frequency channels. Local-Coord promises increased throughputs and mitigation of interference by performing frequency allocation irrespective of network topology, AP activity level, number of APs, rogue interferers, or available channels. The weighted interference constitutes the cost function in this approach and it is applied for every BSS. Metrics like the average traffic volume and average RSSI of the clients within a BSS can be used as weights, thus guiding the protocol to tolerate different metrics accordingly. Correspondingly, a proposed global coordination scheme, Global-Coord, applied centrally, performs overall coordination and spectrum allocation of a network only if a new channel assignment results in lower co-channel weighted interference.

Leung and Kim propose MinMax [20], focusing on interactions among APs and aiming to minimize the maximum effective channel utilization at the network's bottleneck AP, so that its throughput is improved and the overall network escapes congestion. The effective channel utilization is defined as the time a channel assigned to an AP is used for transmission, or is sensed busy because of interference from its co-channel neighbors. More specifically, initially random channel assignment is performed to all the APs in the network. Next, the bottleneck AP's interfering neighbors channels are readjusted so that the effective channel utilization of the bottleneck AP is minimized [8,20].

Yu et al. propose a dynamic radio resource management scheme in [31], where again the maximum effective channel utilization at the network's bottleneck AP, is to be minimized but without interactions among APs. In their work, the channel utilization is determined by a real-time algorithm that estimates the number of active stations from an AP's point of view. The real time consideration of active stations before each channel assignment, as well as a post channel assignment QoS check reinforces the dynamic nature of this approach. However, this scheme does not scale to large networks [8].

An added degree of freedom arises from power management. Power management of APs accounting for traffic load distribution and spectrum allocation is proposed in [11] by considering the notion of a "bottleneck" AP in a network and suggest its transmission power readjustment (reduction) takes place together with channel assignment in order to reduce the total interference over the network. The power reduction only applies to beacon packets and not data packets, so that clients that cannot longer be supported turn to another less congested AP. In this work, the total data rate required by an AP's clients over the AP's available bandwidth forms the, so–called, congestion indicator [8,11]. Their proposed optimization algorithm is claimed to be able to redistribute the load in a network, reduce the AP congestion and finally perform spectrum allocation so that the interference is minimized.

Concluding, we also note that whereas the coloring-based channel allocation problem is a useful abstraction, there has been evidence that a potential overlap of the allocated channels (hence abandoning a strict vertex coloring interpretation) is not as harmful as suggested by most of the literature, while it is almost unavoidable in high density network deployments anyway. Specifically, Mishra et al. [23], by explicitly modelling an interference factor (I-factor), derive capacity improvements. We conjecture a similar model, which explicitly accounts for interference factors, may be the way of the future, in particular if the I-factor is defined for cross-technology interference as well. Moreover, in certain instances, accepting interference as unavoidable, may be the only available strategy, e.g., with l-type network elements.

5   Related Work: Characterizing and (Re-)Acting

5.1   Channel Characterization

A crucial element towards performing dynamic capacity management in the face of changing conditions, such as interference, is the collection of suitable measurements. We can collectively call the various facets of this problem as the Channel State Information (CSI) problem, but recognize that, in the same network, different devices can acquire completely different types of CSI metrics, from fine-grained to coarse-grained and for different bandwidths. Our tacit assumption is that similar measurements along with the user behavior predictability imply that similar actions need to be taken on a regular, e.g., daily, basis.

A type of detection that can be performed with p type devices using simple measurements is energy-based signal detection [19] which involves a low computational overhead but generally performs poorly in low SINR environments. Cyclostationarity [10], on the other hand, has been a potent technique but at a high computational cost. Clearly, if the sampled channel data can be shipped quickly and in large volumes to a cloud-based detection algorithm, cyclostationarity computation is a viable option. Nevertheless, cyclostationarity is not exhibited by non-communication (hence not modulated) interference sources. There, a form of feature detection can be used instead. An example are on-line classification mechanisms of interference such as the ones proposed by Boers et al. [4].

Inspired by Boers et al.'s work [3] and in [4,5], where they explore and characterize noise and interference patterns found on 256 frequencies in an indoor urban environment's 900 MHz ISM and non-ISM bands, as well as based on our previous work in [29] where we studied whether the agreement of the same interference patterns on the class of a channel, can be linked to the cross correlation statistic, we envision the next step to be the exploration of the whole spectrum in the ISM band for further characterization of noise and interference patterns.

We anticipate that the perceived interference is location-specific. Indeed, the cross-correlation results of the time series [29] is suggesting that nearby nodes are experiencing similar interference. Seen differently, it is not necessary that all nodes in an area, e.g., not all APs, have to be p-type devices since very similar measurements would end up being collected. However, it is still not known how dense should the spatial sampling be to derive reliable channel state metrics for the various areas of the network. We note that similarities exist with the case of sampling strategies for optimal monitoring [18] as they express the situations where limited sampling resources are to handle a large network. Yet another, natural facet of the same problem is how high the sampling rate should be and how to convey the sampled data for processing in the cloud. Our early results show that wavelet compression of the sampled time series may produce significant data volume savings [28].

5.2   Interference Reaction/Mitigation

Current strategies to handle interference include *interference alignment* which is a transmission strategy relying on the coordination of multiple transmitters so that their mutual interference aligns at the receivers, and promises to improve the network's throughput [2]. Such techniques set as their objective to approximate the maximum degrees of freedom, also known as the channel's maximum multiplexing gain. Nevertheless, there are still open issues that need to be considered, such as realistic propagation environments, and the role of channel state information at the transmitter, and most importantly for our approach, the practicality of interference alignment in large networks [2].

Krishnamachari and Varanasi [17] study systems characterized by multi-user interference channels with single or multiple antennas at each node and develop an interference alignment scheme where there is no global channel knowledge in the network, but where each receiver knows its channels from all the transmitters and broadcasts a quantized version of it to all the other terminals, at a rate that scales sufficiently fast with the power constraint on the nodes. The quantized channel estimates are treated as being perfect and it is shown that they are indeed sufficient to attain the same sum degrees of freedom as the interference alignment implementation utilizing perfect channel state information at all the nodes. Significant is also the observation by Jafar [13], that statistical knowledge of channel autocorrelation structure alone is sufficient for interference alignment and, to this end, an alternative to CSI.

Another philosophy is that of embracing and exploiting interference. For example, Katti et al. [15] show that by combining physical-layer and network-layer information, network capacity can be increased. Their analog network coding scheme actually encourages strategically picked senders to interfere. Instead of forwarding packets, it suggests that routers forward the interfering signals, so that the destination leverages network-level information to cancel the interference and recover the signal destined to it. On a different tangent of embracing interference is the work by Boers et al. [3] which essentially proposes a MAC coordination scheme that takes into consideration the temporal behavior of interference patterns and aims to steer transmissions around them. Their approach simulates a pattern-aware MAC (PA-MAC) and their results include improved packet reception rates in both single and multi-hop environments at the cost of increased latency.

6   Conclusions and Research Directions

Given the reviewed literature, we identify a relative lack of work and, hence, a need to address the following technical issues:

– developing techniques to determine when, and for how long, nodes can take time away from their "regular business" of handling traffic and instead sensing the spectrum, i.e., a form of scheduling the spectrum analysis tasks with minimal impact on the regular operation of the networks,
– developing techniques that allow spectrum sensing data acquired from different nodes to be temporally aligned correctly and referenced back to natural time, despite the lack of strongly synchronized clocks,
– developing tools that will allow us to quickly identify correlations in interference patterns, both for determining the origin of the interference, and for guiding nodes to follow similar mitigation strategies,
– developing simple metrics and models for quantifying the CTI over a broad set of protocols, such that they can be used to capture "cost" metrics useful to resource management optimization decisions.

Acknowledgments. This research has been partly supported by the National Sciences and Engineering Research Council (NSERC) Discovery Grant program. Additionally, Aikaterini Vlachaki's work has been supported by an Alberta Innovates - Technology Futures (AITF) Graduate Student Scholarship.

References

1. Akyildiz, I.F., Lee, W.Y., Vuran, M.C., Mohanty, S.: Next generation/dynamic spectrum access/cognitive radio wireless networks: a survey. Comput. Netw. **50**(13), 2127–2159 (2006)
2. Ayach, O.E., Peters, S.W., Heath, R.W.: The practical challenges of interference alignment. IEEE Wirel. Commun. **20**(1), 35–42 (2013)
3. Boers, N.M., Nikolaidis, I., Gburzynski, P.: Impulsive interference avoidance in dense wireless sensor networks. In: Li, X.-Y., Papavassiliou, S., Ruehrup, S. (eds.) ADHOC-NOW 2012. LNCS, vol. 7363, pp. 167–180. Springer Berlin Heidelberg, Berlin, Heidelberg (2012). doi:10.1007/978-3-642-31638-8_13
4. Boers, N.M., Nikolaidis, I., Gburzynski, P.: Sampling and classifying interference patterns in a wireless sensor network. ACM Trans. Sen. Netw. **9**(1), 2:1–2:19 (2012)

5. Boers, N., Nikolaidis, I., Gburzynski, P.: Patterns in the RSSI traces from an indoor urban environment. In: 2010 15th IEEE International Workshop on Computer Aided Modeling, Analysis and Design of Communication Links and Networks (CAMAD), pp. 61–65 (2010)
6. Centenaro, M., Vangelista, L., Zanella, A., Zorzi, M.: Long-range communications in unlicensed bands: the rising stars in the iot and smart city scenarios. CoRR abs/1510.00620 (2015). http://arxiv.org/abs/1510.00620
7. Chen, J.K., de Veciana, G., Rappaport, T.S.: Improved measurement-based frequency allocation algorithms for wireless networks. In: IEEE GLOBECOM 2007, pp. 4790–4795 (2007)
8. Chieochan, S., Hossain, E., Diamond, J.: Channel assignment schemes for infrastructure-based 802.11 WLANs: a survey. IEEE Commun. Surv. Tutorials 12(1), 124–136 (2010)
9. Cisco: Radio resource management under unified wireless networks. Technical report 71113, Cisco Systems, May 2010
10. Gardner, W.A., Spooner, C.M.: Signal interception: performance advantages of cyclic-feature detectors. IEEE Trans. Commun. 40(1), 149–159 (1992)
11. Haidar, M., Akl, R., Al-Rizzo, H., Chan, Y.: Channel assignment and load distribution in a power-managed WLAN. In: 2007 IEEE 18th International Symposium on Personal, Indoor and Mobile Radio Communications, pp. 1–5, September 2007
12. Hithnawi, A., Li, S., Shafagh, H., Gross, J., Duquennoy, S.: Crosszig: combating cross-technology interference in low-power wireless networks. In: 2016 15th ACM/IEEE International Conference on Information Processing in Sensor Networks (IPSN), pp. 1–12, April 2016
13. Jafar, S.A.: Exploiting channel correlations - simple interference alignment schemes with no CSIT. In: GLOBECOM (2010)
14. Joshi, K., Hong, S., Katti, S.: Pinpoint: Localizing interfering radios. In: 10th USENIX Symposium on Networked Systems Design and Implementation (NSDI 2013), pp. 241–253. USENIX, Lombard, IL (2013)
15. Katti, S., Gollakota, S., Katabi, D.: Embracing wireless interference: analog network coding. Comput. Commun. Rev. 37(4), 397–408 (2007)
16. Katzela, I., Naghshineh, M.: Channel assignment schemes for cellular mobile telecommunication systems: a comprehensive survey. IEEE Pers. Commun. 3(3), 10–31 (1996)
17. Krishnamachari, R.T., Varanasi, M.K.: Interference alignment under limited feedback for MIMO interference channels. In: 2010 IEEE International Symposium on Information Theory, pp. 619–623, June 2010
18. Le, T., Szepesvári, C., Zheng, R.: Sequential learning for multi-channel wireless network monitoring with channel switching costs. IEEE Trans. Sig. Process. 62(22), 5919–5929 (2014)
19. Lehtomäki, J.: Analysis of energy based signal detection. Ph.D. thesis, University of Oulu, Faculty of Technology, Department of Electrical and Information Engineering, Oulu, Finland, November 2005
20. Leung, K.K., Kim, B.J.: Frequency assignment for IEEE 802.11 wireless networks. In: 2003 IEEE 58th Vehicular Technology Conference, VTC 2003-Fall, vol. 3, pp. 1422–1426, October 2003
21. Mahonen, P., Riihijarvi, J., Petrova, M.: Automatic channel allocation for small wireless local area networks using graph colouring algorithm approach. In: PIMRC 2004, vol. 1, pp. 536–539, September 2004

22. Mishra, A., Brik, V., Banerjee, S., Srinivasan, A., Arbaugh, W.: A client-driven approach for channel management in wireless LANs. In: Proceedings of IEEE INFOCOM 2006, 25TH IEEE International Conference on Computer Communications, pp. 1–12, April 2006

23. Mishra, A., Shrivastava, V., Banerjee, S., Arbaugh, W.: Partially overlapped channels not considered harmful. SIGMETRICS Perform. Eval. Rev. **34**(1), 63–74 (2006)

24. Murty, R., Chandra, R., Moscibroda, T., Bahl, P.V.: SenseLess: a database-driven white spaces network. IEEE Trans. Mob. Comput. **11**(2), 189–203 (2012)

25. Naganawa, J., Kim, H., Saruwatari, S., Onaga, H., Morikawa, H.: Distributed spectrum sensing utilizing heterogeneous wireless devices and measurement equipment. In: 2011 IEEE Symposium on New Frontiers in Dynamic Spectrum Access Networks (DySPAN), pp. 173–184, May 2011

26. Pefkianakis, I., Lundgren, H., Soule, A., Chandrashekar, J., Guyadec, P.L., Diot, C., May, M., Doorselaer, K.V., Oost, K.V.: Characterizing home wireless performance: the gateway view. In: 2015 IEEE Conference on Computer Communications (INFOCOM), pp. 2713–2731, April 2015

27. Rost, P., Bernardos, C.J., Domenico, A.D., Girolamo, M.D., Lalam, M., Maeder, A., Sabella, D., Wbben, D.: Cloud technologies for flexible 5G radio access networks. IEEE Comm. Mag. **52**(5), 68–76 (2014)

28. Vlachaki, A., Nikolaidis, I., Harms, J.: Wavelet-based analysis of interference in WSNs. In: 41st IEEE Conference on Local Computer Networks (LCN), November 2016 (to appear)

29. Vlachaki, A., Nikolaidis, I., Harms, J.J.: A study of channel classification agreement in urban wireless sensor network environments. In: Postolache, O., van Sinderen, M., Ali, F.H., Benavente-Peces, C. (eds.) SENSORNETS 2014, pp. 249–259. SciTePress (2014)

30. Yeo, J., Youssef, M., Agrawala, A.K.: A framework for wireless LAN monitoring and its applications. In: Proceedings of the 2004 ACM Workshop on Wireless Security, Philadelphia, PA, USA, 1 October 2004, pp. 70–79 (2004)

31. Yu, M., Luo, H., Leung, K.K.: A dynamic radio resource management technique for multiple APs in WLANs. IEEE Trans. Wirel. Commun. **5**(7), 1910–1919 (2006)

Distance Distributions in Finite Ad Hoc Networks: Approaches, Applications, and Directions

Fei Tong[1]([✉]), Jianping Pan[1], and Ruonan Zhang[2]([✉])

[1] University of Victoria, Victoria, BC, Canada
{tongfei,pan}@uvic.ca
[2] Northwestern Polytechnical University, Xi'an, Shaanxi, China
rzhang@nwpu.edu.cn

Abstract. Most performance metrics in wireless ad hoc networks, such as interference, Signal-to-Interference-plus-Noise Ratio, path loss, outage probability, link capacity, node degree, hop count, network coverage, and connectivity, are nonlinear functions of the distances among communicating, relaying, and interfering nodes. A probabilistic distance-based model is definitely needed in quantifying these metrics, which eventually involves the Nodal Distance Distribution (NDD) in a finite network intrinsically depending on the network coverage and nodal spatial distribution. In general, there are two types of NDD, i.e., (1) Ref2Ran: the distribution of the distance between a given reference node and a node uniformly distributed at random, and (2) Ran2Ran: the distribution of the distance between two nodes uniformly distributed at random. Traditionally, ad hoc networks were modeled as rectangles or disks. Recently, both types of NDD have been extended to the networks in the shape of one or multiple arbitrary polygons, such as convex, concave, disjoint, or tiered networks. In this paper, we survey the state-of-the-art approaches to the two types of NDD with uniform or nonuniform node distributions and their applications in wireless ad hoc networks, as well as discussing the open issues, challenges, and future research directions.

Keywords: Wireless ad hoc networks · Performance metrics · Distance distributions

1 Introduction

As one of the most significant applications of wireless communication technologies, a wireless ad hoc network consists of autonomous or mobile nodes which communicate with each other without a centralized control or assistance. All the nodes in the network can transmit, receive and forward messages, and thus require no support of backbone networks. Therefore, ad hoc networks provide more robustness and flexibility in the presence of node failures than those requiring infrastructure supports and are quite useful in environmental monitoring, infrastructure surveillance, disaster relief, battlefield, and scientific exploration.

© ICST Institute for Computer Sciences, Social Informatics and Telecommunications Engineering 2017
Y. Zhou and T. Kunz (Eds.): ADHOCNETS 2016, LNICST 184, pp. 167–179, 2017.
DOI: 10.1007/978-3-319-51204-4_14

To reduce the computational complexity and cost of studies and designs in wireless ad hoc networks, in addition to using powerful computer-aided design and analysis tools, the past decades have seen an increasing amount of attempts focusing on the analytical description of system characteristics and performance metrics through the modeling and analysis of wireless networks. As one of the promising tools, stochastic geometry [1] has been widely adopted, where the node distribution is assumed to follow a Poisson Point Process (PPP) [2–4] or a Binomial Point Process (BPP) [5–7]. However, the PPP model is inadequate/inaccurate in many practical wireless networks where a finite number of nodes are randomly distributed in a finite area, because it assumes an unbounded number of nodes and does not take into account the effects of the network boundaries; and the BPP model can analyze neither the exterior interference at a reference receiver nor the average performance metrics at any node but only at a specific reference node. Both models provide us with average results over time and space, but cannot present performance metrics for a specific network deployment and/or a time instance [8].

Since most of the performance metrics in finite wireless ad hoc networks are nonlinear functions of the distances among communicating, relaying, and interfering nodes, probabilistic distance-based model has been extensively studied and applied as a significant complementary tool to the PPP and BPP models to quantify these metrics, such as interference [9], Signal-to-Interference-plus-Noise Ratio (SINR) [10], path loss [11], node degree [10], link/hop distance [12,13], outage probability [11], link capacity [10], localization [14], energy consumption [15], stochastic properties of a random mobility model [16,17], etc. As a result, nodal distance distributions (NDDs) are eventually involved in such quantifications, which intrinsically depend on the network coverage and nodal spatial distribution. Depending on whether one of the communicating nodes in a node pair is fixed or random, there are generally two types of NDD, i.e., Ref2Ran NDD from a given reference node to a random node and Ran2Ran NDD between two random nodes. In addition to the utilization in wireless ad hoc networks, NDD-based models have also been widely adopted for modeling and analyzing cellular networks [8,18–23]. Therefore, how to effectively obtain the relevant NDDs is definitely significant to accurately quantify the distance-dependent performance metrics when modeling and analyzing finite wireless networks, which has attracted plenty of attention in the current literature [11,18–21,24–37].

This paper focuses on the survey of the state-of-the-art approaches to the two types of NDD as shown in Sect. 2 and their applications in wireless ad hoc networks as shown in Sect. 3. The relevant open issues, challenges, and directions are also discussed in Sect. 4. Finally, Sect. 5 concludes the paper.

2   NDDs Associated with Finite Regions

In this section, we briefly survey the state-of-the-art approaches to both Ref2Ran and Ran2Ran NDDs associated with arbitrary polygons.

2.1 Ref2Ran NDD

In our technical report [35], we have proposed a systematic recursive approach to obtain the Ref2Ran NDD from an arbitrary reference node (i.e., anywhere in the plane) to a random node inside an arbitrary polygon, which eliminates the inappropriate assumptions and limitations in the previous work that the network area has to be in certain specific shapes (including squares [18], disks/circles [11], hexagons [19,20], regular polygons [31], and convex n-gons [30]), and the reference node has to be inside or on the boundary of the network. Specifically, we first obtain the Ref2Ran NDD from a vertex of an arbitrary triangle to the triangle using the area-ratio approach, and then based on which the Ref2Ran NDD from an exterior or interior reference point to the triangle can be obtained through decomposition and recursion (D&R) methods. Such NDDs from an arbitrary reference point to an arbitrary triangle are called Ref2Ran triangle-NDDs.

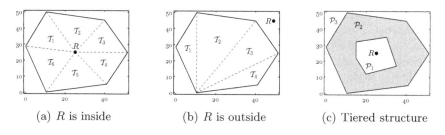

(a) R is inside   (b) R is outside   (c) Tiered structure

Fig. 1. An arbitrary reference point R and arbitrary polygons (unit: m).

Based on the obtained Ref2Ran triangle-NDDs, the distribution of the distance from an arbitrary reference point to an arbitrary polygon can also be obtained through a D&R method, since any polygon can be triangulated. If a reference point R is inside a polygon, the polygon can be triangulated from R as shown in Fig. 1(a). On the other hand, if R is outside, we can just triangulate the polygon as shown in Fig. 1(b). Then with a weighted probabilistic sum, the CDF of the distance distribution from R to the polygon is

$$F(d) = \sum_{i}^{K} \frac{S_i}{S} F_i(d), \tag{1}$$

where K is the number of triangles generated after the triangulation ($K = 6$ and 4 for the examples shown in Fig. 1(a) and (b), respectively), S_i is the area of triangle $\mathcal{T}_i$, S is the area of the polygon, and $F_i(d)$ is the CDF of the distance distribution from R to $\mathcal{T}_i$, which has been obtained already.

It is possible that a subarea of a network area has a higher node density than other subareas, which is referred to as tiered network structure in this paper, due to the reasons such as the bottleneck nodes near the hotspot under heavy loads run out of energy, some nodes are physically damaged in a hostile environment, or there are no nodes being deployed in a specific area which needs not to be monitored. In this case, the above weighted probabilistic sum is still applicable, but with the weights modified correspondingly due to the node density difference. Take the tiered structure shown in Fig. 1(c) for example. Assuming the node density ratio between $\mathcal{P}_1$ (the white area) and $\mathcal{P}_2$ (the grey area) is $\lambda_1 : \lambda_2$ (λ_1 and λ_2 are not 0 at the same time), then the distance distribution from an arbitrary reference point R to the whole area $\mathcal{P}_3$ ($\mathcal{P}_1$ plus $\mathcal{P}_2$) is

$$F_3(d) = \sum_i^2 \frac{S_i \lambda_i}{\sum_j^2 S_j \lambda_j} F_i(d), \tag{2}$$

where S_i is the area of $\mathcal{P}_i$, and $F_i(d)$ is the distance distribution from R to $\mathcal{P}_i$, which can be obtained by (1). The obtained results have been applied in our recent work [8] to analyze the outage probability of the macro and femto BSs in arbitrarily-shaped cells for tiered cellular networks.

The authors in [36] made an algorithmic implementation based on our proposed approach. Instead of using D&R methods, they modified the shoelace formula to calculate the area of the intersection between the polygon and the circle centered at an arbitrary reference point R with a radius of d. Then the probability that the distance from R to a point uniformly distributed at random within the polygon is no longer than d is the area of the intersection divided by the area of the polygon. In the modified shoelace formula, the area of the intersection between the circle and each triangle generated by triangulating the polygon from R is obtained based on our approach.

2.2   Ran2Ran NDD

For obtaining Ran2Ran NDDs, previous work had to assume that the networks are in certain specific shapes, including disks [9,26,28], triangles [29,32], rectangles [10,12,13,28], rhombuses [24], trapezoids [27], and regular polygons [10,15,21,25,33], which considerably limits the applicability of these approaches in modeling and analyzing wireless networks. Our recently proposed approach in [37] can handle the networks in the shape of arbitrary polygons as well as the polygons with different node densities in different subareas. The D&R method is also applied to this end by triangulating polygons. Specifically, we first obtain the Ran2Ran NDDs associated with arbitrary triangles (i.e., Ran2Ran triangle-NDDs), including the Ran2Ran NDD within an arbitrary triangle, and that between two arbitrary triangles which can be disjoint or share a common vertex or side. Then the Ran2Ran NDDs associated with arbitrary polygons can be obtained through D&R methods, since any polygon can be triangulated. Therefore, the Ran2Ran NDD-based performance metrics of wireless ad hoc networks associated with arbitrary polygons can be quantified properly.

Nonuniform node distribution can also be considered. Take the irregular polygon with the triangulation shown in Fig. 1(b) for example. Assuming the node density ratio among $\mathcal{T}_1$, $\mathcal{T}_2$, $\mathcal{T}_3$, and $\mathcal{T}_4$ is $\lambda_1{:}\lambda_2{:}\lambda_3{:}\lambda_4$ (λ_1, λ_2, λ_3, and λ_4 are not

0 at the same time), through D&R, the CDF of the Ran2Ran NDD within the polygon is given by a weighted probabilistic sum,

$$F(d) = \sum_{i=1}^{4} \sum_{j=1}^{4} \frac{S_i \lambda_i S_j \lambda_j}{\left(\sum_{k=1}^{4} S_k \lambda_k\right)^2} F_{ij}(d), \tag{3}$$

where S_x is the area of triangle $\mathcal{T}_x$, and $F_{ij}(d)$ or $F_{ji}(d)$ is the CDF of the Ran2Ran NDD within triangle $\mathcal{T}_i$ if $i = j$, or between two triangles $\mathcal{T}_i$ and $\mathcal{T}_j$ if $i \neq j$, which have been obtained above as Ran2Ran triangle-NDDs.

For tiered polygons shown as in Fig. 1(c), with the Ran2Ran triangle-NDDs obtained based on the above approach, the CDFs of the Ran2Ran NDDs within $\mathcal{P}_1$ and $\mathcal{P}_2$, i.e., $F_{11}(d)$ and $F_{22}(d)$, can be obtained. Assuming the node density ratio among $\mathcal{P}_1$ and $\mathcal{P}_2$ is $\lambda_1{:}\lambda_2$ (λ_1 and λ_2 are not 0 at the same time), and with a weighted probabilistic sum, we have

$$F_{33}(d) = \sum_{i=1}^{2} \sum_{j=1}^{2} \frac{S_i d_i S_j d_j}{\left(\sum_{k=1}^{2} S_k \lambda_k\right)^2} F_{ij}(d),$$

$$F_{3i}(d) = \sum_{j=1}^{2} \frac{S_j \lambda_j}{\sum_{k=1}^{2} S_k \lambda_k} F_{ij}(d), \qquad (i \in \{1, 2\}) \tag{4}$$

where S_x is the area of $\mathcal{P}_x$, and $F_{ij}(d)$ or $F_{ji}(d)$ is the CDF of the Ran2Ran NDD within $\mathcal{P}_i$ if $i = j$, or between $\mathcal{P}_i$ and $\mathcal{P}_j$ if $i \neq j$. With $F_{11}(d)$, $F_{22}(d)$, and $F_{12}(d)$ obtained by (3), $F_{33}(d)$, $F_{13}(d)$, and $F_{23}(d)$ can be obtained by (4).

3   Applications in Finite Wireless Ad Hoc Networks

As mentioned before, NDD can be utilized to characterize most of the performance metrics in finite wireless ad hoc networks due to their nonlinear relationships with the distances among nodes. In this section, we categorize the existing applications in the current literature into different levels, including graph, transceiver, link, path, and network levels. We also show the efficacy of the new approaches highlighted in Sect. 2 on some selected performance metrics with arbitrary shapes/densities, in comparison with previous approaches/approximations such as using average density, ignoring border effect (e.g., in PPP), and so on.

3.1   Graph Level

There are several representative performance metrics at the graph level, such as kth nearest neighbor (k–NN) distance, node degree (just k–NN below a certain threshold), etc. Especially, 1–NN and $(n - 1)$–NN (n is the total number of nodes in the network) correspond to the nearest and farthest neighbor distances, respectively, which are useful for routing protocol design in ad hoc networks [21,30]. For example, in a sparse network where the network size is much larger than the communication range of the nodes, a nearest-neighbor routing is beneficial to reducing energy consumption and increasing network throughput. On the other hand, in a small dense network, choosing the farthest node for packet relay, the routing overhead can be alleviated by reducing the number of transmissions. In addition, the nearest neighbor distance distribution was also utilized in [16,17] to evaluate the nearest-job-next service discipline for mobile collectors or chargers (known as mobile elements).

Suppose there are n nodes uniformly distributed at random in a network area. For a node i (either a random or reference node), the distances from the other $n-1$ nodes to node i are ordered as $d_1 \leq d_2 \leq \cdots \leq d_{n-1}$. Let Δ_k denote the random variable which represents the k–NN distance to node i. The PDF of Δ_k according to order statistic is

$$f_{\Delta_k}(d) = \frac{(n-1)!}{(k-1)!(n-1-k)!}[F(d)]^{k-1}[1-F(d)]^{n-1-k}f(d), \qquad (5)$$

where $F(d)$ and $f(d)$ are the CDF and PDF of any NDD introduced in Sect. 2, respectively.

Suppose there are $n = 10$ nodes randomly distributed in $\mathcal{P}_3$ shown in Fig. 1(c). For $\lambda_1{:}\lambda_2 = 1{:}1$ (i.e., uniform distribution) and 10:1 (nonuniform distribution), Fig. 2 shows the corresponding Ran2Ran nearest neighbor distance distributions, compared with the result obtained based on the PPP model. Due to the ignoring of the network border effect in PPP and the different node density ratios, there exist gaps among the comparisons. Also in the nonuniform case where $\mathcal{P}_1$ has a higher density, surrounded by $\mathcal{P}_2$ with a lower density, nodes are more likely closer to each other in $\mathcal{P}_1$, with nearer nearest-neighbors than the uniform case, as shown in Fig. 2.

3.2   Transceiver Level

The performance metrics at the transceiver level include path loss [11], received signal strength for a given transmission power, transmission energy consumption to ensure a certain received power [15,30,38], etc.

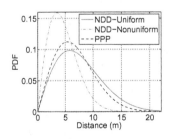

Fig. 2. Nearest neighbor distance distribution.

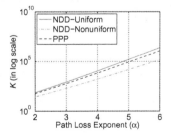

Fig. 3. Nearest neighbor energy consumption (K vs. α).

Path Loss and Signal Strength. Let us assume a general path-loss model, where the path loss of the transmission power at distance d is

$$L(d) = \beta d_0^\alpha d^{-\alpha}, \tag{6}$$

where β is a path-loss constant determined by the hardware features of transceivers, d_0 is a given reference distance, and α is the path-loss exponent. As a result, the received signal strength at distance d is just

$$P_r(d) = L(d)P_t, \tag{7}$$

where P_t is the transmission power. Therefore, given a distance distribution, the distributions of path loss and received signal strength can also be obtained by using the change-of-variable technique. The model can be readily extended to include the shadowing and fading effects of wireless channels. For example, log-normal shadowing and Rayleigh fading can be considered. For the Rayleigh fading channel, we have the PDF of the channel power gain as

$$f_X(x|d) = \frac{1}{P_r(d)} e^{\frac{-x}{P_r(d)}}. \tag{8}$$

Then the PDF of the signal strength at the receiver is

$$f_X(x) = \int_{d_{\min}}^{d_{\max}} f_X(x|d) f(d) \mathrm{d}d, \tag{9}$$

where $f(d)$ is the PDF of any NDD, and $d_{\min}$ and $d_{\max}$ are the minimum and maximum distances between the transmitter and receiver, respectively. In addition, log-normal shadowing and Rayleigh fading can also be modeled as independent random variables that are not related to inter-node distances. As shown in [39], the shadowing effect follows a log-normal distribution with standard deviation σ (typically between 0 and 8 dB), and Rayleigh fading follows an exponential distribution of mean 1. Therefore, the extension of the path-loss model along with shadowing and fading is the multiplication of independent random variables and can still be analyzed based on the NDD-based model.

Transmission Energy Consumption. The energy consumed by a radio transmitter is proportional to the αth power of the distance to the receiver. In a simplistic model with wide applicability [15,38], the average one-hop energy consumption of the radio transmitter can be formulated as

$$E_{\mathrm{Tx}} = \epsilon \int d^\alpha f(d) \mathrm{d}d = \epsilon K, \tag{10}$$

where ϵ is a constant related to the environment, $f(d)$ is the PDF of any relevant NDD, and K can be viewed as the normalized average bit-energy consumption. Based on the obtained NDDs shown in Fig. 2, Fig. 3 shows the variations of K as α increases. Due to the nonlinear effect of the path loss exponent, even a small difference in distance distributions can lead to a big difference in energy consumption. Again, PPP-based model differs from the reality due to the ignored border effect, and nonuniform node distribution also has a great effect.

3.3   Link Level

The interference [9,10], SINR [10], outage (just SINR below a certain thresh-old) [11], link capacity [10], etc., achieved at either a random or fixed receiver are link-level performance metrics. Assuming that all the transmitters in the network have the same transmission power P_t, the cumulative interference at the receiver from all its interfering nodes is

$$I = P_t \sum_i L(d_i), \tag{11}$$

where $L(d_i)$ is given in (6), and d_i is the distance from the receiver to the ith interfering node. So the SINR achieved at the receiver is

$$SINR = \frac{P_t L(d)}{N_o W + I}, \tag{12}$$

where d is the distance from the receiver to its transmitter, W is the commu-nication bandwidth, and N_0 is the one-sided spectral density of additive white Gaussian noise. Given a modulation and coding scheme, outage probability rep-resents the chance that the SINR achieved at a receiver is no larger than a spec-ified threshold so that the reception is considered unsuccessful. Therefore, the CDF of the received SINR is significant to determine the link outage probability. Meanwhile, according to Shannon's theory, the capacity of the link between the transmitter and receiver is

$$C = W \log_2(1 + SINR). \tag{13}$$

Since I, $SINR$, and C are all functions of distance, given the corresponding NDD, their distributions can also be obtained, which are significant for statistically analyzing the performance of ad hoc networks. For the network shown in Fig. 1(c) with $n = 10$ nodes randomly distributed in $\mathcal{P}_3$, Figs. 4 and 5 show the cumulative interference from the other 9 interferers and SINR at R, respectively, for both $\lambda_1{:}\lambda_2 = 1{:}1$ and 10:1, in comparison with the result obtained based on the PPP model ($P_t = 2$ mWatt, $L(d) = -38 - 20\lg(d)$ (dB), $W = 5$ MHz, and $N_0 = -174$ dBm/Hz). Higher density in $\mathcal{P}_1$, surrounding R, thus causes a higher interference and yields a lower SINR at R, as shown in the figures.

3.4   Path Level

The metrics at the link level shown above are utilized to investigate the per-formance of single-hop communications (i.e., via a direct link). For analyzing multi-hop transmissions at the path level, NDD can still be utilized. For exam-ple, hop distance is crucial to the route discovery delay, the reliability of message delivery, and the minimization of multi-hop energy consumption. The authors in [13] investigated the distribution of the minimum hop distance H between a random source and destination pair based on NDD. The closed-form expres-sions for the probability that two nodes can communicate within $H = 1$ hop or

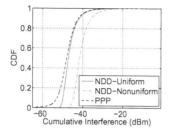

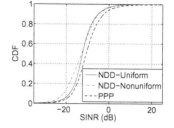

Fig. 4. Cumulative interference at R shown in Fig. 1(c).

Fig. 5. SINR achieved at R shown in Fig. 1(c).

$H = 2$ hops were derived. Analytical bounds were provided for the paths with $H > 2$ hops. In [40], the NDD between a fixed source and destination pair with a single relay uniformly distributed at random in between is utilized to obtain the distribution of the capacity of the two-hop relay communication.

3.5   Network Level

The analysis on the network capacity belongs to this level. Network capacity can be investigated from the perspective of either concurrent links or flows. For example, in a clustered ad hoc network, there are concurrent single-hop communications between cluster members and their heads in several clusters, where the network capacity can be obtained based on the link capacity. On the other hand, in an ad hoc mesh network, there might be several multi-hop communications (referred to as flows) happening concurrently. The network transport capacity in this case can be investigated based on the capacity studied at the path level.

   As shown above, ignoring border effect or using average density often in PPP and existing work skews results greatly, which highlights the need for NDDs to analyze performance metrics accurately in finite ad hoc networks.

4   Issues, Challenges, and Directions

Although the relevant research on both types of NDDs has achieved remarkable breakthroughs, there are still open issues which are challenging to be solved.

Nonuniform Distance Distributions. Most of the existing probabilistic distance-based models and the tools from stochastic geometry assume uniform node distribution. However, in many realistic ad hoc networks, nodes are not always uniformly distributed, either initially or due to node mobility. For both types of NDDs, the approaches based on D&R methods can handle the case where nodes are uniformly distributed with different node densities in different subareas of a network, which leads to a discrete nonuniform node distribution. It is necessary to consider a more general, continuous nonuniform node distribution.

3D Distance Distributions. The Ref2Ran NDDs can be easily extended to consider a reference point with height by the Pythagorean theorem, while for a more general 3D scenario and the Ran2Ran NDDs, there is little work on the approaches in the current literature.

Non-Euclidean Distance Distributions. Most of our existing work, and the existing literature, focused on Euclidean (2-norm) distances, which indeed have wide applications in wireless communication systems as radio signals propagate in the same, uniform medium. However, other distance norms, e.g., Manhattan (1-norm), Chebyshev (∞-norm) and generic l-norms, also have wide application in natural sciences and engineering disciplines, including computer networking. For example, when vehicles travel in a downtown urban scenario, Manhattan distance is more appropriate for travel distance calculation (or carry-and-forward delay in VANETs), which is in fact called taxicab geometry. There are only a few isolated distance results on taxicab geometry and other non-Euclidean geometries, including non-planar geometries such as Lobachevskian (hyperbolic) and Riemannian (spherical) geometries, which are increasingly used for modeling logical and physical networks. We plan to extend our planar geometry results to non-planar ones, with new results and new approaches.

High-Order and Multi-hop Distance Distributions. So far, our work focused on the distance distributions between two random points, or between a random point and a reference point. A high-order distance distribution involves more than two points. For example, in relay communications, the relay can choose different forwarding schemes (amplify-and-forward, decode-and-forward, and so on) and the destination can select or combine the signal from the source or relay (selective combination, maximum rate combination, and so on). For amplify-and-forward and selective combination, the signal strength received at the destination depends on the path loss of both the source-to-relay and relay-to-destination channels, eventually the product of the distances among the source, relays and destination. Similarly there is a need for the sum, difference and ratio of two distances. Furthermore, more than two distances can be involved, e.g., in the distribution of hop distances between source and destination. This is a very hard problem and there are only a few results for two or three hops [40].

Joint and Conditional Distance Distributions. The ultimate goal is to address the joint and conditional distance distributions, with potentially correlated distances due to the triangular inequality. For engineering problems, although closed-form explicit expressions with elementary functions are our target, as what we have achieved for rhombuses, hexagons and triangles, we also have the freedom to develop algorithmic approaches that produce symbolic results in a parametrized way. For example, for an arbitrary geometry without symbolic expressions, it is unlikely to derive its properties symbolically, but given an arbitrary geometry with parameters, the algorithm and thus computer program can output parametrized expressions. This can further guide and speed up numerical calculation and error or bound analysis, determining expression truncation and numerical precision, for practical usage purposes.

5 Conclusion

As a significant complementary tool to the PPP/BPP models from stochastic geometry, probabilistic distance-based model has been extensively studied and applied in finite wireless networks. In this paper, we surveyed the state-of-the-art approaches to both the Ref2Ran and Ran2Ran NDDs with arbitrary shapes and nonuniform densities, and their applications in ad hoc networks. The still-open issues were also discussed for the future research directions in this field.

Acknowledgment. This work is supported in part by NSERC, CFI, and BCKDF, and by National Natural Science Foundation of China (61571370) and National Civil Aircraft Major Project of China (MIZ-2015-F-009), and Fei Tong is also supported in part by the Ministry of Educations Key Lab for Computer Network and Information Integration at Southeast University, China.

References

1. Baccelli, F., Błaszczyszyn, B.: Stochastic Geometry and Wireless Networks, Volume I: Theory; Volume II: Applications. NOW Publisher, Delft (2009)
2. Weber, S., Andrews, J., Jindal, N.: An overview of the transmission capacity of wireless networks. IEEE Trans. Commun. **58**(12), 3593–3604 (2010)
3. Zhao, S., Fu, L., Wang, X., Zhang, Q.: Fundamental relationship between node density and delay in wireless ad hoc networks with unreliable links. In: ACM MobiCom, pp. 337–348 (2011)
4. Ren, W., Zhao, Q., Swami, A.: Temporal traffic dynamics improve the connectivity of ad hoc cognitive radio networks. IEEE/ACM Trans. Netw. **22**(1), 124–136 (2014)
5. Srinivasa, S., Haenggi, M.: Distance distributions in finite uniformly random networks: theory and applications. IEEE TVT **59**(2), 940–949 (2010)
6. Torrieri, D., Valenti, M.: The outage probability of a finite ad hoc network in Nakagami fading. IEEE Trans. Commun. **60**(11), 3509–3518 (2012)
7. Valenti, M., Torrieri, D., Talarico, S.: A direct approach to computing spatially averaged outage probability. IEEE Commun. Lett. **18**(7), 1103–1106 (2014)
8. Ahmadi, M., Tong, F., Zheng, L., et al.: Performance analysis for two-tier cellular systems based on probabilistic distance models. In: INFOCOM, pp. 352–360 (2015)
9. Naghshin, V., Rabiei, A., Beaulieu, N., et al.: Accurate statistical analysis of a single interference in random networks with uniformly distributed nodes. IEEE Commun. Lett. **18**(2), 197–200 (2014)
10. Fan, P., Li, G., Cai, K., Letaief, K.: On the geometrical characteristic of wireless ad-hoc networks and its application in network performance analysis. IEEE Trans. Wirel. Commun. **6**(4), 1256–1265 (2007)
11. Baltzis, K.B.: The distribution of path losses for uniformly distributed nodes in a circle. Res. Lett. Commun. **2008**(4), 1–4 (2008)
12. Miller, L.E.: Distribution of link distances in a wireless network. J. Res. Natl. Inst. Stand. Tech. **106**(2), 401–412 (2001)
13. Bettstetter, C., Eberspacher, J.: Hop distances in homogeneous ad hoc networks. In: IEEE VTC, pp. 2286–2290 (2003)
14. Leão, R.S., Barbosa, V.C.: Exploiting the distribution of distances between nodes to efficiently solve the localization problem in wireless sensor networks. In: ACM PM^2HW^2N, pp. 9–16 (2010)

15. Zhuang, Y., Pan, J., Cai, L.: Minimizing energy consumption with probabilistic distance models in wireless sensor networks. In: IEEE INFOCOM, pp. 1–9 (2010)
16. He, L., Yang, Z., Pan, J., et al.: Evaluating service disciplines for mobile elements in wireless ad hoc sensor networks. In: IEEE INFOCOM, pp. 576–584 (2012)
17. He, L., Yang, Z., Pan, J., et al.: Evaluating service disciplines for on-demand mobile data collection in sensor networks. IEEE TMC **13**(4), 797–810 (2014)
18. Pirinen, P.: Outage analysis of ultra-wideband system in lognormal multipath fading and square-shaped cellular configurations. EURASIP J. Wirel. Commun. Netw. **2006**, 1–10 (2006)
19. Zhuang, Y., Luo, Y., Cai, L., Pan, J.: A geometric probability model for capacity analysis and interference estimation in wireless mobile cellular systems. In: IEEE GLOBECOM, pp. 1–6 (2011)
20. Baltzis, K.B.: Analytical and closed-form expressions for the distribution of path loss in hexagonal cellular networks. Wirel. Personal Commun. **60**(4), 599–610 (2011)
21. Zhuang, Y., Pan, J.: A geometrical probability approach to location-critical network performance metrics. In: IEEE INFOCOM, pp. 1817–1825 (2012)
22. Apilo, O., Lasanen, M., Boumard, S., Mammela, A.: The distribution of link distances in distributed multiple-input multiple-output cellular systems. In: IEEE VTC, pp. 1–5 (2013)
23. Baltzis, K.B.: Spatial characterization of the uplink inter-cell interference in polygonal-shaped wireless networks. Radioengineering **22**(1), 363–370 (2013)
24. Zhuang, Y., Pan, J.: Random distances associated with rhombuses. arXiv:1106.1257 (2011)
25. Zhuang, Y., Pan, J.: Random distances associated with hexagons. arXiv:1106.2200 (2011)
26. Baltzis, K.B.: A geometric method for computing the nodal distance distribution in mobile networks. Prog. Electr. Res. **114**, 159–175 (2011)
27. Ahmadi, M., Pan, J.: Random distances associated with trapezoids. arXiv:1307.1444 (2013)
28. Moltchanov, D.: Distance distributions in random networks. Ad Hoc Netw. **10**(6), 1146–1166 (2012)
29. Bäsel, U.: The distribution function of the distance between two random points in a right-angled triangle. arXiv:1208.6228 (2012)
30. Baltzis, K.B.: Distance distribution in convex n-gons: mathematical framework and wireless networking applications. Wirel. Personal Comm. **71**(2), 1487–1503 (2013)
31. Khalid, Z., Durrani, S.: Distance distributions in regular polygons. IEEE Trans. Vech. Tech. **62**(5), 2363–2368 (2013)
32. Tong, F., Ahmadi, M., Pan, J.: Random distances associated with arbitrary triangles: a systematic approach between two random points. arXiv:1312.2498 (2013)
33. Bäsel, U.: Random chords and point distances in regular polygons. Acta Mathematica Universitatis Comenianae **83**(1), 1–18 (2014)
34. Tong, F., Ahmadi, M., Pan, J., Zheng, L., Cai, L.: Poster: geometrical distance distribution for modeling performance metrics in wireless communication networks. In: ACM MobiCom, pp. 341–343 (2014)
35. Ahmadi, M., Pan, J.: Random distances associated with arbitrary triangles: a recursive approach with an arbitrary reference point. UVicSpace (2014)
36. Pure, R., Durrani, S.: Computing exact closed-form distance distributions in arbitrarily shaped polygons with arbitrary reference point. Math. J. **17**, 1–27 (2015)

37. Tong, F., Pan, J.: Random distances associated with arbitrary polygons: An algorithmic approach between two random points. arXiv:1602.03407 (2016)
38. Heinzelman, W.B., Chandrakasan, A.P., Balakrishnan, H.: An application-specific protocol architecture for wireless microsensor networks. IEEE Trans. Wirel. Commun. $1(4)$, 660–670 (2002)
39. Cheikh, D.B., Kelif, J.-M., Coupechoux, M., et al.: SIR distribution analysis in cellular networks considering the joint impact of path-loss, shadowing and fast fading. EURASIP J. Wirel. Commun. Network. $2011(1)$, 1–10 (2011)
40. Song, X., Zhang, R., Pan, J., et al.: A statistical geometric approach for capacity analysis in two-hop relay communications. In: GLOBECOM, pp. 4823–4829 (2013)

Towards More Realistic Network Simulations: Leveraging the System-Call Barrier

Roman Naumann$^{(\boxtimes)}$, Stefan Dietzel, and Björn Scheuermann

Humboldt-Universität zu Berlin, Berlin, Germany
{roman.naumann,stefan.dietzel}@hu-berlin.de,
scheuermann@informatik.hu-berlin.de

Abstract. Network simulations play a substantial role in evaluating network protocols. Simulations facilitate large-scale network topologies and experiment reproducibility by bridging the gap between analytical evaluation and real-world measurements. A recent trend in discrete event network simulations is to enhance simulation realism and reduce duplicate implementation efforts by maximizing code reuse. Despite such efforts, it is not yet possible to run arbitrary network applications in state-of-the-art network simulators. As a consequence, researchers are required to maintain separate protocol implementations: one for real-world measurements and one for simulations. We review existing approaches that maximize code reuse in simulations, compare their limitations, and propose a novel architecture for protocol simulation that overcomes those restrictions.

Keywords: Simulation · Virtualization · Code reuse · Simulation realism

1 Introduction

When developing novel ad hoc networking protocols, extensive evaluation to gauge their performance and fitness to fulfill use case requirements is an integral part of the protocol design. The same is true when existing protocols are to be revised or improved. Arguably, network simulations are among the most widespread tools used to evaluate protocols for ad hoc networks. Network simulators strike a balance between the fundamental and asymptotic results that formal, analytical protocol evaluations can provide and the realism that testbed implementations on real hardware can provide.

To implement network simulations, the lower layers of the protocol stack are typically approximated by more or less simplified models, whereas higher protocol layers are re-implemented to mimic real-world protocol stacks. Clearly, the physical layer needs to be simulated, and state-of-the-art physical network models have largely been confirmed by empirical measurements [1,2]. For the protocol under evaluation, it is desirable that it is evaluated using the same code that would be used in real deployments; only then can we draw meaningful conclusions from simulative evaluation results. Designing the protocol layers in

© ICST Institute for Computer Sciences, Social Informatics and Telecommunications Engineering 2017
Y. Zhou and T. Kunz (Eds.): ADHOCNETS 2016, LNICST 184, pp. 180–191, 2017.
DOI: 10.1007/978-3-319-51204-4_15

between is challenging, because they need to operate with the simulated physical layer while at the same time allowing the protocol under evaluation to be implemented under realistic conditions.

Today, simulator implementations for intermediate layers are often based on standards or available specifications, whereas real-world implementations contain further optimizations and extensions that affect performance. For instance, real-world TCP variants are a complex interaction of several optimizations performed by the operating system's implementation [3]. Likewise, the most widely used implementation [4] of Optimized Link State Routing [5], a routing daemon for ad hoc networks, implements non-standard link quality extensions that drastically improve performance in wireless mesh networks, whereas a state-of-the-art network simulator's version is based on the official specification only.

In the worst case, the network system – be it a protocol or an application – has to be developed twice: once in the simulator and once for real-world deployment. Such duplicate implementations have several negative implications. First, the implementations' behavior may diverge due to implementation differences. In addition to increased development effort, the differences between the implementations may invalidate simulation results, since they no longer match real-world behavior. Second, the choice of a network simulator may require to use a specific programming language that the development team would not use for real deployments. Finally, the additional effort slows down the development and evaluation of network systems. These and other issues with current simulators have been acknowledged by the simulation community, spawning a trend to increase protocol code reuse [3,6–8].

Code-reuse issues are emphasized by the abstraction level of widely used network simulators, such as OMNeT++ [9] and ns-3 [10]. When developing network protocols for ad hoc networks, researchers interact with artificial interfaces towards the network, medium access control, and physical layers. We argue that shifting the abstraction between actual implementation parts and modeled parts towards the lower layers will provide more realistic simulation results and facilitate more widespread code reuse. To do so, we propose to utilize the system call interface, which is well established in Unix-like operating systems. As ad hoc network systems often use Unix-like operating systems, the system call (short "syscall") interface provides a clear interface to separate real protocol implementations from simulated parts of the network.

In this paper, we contribute a taxonomy of different abstraction levels for network simulations, which we use to survey existing approaches to achieve more realistic network simulations. Moreover, we discuss a novel syscall-level approach to combine the benefits of realistic protocol implementations with those of discrete event network simulations.

We discuss the general role of simulation in network evaluation and introduce discrete event network simulation in Sect. 2. Section 3 then structures options for code reuse in network simulations and discusses existing approaches' advantages and limitations. We present a less restrictive simulation architecture that maximizes code reuse in Sect. 4. In Sect. 5 we summarize and conclude the paper.

2   Discrete Event Network Simulation

When designing or implementing a new network system, verifying correctness and efficiency is an important but difficult task. A network system is generally developed with a use-case environment in mind. In an ideal world, the system's designer could test her protocol in this exact environment as long and as often as necessary. In reality, the exact environment is often not available due to practical considerations such as cost of components or the time it takes to perform measurements, which is especially true when a large number of systems are involved. Likewise, real-world measurements are not easily reproducible, since external influences often cannot be controlled.

For these reasons, several accepted techniques for network system evaluation aim to increase scalability and reproducibility over real-world measurements by controlling external influences to different degrees. The system designer can use these techniques to evaluate a network system without requiring the exact deployment environment. Common approaches to evaluation fall in the categories *analytical evaluation*, *simulation*, and measurements using a *testbed*. As sketched in Fig. 1, evaluation by analytical evaluation, simulation, and testbeds typically offer scalability and reproducibility in decreasing order, and they offer closeness to real-world measurement results, i.e., "realism," in increasing order.

The testbed is the method closest to real-world measurements; protocols are evaluated on real hardware. Testbeds provide partially controlled environments where measurement time and topology of the nodes are pre-determined, whereas external interference, for instance, cannot be predicted in the general case. Scalability (i.e., number of nodes, size of topology, number of measurements) is limited by practical considerations, such as cost of hardware, available space, and the time it takes to perform measurements. The benefit of testbeds is that results are close to real-world measurements when the topology resembles the network system's use-case environment.

Analytical evaluation is on the opposite end of the spectrum. It involves finding the right abstractions to formally model a network protocol and its environment, and it allows to mathematically assess their interaction. Once such a model is found, it is generally possible to arbitrarily scale parameters, such as, number of nodes or size of topology. Realism of analytical evaluation results is highly dependent on the choice of abstractions, since it is seldom possible to formalize all facets of a protocol in a tractable analytical model.

From the viewpoint of realism, network simulations fill the middle ground between analytical protocol evaluation and real-world measurements performed in a testbed. In comparison to testbeds, network simulations provide better scalability and reproducibility. Using modern network simulators, it is possible to simulate hundreds or thousands of nodes in arbitrary topologies and repeat experiments with fully controlled randomness.

The most common technique for network simulators is discrete event network simulation. A discrete event network simulator is driven by events, which can be a timer running out or a packet being received by a simulated network card.

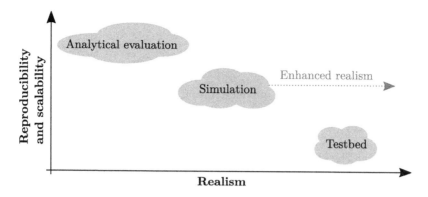

Fig. 1. Evaluation realism

An important property of discrete event network simulators is that events are mere points in time, i.e., no time passes *during* an event. Similarly, no time passes *between* events. Instead, the simulator skips to the next event after processing one event, not simulating anything in between. This approach scales well since it only requires to process what happens during events. Simulated time in a discrete event network simulator is different from system time: simulated time may run faster or slower than system time, depending on the system load of the machine running the simulator and the simulation's complexity.

Discrete event simulators facilitate reproducibility: if properly implemented, running a simulation twice with the same parameters yields the exact same results, enabling precise debugging of rare corner cases. Randomness in simulations is controlled by the simulator's (pseudo-)random number generator, which can be initialized with different seeds to select a statistically meaningful sample size.

From a protocol-implementation perspective, a simulated environment is therefore different from a real-world environment. When considering code reuse, we need to carefully consider the effect of discrete event simulation on real-world implementations. If too many aspects of the implementation are affected by the simulator interface, code reuse is difficult or impossible, limiting meaningfulness of simulation results. If too few aspects are affected, we lose the benefits of discrete event simulations, foremost its reproducibility.

3   Options for Code Reuse

We have established that finding the right level of abstraction between realistic implementations and simulated parts is key to reusable yet scalable and reproducible network simulations. The level of abstraction is determined by the extent to which code can be reused between simulations and real deployments. In this section, we identify different options for code reuse and survey existing approaches within this structure.

3.1   Partial Source Reuse

The simplest form of code reuse is what we term *partial source reuse*. When the protocol implementation's programming language is compatible with the simulator language, it is trivially possible to copy-and-paste chunks of source code to the network simulator implementation and execute them as part of the simulation.

For instance, the state-of-the-art discrete event network simulators OMNeT++ [9] and ns-3 [10] support the C++ programming language for protocol simulations. Therefore, C or C++ protocol implementation source code can be used as part of a simulation. Likewise, existing Java protocol source code can be used in the JiST/SWANS simulator [11].

There are several software components, however, that need porting or re-implementation to work in a discrete event simulator:

- real-world socket APIs cannot be used in a simulator; instead, the network abstractions provided by the simulator need to be used;
- time is different from the system time in a network simulator, so no system time queries must be made;
- random numbers have to stem from the simulators pseudo-random number facility exclusively;
- concurrency is often not supported by discrete event simulators, instead the asynchronous event dispatcher of the simulator has to be used;
- global variables may prevent spawning more than one application instance in a simulator; and
- likewise, file system operations may conflict when more than one application instance is simulated.

We conclude that partial code reuse can help alleviate duplicate implementation efforts, but by no means eliminates them, because all of the above issues have to be addressed manually.

3.2   Full Source Reuse

Recent research [3,6,8,12,13] has investigated how duplicate implementation effort can be minimized by increasing code reuse. Here, we discuss approaches based on sharing the entire source code of a protocol implementation for simulation and real-world deployment. We distinguish two different approaches for full source reuse: employing a software compatibility layer and using alternative compilation methods.

Software Compatibility Layer. A special case of code reuse is the approach taken by Click [14], where network protocols are implemented in a modular fashion in C++ and a domain-specific router configuration language. Click protocol implementations can be deployed on real hardware or integrated in a simulator such as ns-3 [6]. Click's aim is to find suitable programming abstractions for the critical software components listed in Sect. 3.1. The protocol is implemented against this compatibility layer instead of APIs that are specific to the real world or simulation environments.

Of course, the Click approach only works when developing a new protocol, as tight integration with Click is needed. Another restriction is that the compatibility layer can only support those features that *all* supported platform APIs provide.

Mayer *et al.* [13] consider a more lightweight compatibility layer for the OMNeT++ simulator. Instead of compiling a user space protocol's sources into an executable, a shared library is built and dynamically loaded into the simulator. The authors suggest to replace the network functionality with a compatibility wrapper, so that it can quickly be exchanged depending on whether the protocol is built for real-world deployment or for simulation. Likewise, calls that query the current time are replaced by pre-processor macros that switch between simulation time and system time depending on the compilation mode.

Alternative Compilation. Tazaki *et al.* [15] propose a refined shared library approach for the ns-3 simulator that, with some restrictions, allows a protocol implementation's sources to run unmodified in the simulator, i.e., without a compatibility layer. Again, a shared library is built from the implementation's sources, as depicted in Fig. 2, and dynamically loaded into the simulator. However, instead of using a compatibility layer (which requires in-source changes), calls to the operating system's standard library are redirected to a wrapper library. The wrapper library decides whether to pass the call to the operating system (for most calls), or provide an alternative implementation based on simulator facilities. For example, a call to the function that returns the length of a string (`strlen`) can safely be passed through, as it does not perform input or output operations, whereas a call to a function that normally returns the current system time (e.g., `gettimeofday`) is replaced by a wrapper that returns simulation time. The approach can be used for kernel-space protocols in a similar fashion [3,7].

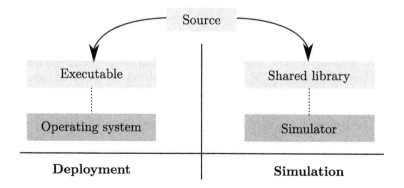

Fig. 2. Shared library approach

The shared library approach is the first to allow running unmodified applications without reducing the reproducibility guarantees provided by discrete event simulation, but it has restrictions on a conceptual level: compilation to a shared library requires that the source code is available and in fact can be compiled into such a shared library. The former is not necessarily true when proprietary implementations are evaluated and the latter does not usually hold for most interpreted programming languages and even many mainstream compiled languages, such as Java or Go.

3.3   Process Reuse

Another approach is to run node processes or even the nodes' operating systems via virtualization and solely exchange network traffic between these real processes and the simulation. In the context of the OMNeT++ simulator, this approach was first briefly discussed in [13] and later implemented by Staub et al. [12].

The advantage of network traffic exchange between simulation and real processes is that full code reuse is trivial, since processes run in the same environment as they would when deployed. No programming language limitations or tool chain restrictions apply when the system is implemented as in [12]. Unfortunately, this approach does not maintain perfect reproducibility, because only network operations are simulated. Processes or operating systems do not run in the simulated time domain but in their respective system time domain; system (pseudo-)random numbers cannot be predicted, i.e., reproduced.

4   Leveraging the System Call Barrier

The shared-library approach that we saw in Sect. 3.2 chose the operating system's standard library as the barrier between simulation and a user's protocol implementation. What we propose here is to use a lower-level abstraction as the border between simulation and real-world applications. Operating systems already provide a natural barrier between user-space and kernel-space that can only be transgressed via so-called system calls (syscalls).

4.1   The Syscall Interface

An obvious property of the system call barrier is that the operating system is agnostic towards programming language details: every process, regardless of whether it is a compiled executable, an interpreter, or a just-in-time compiled program fragment, uses the same syscalls to interface with the operating system's kernel. So the approach supports running all of these protocol implementations, even proprietary ones, with zero modification, thereby maximizing code reuse.

Techniques to capture and modify system calls, often called *syscall wrapping*, have been used before in the security context [16] and for operating system emulation [17]. To make use of the system call barrier for discrete event network simulation, it is necessary to filter and selectively re-implement system calls.

The Linux operating system kernel version 4.5, for instance, supports a total of 385 system calls for file manipulation, signal handling, concurrency, socket operations, and so forth. While this number may appear to be large, most system calls are rarely used and implementing only a subset would already support numerous protocol implementations. For example, our experiments show that a simple web page served by the Nginx web server utilizes 46 distinct syscalls and the olsrd [4] daemon uses 26 unique system calls when running in minimal mode. Both implementations invoke largely the same – frequently used – system calls and jointly require only 51 distinct system calls. Of these commonly used syscalls, only a fraction needs to be modified during execution, whereas most system calls need not be modified to support reproducibility in discrete network simulations. System call groups that can be passed through instead of being re-implemented include security options, memory manipulation, process manipulation, and most concurrency operations, since these do not usually involve network traffic or system time [15].

4.2 Syscall-Barrier Process Simulation

Figure 3 shows an overview of our proposed process simulation architecture: Topmost are user-space processes, and bottommost is the operating system kernel. Two components constitute the simulator process in the middle: the discrete event simulation logic to the right and the syscall wrapper to the left. We propose to use syscall wrapping, as shown in Fig. 3, to selectively redirect syscalls to the simulator logic and emulate them there. Non-emulated system calls are forwarded to the operating system as is. The simulator process thereby implements a secondary system call barrier to run real-world processes within the simulation environment. Previous work that uses syscall wrappers for process virtualization suggests that the syscall barrier's performance is lower than hardware virtualization as in, e.g., XEN [18]. Dike et al. [17] notes that the performance penalty is dominated by additional context switches. This factor can, however, be mitigated on platforms that provide special operating system support for syscall wrapping [19,20].

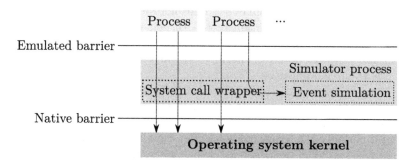

Fig. 3. Syscall-barrier process simulation

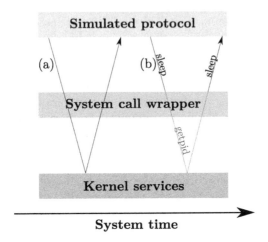

Fig. 4. System call wrapping

To illustrate our approach, we discuss how to wrap two example syscalls. Namely, we discuss two system calls that the OLSR mesh routing daemon issues: the first system call (a) is mprotect, a system call that changes access permissions on a memory region. The second system call (b) is nanosleep, which causes the kernel to suspend the calling thread's execution via a high-resolution timer. As shown in Fig. 4, (a) is an example for a syscall that can be passed through to the actual kernel services, whereas (b) is a syscall that needs to be caught and handled by the wrapper.

Like [17], we assume a Linux system and a syscall wrapper based on the ptrace framework [20]. The wrapper runs solely in user space and leverages the ptrace system call to trace protocol processes. ptrace enables syscall inspection and modification at two points: (1) just before the system call is processed by the kernel and (2) just after the system call was processed by the kernel, but before the protocol process is notified. As soon as the mprotect system call (a) is issued, but before the kernel processed the call, the syscall wrapper would be notified by the ptrace framework. It can inspect the system call and decide, based on a lookup table, that mprotect does not affect the network simulation. The wrapper hands back execution to the kernel, which processes the system call as usual, and is notified again when the system call's processing is finished but before the protocol process is notified. Again, the syscall wrapper continues execution without modification.

The second system call (b) is nanosleep. Time-related system calls need embedding in a discrete event simulation environment, so the wrapper intercepts the call: the syscall wrapper first registers an event with the simulator that notifies the wrapper once the requested *simulation* time has passed. If operating system support is available, the original system call can be skipped altogether [19]. Otherwise, the syscall is replaced by a dummy system call without input or output, such as getpid [17], as indicated in Fig. 4. After the (dummy) system call is processed, but before the protocol process is notified, the original system call's result is emulated by modifying the protocol processes registers. In particular, the syscall's return value is replaced by zero, which indicates success for `nanosleep`. Next, the syscall wrapper waits until the event it has registered with the simulator expires. Once notified that the event has expired, it continues the protocol process. By following these steps, the `nanosleep` system call is transparently emulated by the simulator, replacing system time with simulation time – which is crucial for experiment reproducibility.

Other system calls can be implemented in a similar fashion. Some syscalls can be passed through, because they do not interfere with the simulation time. In some cases, syscalls may be forwarded, but their parameters need to be modified. Examples are file system operations, where potentially path prefixes should be modified by the syscall wrapper. Others, such as timing and network interactions, need to be intercepted entirely and handled internally.

4.3 Syscall-Barrier System Simulation

Pushing the border even further towards the operating system level, we can emulate the whole operating system while maintaining the syscall barrier as the interface to the network simulator. Normally, this approach would require that the simulator emulates hardware on which a node's operating system can run. Dike *et al.* [17] shows that instead it is possible and feasible to port an operating system "to itself" in terms of system calls.

Instead of creating an environment that the virtualized operating system can run on, the virtualized operating system is ported to run on an existing system call environment. In theory, we can utilize these findings to run a node's operating system and all associated protocol implementation's processes via the same interface that we propose in Sect. 4.2 – the system call barrier.

This approach maximizes code reuse: nodes run fully virtualized, running real-world protocol stacks on all layers above the medium access and physical layers, which the network simulator models and simulates.

It is an open research challenge to evaluate how a real-world operating system behaves when running on top of a discrete event network simulator, since time-related syscalls behave differently in a discrete simulation environment. However, due to the successful porting of kernel-space UDP and TCP implementations [3,7], we are positive that this next level of code reuse can be obtained without much modification to the kernel.

5 Conclusion

During the design and implementation phase of a network system, it is important to verify the system's correctness and performance. Simulating a protocol during the design phase allows to carefully tune parameters and quickly assess a proposed modification's performance impact. Unfortunately, with today's tool support, it is often required to maintain two separate implementations for simulation and real-world deployment. This undermines both correctness – as implementation differences question simulation results – and efficiency – as two implementations duplicate development efforts.

We reviewed and structured a number of approaches that maintain the reproducibility and scalability of discrete event network simulation, and at the same time, improve correctness and reduce duplicate effort by increasing code reuse. Among those approaches, the recently proposed shared library approach [3,7] facilitates full source reuse with a state-of-the-art simulator, albeit with a number of restrictions.

We proposed a system-call barrier design as an alternative abstraction level to form the border between simulator and protocol stack, i.e., model and real-world code. Our design has the potential to solve the remaining restrictions that are inherent to the shared-library approach. It is agnostic to programming language, it can run compiled, interpreted, or just-in-time compiled code, and it does not require a modified tool chain nor modified source code. The proposed design is based on a technique called system call wrapping, which has been used for security and virtualization previously. We also describe an extended design that utilizes an operating system's port to itself to simulate nodes' operating systems via the system-call barrier.

We expect that, along with the trend to improve code reuse, the use of simulation in the evaluation of network systems will increase. It remains to be seen whether perfect reproducibility can be upheld when modeling arbitrarily complex systems such as full operating systems without modification, but the direction is promising and we expect more results from this line of research in the near future.

Acknowledgments. This work has received funding from the European Union's Horizon 2020 research and innovation programme under grant agreement No. 636892 in the context of the PREVIEW project [21].

References

1. Hashemi, H.: The indoor radio propagation channel. Proc. IEEE **81**(7), 943–968 (1993)
2. Tanghe, E., Joseph, W., Verloock, L., et al.: The industrial indoor channel: largescale and temporal fading at 900, 2400, and 5200 MHz. IEEE Trans. Wirel. Commun. **7**(7), 2740–2751 (2008)
3. Jansen, S., McGregor, A.: Simulation with real world network stacks. In: Proceedings of the Winter Simulation Conference, December 2005
4. OLSR.org Wiki (2016). http://www.olsr.org/mediawiki/index.php/Main_Page. Accessed 26 June 2016
5. Clausen, T., Jacquet, P.: Optimized Link State Routing Protocol (OLSR). Request for Comments, No. 3626, October 2003
6. Suresh, P.L., Merz, R.: ns-3-click: click modular router integration for ns-3. In: Proceedings of the 4th International ICST Conference on Simulation Tools and Techniques, SIMUTools 2011, pp. 423–430. Institute for Computer Sciences, Social-Informatics and Telecommunications Engineering (ICST) (2011)

7. Tazaki, H., Urbani, F., Turletti, T.: DCE cradle: simulate network protocols with real stacks for better realism. In: Proceedings of the 6th International ICST Conference on Simulation Tools and Techniques, SimuTools 2013. Institute for Computer Sciences, Social-Informatics and Telecommunications Engineering (ICST), pp. 153–158 (2013)

8. Camara, D., Tazaki, H., Mancini, E., et al.: DCE: test the real code of your protocols and applications over simulated networks. IEEE Commun. Mag. **52**(3), 104–110 (2014)

9. Varga, A., Hornig, R.: An overview of the OMNeT++ simulation environment. In: Proceedings of the 1st International Conference on Simulation Tools, Techniques for Communications, Networks, Systems and Workshops, Simutools 2008, pp. 60:1–60:10. Institute for Computer Sciences, Social-Informatics, Telecommunications Engineering (ICST) (2011)

10. Henderson, T.R., Lacage, M., Riley, G.F., et al.: Network simulations with the ns-3 simulator. In: SIGCOMM Demonstration, vol. 14 (2008)

11. Barr, R.: An Efficient, Unifying Approach to Simulation Using Virtual Machnies (2004)

12. Staub, T., Gantenbein, R., Braun, T.: VirtualMesh: an emulation framework for wireless mesh and ad hoc networks in OMNeT++. In: Simulation (2010)

13. Mayer, C.P., Gamer, T.: Integrating real world applications into OMNeT++. Technical report TM-2008-2, Karlsruhe Institute of Technology, Karlsruhe, Germany (2008)

14. Morris, R., Kohler, E., Jannotti, J., et al.: The click modular router. ACM Trans. Comput. Syst. **18**, 263–297 (2000)

15. Tazaki, H., Uarbani, F., Mancini, E., et al.: Direct code execution: revisiting library OS architecture for reproducible network experiments. In: Proceedings of the Ninth ACM Conference on Emerging Networking Experiments and Technologies, CoNEXT 2013, pp. 217–228. ACM (2013)

16. Watson, R.N.: Exploiting concurrency vulnerabilities in system call wrappers. In: WOOT, vol. 7, pp. 1–8 (2007)

17. Dike, J., et al.: User-mode Linux. In: Annual Linux Showcase & Conference (2001)

18. Emeneker, W., Stanzione, D., et al.: HPC cluster readiness of Xen and user mode Linux. In: CLUSTER (2006)

19. Vivier, L.: User-Mode-Linux SYSEMU Patches (2011). http://sysemu.sourceforge.net/. Accessed 26 June 2016

20. Ptrace(2) Linux User's Manual, June 2016

21. Preview Project EU - Predictive System for Injection Mould Process Optimisation. http://www.preview-project.eu/. Accessed 27 Oct 2015

Resource Allocation for Relay-Aided Cooperative Hospital Wireless Networks

Jingxian Liu[1], Ke Xiong[1($\boxtimes$)], Yu Zhang[2], and Zhangdui Zhong[3]

[1] School of Computer and Information Technology, Beijing Jiaotong University, Beijing, China
xiongke.bjtu.iis@gmail.com
[2] School of Computer and Communication Engineering, University of Science and Technology Beijing, Beijing, China
[3] State Key Laboratory of Rail Traffic Control and Safety, Beijing Jiaotong University, Beijing, China

Abstract. In this paper, we investigate the relay-aided hospital wireless systems in cognitive radio environment, where multiple transmitter-relay pairs desire transmit their collected information to a data center. For the system, we propose a transmission framework, which follows IEEE 802.22 WRAN and adopts the listen-before-talk and geo-location/database methods to protect the primary users. The transmission strategy is presented, where in each subsystem, the wireless sensor device (WSD) with the highest signal-to-noise ratio (SNR) is selected to transmit signals at each time and then, a two-hop half-duplex decode-and-forward (DF) relaying transmission is launched among the selected WSD, the corresponding personal wireless hub (PWH) and the data center. To explore the potential system performance, an optimization problem is formulated to maximize the system sum rate via power allocation. We then solve it by using convex optimization theory and KKT condition method and derive a closed-form solution of the optimal power allocation. Simulation results demonstrate the validity of our proposed scheme and also show the effects of the total power, the interference thresholds and the scale of the network on the system performance, which provide some insights for practical hospital wireless system design.

Keywords: Relays · Wireless sensor devices · Hospital environment · Power allocation

1 Introduction

1.1 Background

With the fast development of wireless sensor devices (WSDs), various wireless networks have been developed, which are used for various aspects, including traffic control, healthcare, home automation and habitat monitoring [1]. An emerging paradigm of this kind of network is wireless sensor networks (WSNs), which has

© ICST Institute for Computer Sciences, Social Informatics and Telecommunications Engineering 2017
Y. Zhou and T. Kunz (Eds.): ADHOCNETS 2016, LNICST 184, pp. 192–204, 2017.
DOI: 10.1007/978-3-319-51204-4_16

become a very important role in civilian and industrial applications. Besides, the quality of life is becoming a common focus among people all over the world. As a result, WSDs will be used widely in remote and infrastructure-based healthcare facilities [2]. In practice, it is able to reduce workload of staff if to compose a WSN in a hospital. The data of patients can be monitored by a WSN and then these data can be transmitted to their designated doctor or nurses.

The ballistocardiograph (BCG) device is one of the most widely deloped sensor devices, which transmits cardiac respiratory signals, impulse signals and kinetic signals to the health-care center [3]. Another widely deployed sensor device is the Electroencephalography (EEG) monitoring device, which transmits electrophysiological monitoring data of recording electrical activity of the brain. Besides, the device used to transmit ECG traces, metadata and annotations, is called SCP-ECG [4,5]. These WSDs are low-power devices, which cannot transmit signals over a long distance, so relay nodes are employed to help them transmit signals more reliable over a long distance. In hospital wireless networks, personal wireless hubs (PWHs) are commonly used as helping relays, which help forward the signals from WSDs to the data center. PWHs are capable of enhancing the system performance and improving the reliability of the wireless networks in hospital environment [6].

1.2   Related Work

Wireless resource allocation, which is a very effective way to improve the system performance of wireless networks, has been widely studied in the past few years. Wireless networks are resource-limited systems, including spectrum and power. Due to spectrum scarcity, cognitive radio (CR), which allows the unlicensed users (secondary users) to use the spectrum resource of the licensed users (primary users), was raised for solving this problem and its detailed definition can be found in [7]. In WSNs, CR technology was considered for the information transmission in [8–10].

Moreover, power allocation is also a typical issue in relay-aided WSNs. So far, different relaying protocols (e.g., amplify-and-forward (AF) and decode-and-forward (DF)) have been proposed for wireless cooperative communications [11]. So far, power allocation in AF or DF relaying networks in CR environment have been investigated. For source and relay nodes respectively, a power allocation schemes in CR was analysed in [12–14]. Hence, for hospital wireless networks, resource allocation (e.g., power allocation and bandwidth) is very critical.

1.3   Motivations

As much attention has been paid to health-care facilities and hospital environment recently, resource allocation in hospital wireless networks is attracting more and more interests. In [15], a discrete event system model of operating room (OR) was built on a platform named as SIMIO to allocate resource for health-care networks. The author in [16] formulated a dynamic programming problem to allocate bandwidth to enhance the information capacity of patients.

However, relatively few work has been done on power allocation in cooperative relaying hospital wireless networks. In this paper, we investigate a relay-aided hospital wireless systems, which has n subsystems in CR network environment. Our goal is to maximize the sum rate by power allocation for WSDs (transmitters) and PWHs (relays). Some differences between our work and the similar one in [6] is worthy mentioned. Firstly, in [6] multi-relay assignment was investigated with CR technology in hospital environment, which proposed an iterative joint relay assignment and power allocation algorithm for CR. Moreover, in [6], more than one WSDs are allowed to transmit signals at the same time so that each relay can receive signals from more than one WSD. This may cause multi-user interference and consequently deceases the system performance. To meet the data transmission reliability requirement of each patient in one ward, in our work only one WSD (i.e., the one with highest SNR) is selected to transmit at each time, where the PWHs within the wards are regarded as relays. Secondly, in [6], AF relay protocol was considered, while in this paper we adopt the DF relaying protocol, because compared with AF, DF avoids amplifying the noise by decoding information at the relay. Besides, in our work, the channel gain between the transmitter and the relay is good enough to ensure the relay can decode the signals successfully. Thirdly, we consider the power constraint of each ward (i.e., each subsystem), which may keep the fairness for the patients in different wards and each PWH is only allowed to work over its designated frequency so that the interference among PWHs can be avoided, while in [6], these factors were not considered.

1.4   Contributions

In this paper, we investigate the relay-aided hospital wireless systems in CR environment, where multiple transmitter-relay pairs desire transmit the collected information to the data center. Firstly, we propose a transmission framework for the system and our proposed transmission standard follows IEEE 802.22 WRAN, which has listen-before-talk and geo-location/database schemes to protect the primary users [6]. Secondly, we present the detailed transmission strategy for the system, where in each subsystem, the WSD with the highest SNR is selected to transmit signals at each time. Then, a two-hop half-duplex DF relaying transmission is launched among the selected WSD, the corresponding PWH and the data center, where the PWHs and the WSDs are unlicensed users. Thirdly, to explore the potential system performance, we formulate an optimization problem to maximize the sum rate of the system via power allocation, and then solve it by using convex optimization theory and KKT condition method. By doing so, the closed-form solution of the optimal power allocation is provided. Fourthly, we discuss the effects of the total power, the interference thresholds and the scale of the network on the system performance, which provides some insights for practical hospital wireless system design.

The rest of this paper is organized as follows. The system model and transmission protocol are described in Sect. 2. Section 3 presents the problem formulation and solution and Sect. 4 shows the simulation results. Finally, Sect. 5 concludes the paper.

2   System Model and Transmission Protocol

2.1   System Model

We consider a hospital network, which consists of M WSDs and N PWHs as shown in Fig. 1. Low-power WSDs cannot transmit data only by itself to achieve high reliability, so PWHs of patients in the hospital are considered as relays to help WSDs send data from patients to the data center.

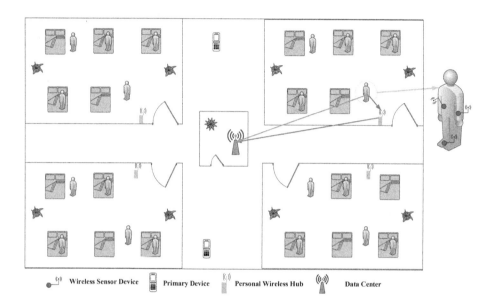

Fig. 1. A part of hospital environment with communication devices.

Considering the limit of spectrum resource, we adopt cognitive radio (CR) to share the spectrum with those licensed wireless devices [6]. In our work, the licensed wireless devices, such as cellphones, are considered as primary wireless devices.

All WSDs and PWHs are unlicensed devices and there are K licensed wireless devices (i.e., primary users) and one data center. There are several patients living together in the same ward, and each of them has several WSDs on his or her body. We assume that one ward only has one PWH. So the number of WSDs are much larger than the related PWHs, i.e., $M > N$. Due to the short distance between the WSDs and their related PWHs, the channel quality between WSDs and their related PWHs are good enough, so the PWH can successfully decode the information transmitted from the sensor devices on the patients. One PWH can only decode signals from one transmitter at some moment. After successfully decoding the information, the PWH re-transmits the data to the data center. p_m and p_n represent the mth WSD's transmission power and the nth PWH's

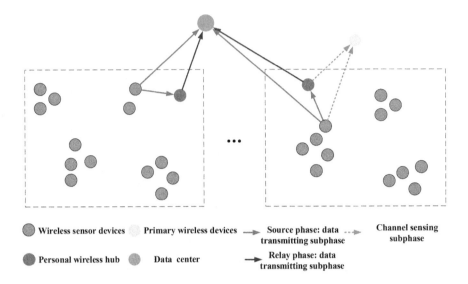

Fig. 2. System communication illustration.

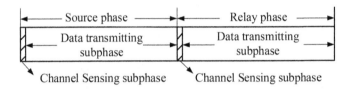

Fig. 3. Transmission protocol structure.

transmission power, respectively. $h_{m,n}$, $h_{m,c}$, $h_{n,c}$ are the channel gains between the mth WSD and the nth PWH, the mth WSD and the data center, and nth PWH and the data center, respectively. The channel gains of the mth WSD and the kth primary wireless device and the nth PWH and the kth primary wireless device are respectively denoted by $g_{m,k}$ and $g_{n,k}$.

2.2   Transmission Protocol

Briefly, the transmission strategy is illustrated as Fig. 2. We consider w WSDs, one PWH and the data center as a subsystem.

We assume that all channels are flat-fading channels. Each PWH only serves for just one WSD and each WSD operates on its separate frequency band, which is not in mutual frequency band with others.

In this paper, time of transmission in every subsystem is divided into two equal parts, which are source phase and relay phase. Source phase is for the WSDs to transmit, while relay phase is for the related PWHs transmitting. Before transmission of the WSDs and the related PWH, they should listen to the primary wireless devices' bands firstly. Secondly, every secondary wireless device should ensure that its transmission power is under a specific threshold. The structure of this protocol is shown in Fig. 3. In the first channel sensing subphase, each WSD should guarantee its interference to the primary devices being under a threshold, i.e.,

$$p_m |g_{m,k}|^2 \leq I_{k,m}, \tag{1}$$

where $I_{k,m}$ represents the interference threshold of mth WSD operating within kth primary device's licensed frequency band. According to the p_m and the channel gain between the mth WSD and related PWH, we let only one WSD in a subsystem transmit its signals and its transmission power is rewritten as $p_{m'}$. m' is in the set $\{1, 2, ..., N\}$. It's easy to see that m' and n are matched. Our selection strategy is a two-stage method. The first stage is, for several WSDs and their related PWH, to sort these WSDs according to their maximum value of achievable transmission power. We assume that a subsystem with w WSDs. According to (1), in a subsystem, for each WSD,

$$p_m = \min \left\{ P_S^{\max}, \frac{I_{1,m}}{|g_{m,1}|^2}, \frac{I_{2,m}}{|g_{m,2}|^2}, ..., \frac{I_{K,m}}{|g_{m,K}|^2} \right\} \tag{2}$$

where $P_S^{\max}$ is the maximal transmission power of the transmission device.

The second stage is, for every subsystem, to calculate each SNR at PWH, and choose the WSD with highest SNR at PWH to transmit its data, and then we write $p_{m'}$ as

$$p_{m'} = \arg\max_{p_i} \left\{ p_1 |h_{1,n}|^2, ..., p_i |h_{i,n}|^2, ..., p_w |h_{w,n}|^2 \right\}, \tag{3}$$

which m' starts from 1 to N.

In the first data transmitting subphase, the WSDs broadcast data to related PWHs and the data center. The channel output at nth PWH is

$$Y_n = \sqrt{p_{m'}} h_{m',n} X_{m'} + Z_n, \tag{4}$$

where $X_{m'}$ is complex-valued transmitted signal and Z_n is complex-valued white Guassian noise at nth PWH, which is zero-mean random variable with variance σ^2. And at the data center is

$$Y_c^{(1)} = B_1 X_1 + Z_{c,1}, \tag{5}$$

where $B_1 = \mathrm{diag}[\sqrt{p_1} h_{1,c}, ..., \sqrt{p_{m'}} h_{m',c}, ..., \sqrt{p_N} h_{N,c}]$ and $Z_{c,1}$ is a Guassian noise vector with convariance matrix $\sigma^2 I_N$ at the data center and the signal X_1 is a vector $[X_1, ... X_{m'}, ..., X_N]^T$. In the second channel sensing subphase, each PWH should guarantee its interference for the primary devices is under a threshold, defined as

$$p_n |g_{n,k}|^2 \leq I_{k,n}, \tag{6}$$

where $I_{k,n}$ represents the interference threshold of nth PWH operating within kth primary device's licensed frequency band. In the second data transmitting subphase, the PWHs transmit signals to the data center. The channel output at the data center is

$$Y_c^{(2)} = B_2 X_2 + Z_{c,2}, \tag{7}$$

where $B_2 = \mathrm{diag}[\sqrt{p_1}h_{1,c}, ..., \sqrt{p_n}h_{n,c}, ..., \sqrt{p_N}h_{N,c}]$ and $Z_{c,2}$ is a Guassian noise vector with convariance matrix $\sigma^2 I_N$ at the data center. In order to save limited energy and guarantee this system will not be strong interference to other important medical devices, in this paper, we assume that the whole transmission power of this system is limited, and from the perspective of nth subsystem, this constraint is written as

$$\sum_{m'=n=1}^{N} (p_{m'} + p_n) \leq P_{\text{total}}, \tag{8}$$

where P_{total} presents the transmission power of the whole system. Besides, for nth subsystem, it has its own power control, which means available power for every subsystem is limited in a proper proportion of the whole system for fairness, and this constraint is written as

$$p_{m'} + p_n \leq \theta_n P_{\text{total}}, \tag{9}$$

where $0 \leq \theta_n \leq 1$ is the proportional factor. Obviously, $\sum_{n=1}^{N} \theta_n = 1$ should be met.

3   Problem Formulation and Solution

In this section, we formulate an optimization problem to allocate power of WSDs and PWHs, and then we get the closed-form of optimal power allocation via two-hop half-duplex DF scheme.

We consider DF strategy to transmit, which means the related PWHs should decode the signals from the WSDs correctly. As shown in Fig. 3, we assume the channel sensing subphase is small enough compared with the data transmitting subphase, so we just consider the data transmitting subphase as the main part of the transmission time and the time of the channel sensing subphase is negligible.

For the nth subsystem, in the first data transmission subphase, the achievable rate $C_{1,n}$ is

$$C_{1,n} = \frac{1}{2N} \log \left(1 + \frac{p_{m'} |h_{m',n}|^2}{\sigma^2/N} \right). \tag{10}$$

In the second data transmission subphase, the achievable rate $C_{2,n}$ is

$$C_{2,n} = \frac{1}{2N} \log \left(1 + \frac{p_{m'} |h_{m',c}|^2}{\sigma^2/N} + \frac{p_n |h_{n,c}|^2}{\sigma^2/N} \right). \tag{11}$$

So the achievable rate of the nth subsystem is the minimum of $C_{1,n}$ and $C_{2,n}$, written as

$$C_n = \min \{C_{1,n}, C_{2,n}\}. \tag{12}$$

And our aim is to achieve maximum of the achieved rate of the whole system

$$\max_{p_{m'},p_n} \sum_{n=1}^{N} \mu_n C_n \tag{13}$$
$$\text{s.t. } (1),(6),(8),(9)$$
$$p_{m'}, p_n > 0.$$

The factor $\mu_n > 0$ ensures the fairness of every subsystem.

Proposition 1. *The optimal power allocation in the problem (12) can be achieved by letting $C_{1,n} = C_{2,n}$.*

Proof. The condition of choosing relay is $h_{m',n} > h_{m',c}$ as well as $h_{n,c} > h_{m',c}$. And for a given $p_{m'}$, we can get a corresponding p_n. $C_{1,n}$ is a monotonically increasing function of $p_{m'}$, while $C_{2,n}$ is a monotonically decreasing function of $p_{m'}$. Because $h_{n,c} > h_{m',c}$ and the sum of $p_{m'}$ and p_n is constant. Only if $C_{1,n} = C_{2,n}$, the subsystem can achieve the maximum achieved rate.

So we can obtain that the relationship between $p_{m'}$ and p_n as following:

$$p_n = \frac{|h_{m',n}|^2 - |h_{m',c}|^2}{|h_{n,c}|^2} p_{m'}, \tag{14}$$

and we set $\alpha_n = \frac{|h_{m',n}|^2 - |h_{m',c}|^2}{|h_{n,c}|^2}$.

Proposition 2. *The problem (13) is a convex optimization.*

Proof. Based on Proposition 1, we already know that $C_{1,n} = C_{2,n}$. Therefore, C_n is equal to $C_{1,n}$ or $C_{2,n}$. For the sake of convenience, we consider C_n is equal to $C_{1,n}$. Obviously, $C_{1,n}$ is a concave function of $p_{m'}$, so the sum of them is also concave. As for (1), (6), (8) and (9), they are all linear functions of $p_{m'}$, which are both convex and concave. So the problem (13) is a convex optimization.

Based on Proposition 2, this problem can be solved by KKT condition after constructing Lagrangian function. The Lagrangian of problem (13) is:

$$L(p_{m'}, \lambda) = \sum_{n=1}^{N} \mu_n C_n - \lambda \left(\sum_{m'=n=1}^{N} (1+\alpha_n) p_{m'} - P_{total} \right). \tag{15}$$

Applying the KKT condition to this problem, we have that:

$$\frac{\partial L(\mathbf{p}_{m'}, \lambda)}{\partial p_{m'}} = 0, \forall m' \in \{1, 2, ..., N\}. \tag{16}$$

And we get the solution as:

$$p_{m'} = \left[\frac{\mu_n}{\lambda(1+\alpha_n)} - \frac{\sigma^2/N}{|h_{m',n}|^2} \right]^+. \tag{17}$$

After that, we can add the other constraints into this solution. We firstly introduce a variable $X = \min\{\frac{I_{k,m'}}{|g_{m',k}|^2}, \frac{I_{k,n}}{\alpha_n|g_{n,k}|^2}, \frac{\theta_n P_{total}}{1+\alpha_n}\}, \forall m', n, k$ and we already knew that m' and n are matched. So we finally get the optimal power allocation is as following:

$$p_{m'}^* = \min\{X, p_{m'}\}. \tag{18}$$

Then, we will demonstrate how these parameters represent the sum of achieved rates of the whole system in the following section.

4   Simulation Results

In this section, we provide some simulation results to discuss the system performance.

In all simulations, channel gain is set to be $h = K_A(\frac{d_0}{d})^\beta \varphi$, seen in [17]. K_A is a constant which describes the antenna characteristic and average channel attenuation. d_0 presents the reference distance for antenna far filed, and the distance between transmitter and receiver is d, which is larger than d_0. β is the path loss constant. The parameter φ is a Rayleigh random variable. And the values of them are set as: $K_A = 20$, $d_0 = 1$, $\beta = 3$. Besides, we set the power of noise is $0.1\,\mathrm{mW}$.

Figure 4 shows an example of WSDs assignment. As shown in Fig. 4, there are four wards existing at one floor of the hospital. Each ward has several patients along with several sensor devices, which are represented by red pots. And the blue pots represent the related PWHs. We use MATLAB to simulate the result of choosing proper WSDs. We chose the WSD with the best channel quality within one subsystem. Instinctively, in the subsystem, the selected WSD is the one that with the shortest distance between it and the PWH.

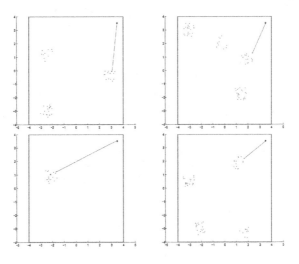

Fig. 4. An example of wireless sensor devices assignment. (Color figure online)

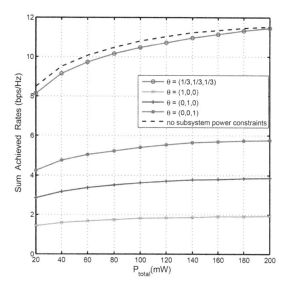

Fig. 5. Sum rate with different P_{total}, when $N = 3$, $K = 2$.

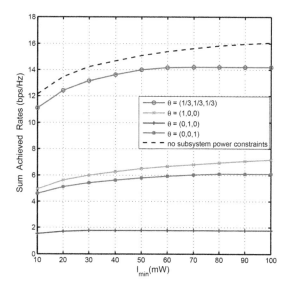

Fig. 6. Sum rates with different I_{min}, when $N = 3$, $K = 2$.

Figures 5 and 6 are simulation results of the system with $N = 3$, $K = 2$. Figure 5 illustrates the sum rate versus θ when P_{total} changes from 20 mW to 200 mW. We set μ randomly and we set $I_1 = 80\,mW$ and $I_2 = 60\,mW$. It shows that, the system with uniformed θ gets higher sum rate than the system with only one subsystem transmitting. It indicates that, to achieve better fairness and spectral efficiency, only one subsystem being permitted to transmit may not

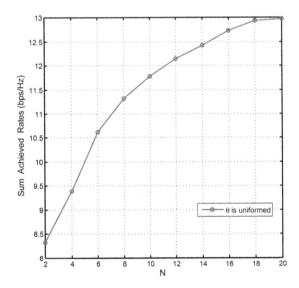

Fig. 7. Sum rates with different N, when $K = 2$, $p_{\text{total}} = 200\,\text{mW}$.

always be the best choice. Moreover, the gap between the system with uniformed θ and the system with no subsystem power constraint is the smallest than the others, and the gap is decreasing when P_{total} increases. It can be concluded that when P_{total} is becoming larger, uniformed θ constrains will not be the main factor on affecting the system sum rate any more. Figure 6 illustrates the sum rate of different θ versus I_{min} changing from $10\,\text{mW}$ to $100\,\text{mW}$. Two different I values limit the optimal values of $p_{m'}$ and the smaller one affects the sum rate more obviously. It can be observed that the gap between the system with uniformed θ and the system with no subsystem power constraint is the smallest than the others, and the gap is increasing when I_{min} increases, because when I_{min} is getting larger, uniformed θ will be the main factor, which limits the system sum rate.

Figure 7 shows that the sum rate is increased when N grows, but the increasing rate is declining with the increment of N.

5   Conclusion

This paper studied the relay-aided hospital wireless systems in cognitive radio environment. We proposed a transmission framework and the corresponding transmission strategy. To explore the potential system performance, an optimization problem was formulated to maximize the system sum rate via power allocation. A closed-form solution was derived for the power allocation. Simulation results demonstrated the validity of our proposed scheme and also showed the effects of the total power, the interference thresholds and the scale of the network on the system performance, which provide some insights for practical hospital wireless system design.

Acknowledgement. This work was supported by the Open Research Fund of National Mobile Communications Research Laboratory, Southeast University (no. 2014D03), the Fundamental Research Funds for the Central Universities (no. 2016JBM015), and in part by the Beijing Natural Science Foundation (no. 4162049).

References

1. Zeb, A., Islam, A., Komaki, S., Baharun, S.: Multi-nodes joining for dynamic cluster-based Wireless Sensor Network. In: Informatics, Electronics & Vision (ICIEV), May 2014
2. Naeem, M., Pareek, U., Lee, D.C., Khwaja, A.S., Anpalagan, A.: Efficient multiple personal wireless hub assignment in next generation healthcare facilities. In: IEEE Conference on Wireless Commununication Network (WCNC), March 2015
3. Kim, J.M., Hong, J.H., Cho, M.C., Cha, E.J., Lee, T.S.: Wireless biomedical signal monitoring device on wheelchair using noncontact electro-mechanical film sensor. In: 29th Annual International Conference on IEEE Engineering in Medicine and Biology Society (EMBS), Lyon, pp. 574–577 (2007)
4. Mandellos, G.J., Koukias, M.N., Styliadis, I.S., Lymberopoulos, D.K.: e-SCP-ECG protocol: an expansion on SCP-ECG protocol for health telemonitoring-pilot implementation. Int. J. Telemed. Appl. **2010**, 137201 (2010)
5. Rubel, P., et al.: Toward personal eHealth in cardiology. IEEE Electrocardiol. J. **38**(4), 100–106 (2005)
6. Naeem, M., Pareek, U., Lee, D.C., Khwaja, A.S., Anpalagan, A.: Wireless resource allocation in next generation healthcare facilities. IEEE Sens. J. **15**(3), 1463–1474 (2015)
7. Akyildiz, I.F., Lee, W.Y., Vuran, M.C., Mohanty, S.: A survey on spectrum management in cognitive radio networks. IEEE Commun. Mag. **46**(4), 40–48 (2008)
8. Vijay, G., Bdira, E.B.A., Ibnkahla, M.: Cognition in wireless sensor networks: a perspective. IEEE Sensors J. **11**(3), 582–592 (2011)
9. Rao, R., Cheng, Q., Varshney, P.K.: Subspace-based cooperative spectrum sensing for cognitive radios. IEEE Sensors J. **11**(3), 611–620 (2011)
10. Lee, K.D., Vasilakos, A.V.: Access stratum resource management for reliable u-healthcare service in LTE networks. Wirel. Netw. **17**(7), 1667–1678 (2011)
11. Moubagou, B.D., Chang, Y.Y.: Performance of AF and DF relays by employing Half Duplex and Full Duplex mode. In: International Conference on Measurement, Information and Control (ICMIC), Harbin (2013)
12. Li, Y., Chen, Z., Gong, Y.: Optimal power allocation for coordinated transmission in cognitive radio networks. In: IEEE 81st Vehicular Technology Conference (VTC Spring), Glasgow, pp. 1–5 (2015)
13. Zhang, C., Fan, P., Xiong, K.: Optimal power allocation with delay constraint for signal transmission from moving train to base stations in high-speed railway scenarios. IEEE Trans. Veh. Technol. **64**, 5775–5788 (2015)
14. Xiong, K., Shi, Q., Fan, P., Letaief, K.B.: Resource allocation for two-way relay networks with symmetric data rates: an information theoretic approach. In: IEEE International Conference on Communications (ICC), pp. 6060–6064, June 2013
15. Zheng, Q., Shen, J., Liu, Z.Q., Fang, K., Xiang, W.: Resource allocation simulation on operating rooms of hospital. In: IEEE 18th International Conference on Industrial Engineering and Engineering Management (IE&EM), Changchun, pp. 1744–1748 (2011)

16. Lin, D., Labeau, F.: Bandwidth allocation in view of EMI on medical equipments in healthcare monitoring systems. In: 13th IEEE International Conference on Healthcom, pp. 209–212, June 2011
17. Goldsmith, A.: Wireless Communications. Cambridge University Press, Cambridge (2005)

Improving the Performance of Challenged Networks with Controlled Mobility

Laurent Reynaud[1,2(✉)] and Isabelle Guérin-Lassous[2]

[1] Orange Labs, Lannion, France
laurent.reynaud@orange.com
[2] Université de Lyon/LIP (ENS Lyon, CNRS, UCBL, Inria), Lyon, France
isabelle.guerin-lassous@ens-lyon.fr

Abstract. In this work, we investigate the application of an adapted controlled mobility strategy on self-propelling nodes, which could efficiently provide network resource to users scattered on a designated area. We design a virtual force-based controlled mobility scheme, named VFPc, and evaluate its ability to be jointly used with a dual packet-forwarding and epidemic routing protocol. In particular, we study the possibility for end-users to achieve synchronous communications at given times of the considered scenarios. On this basis, we study the delay distribution for such user traffic and show the advantages of VFPc compared to other packet-forwarding and packet-replication schemes, and highlight that VFPc-enabled applications could take benefit of both schemes to yield a better user experience, despite challenging network conditions.

Keywords: Controlled mobility · Virtual forces · MANET · Challenged networks · DTN · Unmanned aerial vehicles · Disaster communications

1 Introduction

Over the last 15 years, the notion of ubiquitous network access got closer to reality. As an illustration, by that time, the worldwide Internet penetration rate has grown 7 times, reaching 43% in 2015 [1]. Yet, this encouraging key performance indicator should not conceal the acute challenges still posed by the current need to greatly improve access to network infrastructure in many unconnected or ill-connected territories. Although the reasons for this imperfect network coverage may differ, with various issues and constraints met in either rural areas, remote zones or emerging countries, alternate communication resources need to be deployed on site to grant network access. A similar problem arises in the case of disasters which may leave the existing networks impaired at a time when communications are greatly needed by the rescue, response and restoration teams. To this end, various rapid deployment communication systems were proposed, often relying on different types of terrestrial, aerial and satellite network segments [2]. Moreover, the principles of Mobile Ad Hoc Networks (MANETs) [3] have often been adopted in these contexts [4], since they

© ICST Institute for Computer Sciences, Social Informatics and Telecommunications Engineering 2017
Y. Zhou and T. Kunz (Eds.): ADHOCNETS 2016, LNICST 184, pp. 205–217, 2017.
DOI: 10.1007/978-3-319-51204-4_17

can forgo the time-consuming, staff-demanding and potentially costly roll-out of a surrogate cellular network infrastructure. Further, MANETs allow devices to form temporary and self-organized networks in dynamic topologies, where multi-hop communications are used to extend the inherently limited range of wireless transmissions. Yet, in the context of challenged networks [6] with high node mobility, low node density and other detrimental issues, the performance of MANETs can be severely hindered by the scarcity of network connectivity and subsequent link disruptions, which in turn increase packet losses [5]. In contrast, Delay/Disruption Tolerant Network (DTN) techniques were designed to handle packet delivery in case of intermittent connectivity found in challenged networks. Moreover, while MANETs use synchronous routing schemes based on the determination of an end-to-end path, DTN schemes on the other hand rely on the asynchronous store-carry-and-forward principles [7] wherein a network node buffers and carries incoming packets as it moves. Further, among the proposed DTN routing strategies, two specific directions were abundantly explored:

- *Packet-forwarding*, which, often combined with modified synchronous protocols to support longer delays (e.g. Deep-Space Transport Protocol (DS-TP) or TP-Planet [5]), allows better packet delivery with respect to MANET performance.
- *Epidemic approaches* enable a node to transmit copies of incoming packets to nodes it gets in contact with. As a result, multiple replications of a specific packet may exist in the network at the same time, increasing the chances for this packet to reach its destination. Yet, a systematic packet replication at each contact opportunity incurs a significant resource consumption. As a result, several solutions (e.g. MaxProp [9], RAPID [10] and Spray and Wait [8]) were proposed to keep packet replication as low as possible.

In this regard, due to the unpredictable nature of most intermittently connected networks, traditional DTN schemes fail to fully ensure consistent network performance gains with respect to multiple key routing metrics such as packet delivery, delay, overhead and resource consumption. This observation has been referred to as the incidental effect [10] of existing DTN schemes. To overcome this limitation, the concept of controlled mobility [16] has recently been explored, bringing a new perspective on network node mobility, which was until then mainly considered as an unavoidable nuisance requiring mitigation. In contrast, controlled mobility enforces deployed network protocols with the ability to put nodes in motion and direct them where they can help increase the overall network performance. Some forms of controlled mobility mechanisms have been notably studied in the context of DTNs, where specific nodes may be used as message ferries to enhance connectivity in networks with sparsely deployed nodes [16]. Likewise, wireless sensor networks may benefit from data sinks with controlled mobility for various performance aspects, such as network lifetime increase [17]. In this work, we particularly focus on swarming principles, a distributed form of controlled mobility for which local node interaction can engender desirable emergent behaviors [18]. In effect, swarming mechanisms give network nodes the ability to cooperatively readjust their movements thanks to the exchange of local information and to collectively achieve a given spatial organization. In this regard, although swarming mechanisms generally require a large number of mobile nodes to complete pattern formations such as grids and lattices [13], specific strategies can achieve chain formations with a limited number of nodes, which is a useful property in network deployments, where the number of nodes is often constrained. Thus, several works studied chain formations with probabilistic finite state machine [19] and evolutionary robotics techniques [20].

In our previous studies, we investigated a third approach, for which local information exchange is based on virtual force principles [14,15]. We notably presented in [15] the Virtual Force Protocol (VFP), allowing mobile nodes, and in particular unmanned aerial vehicles (UAVs), to form communication chains and provide network connectivity in the context of disaster relief operations. We assessed VFP performance and notably showed that a peak efficiency was obtained with a limited set of nodes, which confirmed its interest in network deployments where the number of nodes is constrained. Yet, the performance gain from the use of a simple MANET protocol to a joint use of MANET and VFP, however significant, could not exceed a relatively low threshold of about 40%, in terms of Packet Delivery Ratio (PDR).

In this work, we seek to overcome this PDR limitation while keeping end-to-end delays as low as possible. To this end, our main contribution can be summarized as follows: we give our VFP-based strategy, which we here name VFPc, the ability to work jointly with a DTN epidemic scheme, with the objective to thoroughly improve packet delivery. We also design a cross-layer framework that allows switching from packet-replication to packet forwarding (and reciprocally), based on the context given by VFPc to the upper layer routing components, in terms of whether a VFP communication chain is established or not. That way, end-users can benefit from synchronous communications when a VFP chain connects the traffic endpoints. Otherwise, the network autonomously falls back to asynchronous communications. The rest of this paper is organized as follows: Sect. 2 presents the disaster relief scenario which gives the context of this study. Section 3 details the design choices made for our VFPc strategy and other schemes used as comparison references. The performance of VFPc is then evaluated along with the other schemes, in terms of PDR and end-to-end delay, in Sect. 4, and we finally conclude in Sect. 5.

2 Scenario

In the context of this study, we envision a scenario where a rapid deployment communication system is required to provide network coverage on a zone Z_e where network access is non-existent or temporarily impaired. To this end, our system encompasses the following nodes, as illustrated by Fig. 1:

– *Traffic nodes* are regular end-user devices which, like the other nodes, support the VFPc scheme in order to cooperate with the rest of the network. Two such

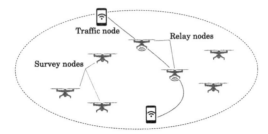

Fig. 1. A representative network deployment in the considered scenario.

nodes are assumed to move randomly within Z_e, respectively acting as source and destination of all user traffic during the considered case flow.

- *Survey nodes* explore Z_e, exchange information with other nodes in radio range and store their location for future use and dissemination in the network.
- *Relay nodes* are initially survey nodes which change their function at given times to become part the set of intermediate nodes in the multihop communication chain between the traffic nodes. When not needed anymore in the chain, relay nodes revert to their former survey type.

3   VFP Protocol Design

3.1   A Force-Based System

Our VFPc scheme implements a virtual force-based distributed system, VFP [14,15], which is used to control node mobility so as to create and maintain a wireless multi-hop communication chain between any traffic (source, destination) pair. Further, VFP defines a beacon message which is regularly broadcast 1-hop away by each node in the network. VFP beacons contain various information such as the emitting node coordinates and velocity vector, whether it belongs to a communication chain and in that case which intermediate nodes are preceding and following in the chain. It additionally encompasses a list of nodes previously discovered, also with relevant information. This local distribution of information is a pivotal mechanism for network nodes to quantify the forces exerted by neighboring nodes, reassess their own subsequent acceleration and velocity vectors, move accordingly, and take part to the relay node election process [15]. As illustrated by Fig. 2 (left), relay nodes in a communication chain (nodes P and N in the given example) are subjected to interaction and alignment forces. To calculate both forces, nodes use received VFP beacon information differently:

- Interaction forces [14], which encompass three repulsive, friction and attractive components, are exerted by the node's predecessor in the chain (i.e. N and P are subjected to interaction forces from respectively P and S in the given example). Depending on the distance with its predecessor, a node can be

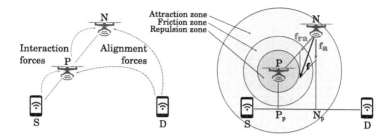

Fig. 2. (left) Principles of virtual forces applied on nodes P and N in a communication chain, (right) detailed forces on N (note that friction $\mathbf{f_{fr}}$ is non-existent in this case). (Color figure online)

located in a virtual repulsion, friction or attraction zone, delineated by the red, grey and green areas in Fig. 2 (right). In this study, we select a repulsion-attraction intensity profile such that $f_{ra} = I$ in the repulsion zone, $f_{ra} = -I$ in the attraction zone and $f_{ra} = 0$ elsewhere. Further, the repulsion-attraction force is used to move the node within a given relative distance from its predecessor, while the friction force $\mathbf{f_{fr}}$ enables smooth decelerations and allows avoiding undesirable oscillating movement effects [14]. These forces are calculated on the basis of 1-hop information received from the predecessor's beacons.

– In contrast, alignment forces [15] can be calculated by a relay node as soon as it learns the position of the traffic source and destination nodes (S and D in the example of Fig. 2). This force steers relay nodes towards line (SD), and ultimately tends to generate a straight line topology for the communication chain. In the example of Fig. 2 (right), the alignment force $\mathbf{f_a}$ tends to steer N towards its projection N_p on line (SD). If N_p was closer from P_p to S, $\mathbf{f_a}$ would be directed towards the symmetric point of N_p about P_p on (SD). That way, the alignment force also helps reordering chains if a relay node is not correctly located with respect to its predecessor and successor in the chain.

3.2 A Cross-Layer Framework

Figure 3 gives an architectural representation of the dual routing stack made of a packet-forwarding scheme (the synchronous stack) and a packet-replication scheme (the asynchronous stack) hosted on each node. It also outlines how three VFP entities interact with the other main components involved both in network traffic transmission as well as in node mobility control:

– The *mobility controller* retrieves information such as coordinates and velocity vectors from the node Inertial Measurement Unit (IMU) and various sensors via the Autopilot application. This entity also performs the calculation of the force system exerted on the node based on available local information and sends the Autopilot the updated measures (e.g. under the form of a velocity vector or a waypoint).

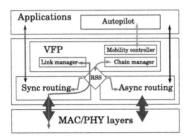

Fig. 3. Representation of the cross-layer framework. (Color figure online)

- The *link manager* is in charge to periodically build and emit VFP beacons as well as receive and store beacons from neighboring nodes. As shown by the blue arrows, VFP beacons only rely on 1-hop broadcasts provided by the synchronous scheme and don't require any DTN persistence.
- The *chain manager* works together with a selection mechanism (illustrated by the red arrows in Fig. 3) which allows triggering the relevant routing stack depending on the VFP status of the node: if this later does not belong to an established VFP chain whose destination matches the considered user traffic destination, then the corresponding packets are handled by the DTN-based asynchronous stack. Otherwise, it is known the node has an available multi-hop route via its successor in the chain to the destination and the selection mechanism lets the synchronous stack handle packet forwarding on a hop-by-hop basis.

3.3   Implementation Aspects

Although the outlined framework may allow the use of multiple routing schemes, specific deployment choices were required to allow a rigorous performance evaluation of VFPc. Because of its well-documented properties, the epidemic protocol [12] was therefore selected for deployment as packet-replication scheme within the cross-layer framework. This DTN protocol exhibits a simple opportunistic flooding-based design, with the use of a dedicated beacon to inform nodes of contact opportunities with neighboring nodes, as well as of another specific mechanism, the *summary vector exchange*, which allows two nodes to exchange their disjoint packets during contacts. Moreover, we implemented the Routing Stack Selector (RSS) shown by Fig. 3 as well as a simple forwarding scheme in the same component. Basically, when an incoming packet needs processing, the component requests the VFP status of the node, and if applicable, the identifier of its successor in the chain. If applicable, the packet is immediately forwarded to the successor. Otherwise, the packet is passed to the epidemic protocol and will be kept into persistent storage for further transmission, when a contact opportunity with a neighboring node arises.

On this basis, we designed seven routing strategies, as Fig. 4 shows. The first six schemes only partially use our framework components. The **RWP1** scheme

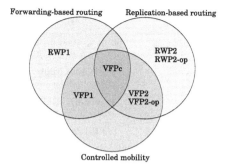

Forwarding-based routing   Replication-based routing

RWP1

RWP2
RWP2-op

VFPc

VFP1

VFP2
VFP2-op

Controlled mobility

Fig. 4. Overview of the implemented schemes, ordered by function support.

exclusively relies on Random Waypoint Mobility (RWP) and not on virtual force-based controlled mobility. Besides, it only supports a MANET forwarding-based routing. We also study two other RWP-based schemes which on the opposite only support the epidemic protocol: while **RWP2** uses a regular epidemic stack with default values, **RWP2-op** lowers the period of packet list exchange between two neighboring nodes by setting $HostRecentPeriod = 1$ s (instead of 10 s by default) [12]. Hence, the use of this later scheme should incur a faster transmission of epidemic packets during contacts (i.e. when nodes are in direct radio range and able to exchange their list of stored epidemic packets to determine which packets should be transmitted). We also consider the **VFP1** scheme, which supports VFP controlled mobility and which relies on a MANET forwarding-based routing, but never on the epidemic routing. As a result, the user traffic in only transmitted when valid end-to-end routes are established and is dropped otherwise. We then implemented two VFP-based schemes which only rely on the epidemic stack: whereas **VFP2** uses the epidemic routing with default values, **VFP2-op** employs optimized values, with the aforementioned expected benefits. The last strategy, **VFPc**, supports all the features offered by our framework: the VFP component controls node mobility when applicable, and the user traffic is contextually passed to the epidemic or the forwarding scheme, depending on whether the considered node belongs to an established communication chain. It is worth noting that VFPc uses the default epidemic parameter valuation.

4 Performance Evaluation

4.1 Simulation Parameters

The simulation parameters were chosen as closely as possible as those described in our previous work [15]. Table 1 summarizes the values of the key parameters of our simulation setup, which uses the network simulator ns-3.23.

The node number, initial positions and mobility patterns are given in Table 1. All nodes use IEEE 802.11b/g communication links with High-Rate

Table 1. Main simulation parameters

Nodes	Exploration zone Z_e = 1000 m × 1000 m, 2 traffic nodes, N (survey + relay) nodes, N = 15 or $1 \leq N \leq 30$	
Mobility patterns	Traffic nodes	Position initially uniformly distributed on Z_e, RWP, velocity $\in [0.25, 1]$ m/s
	Survey nodes	Position initially at center of Z_e, RWP, velocity $\in [5, 10]$ m/s
	Relay nodes	VFP-based mobility, velocity $\in [0, 10]$ m/s
Network	802.11b/g, HR-DSSS at 11 Mb/s, radio range = 100 m constant speed propagation delay model	
VFPc protocol	Beacon emission interval = 1 s, interaction forces $\mathbf{f_{ra}}$ and $\mathbf{f_{fr}}$ configured as in [14], Alignment force $\mathbf{f_a}$ valued as in [15]	
Routing	MANET	OLSR with default values [11]
	DTN	Epidemic protocol [12]
User traffic	CBR bitrate = 10 Kb/s, CBR packet size = 512 B	

Direct Sequence Spread Spectrum (HR-DSSS) at 11 Mb/s, whose communication range is set to 100 m and radio propagation is assumed lossless. Moreover, the MANET routing is supported by the Optimized Link State Routing protocol (OLSR) [11] for the non-crosslayer schemes that use the packet-forwarding stack (i.e. the RWP1 and VFP1 strategies). In contrast and as previously mentioned in Sect. 3.3, the packet-forwarding component of VFPc relies on a VFP-based simplified hop-by-hop routing. With respect to the force-based controlled mobility, Table 1 refers to our previous works [14,15] which provide a detailed justification of the chosen values for respectively the interaction forces $\mathbf{f_i}$ and the alignment forces $\mathbf{f_a}$. Moreover, the user traffic is modelled with a Constant Bitrate (CBR) flow at 10 Kb/s, which is assumed sufficient in the context of the considered scenario to convey important and potentially delay-tolerant traffic, such as UAV telemetry or payload sensor data. Note that the case of larger bitrates was studied in [14]. Further, each point of Figs. 5, 6 and 7 are respectively averaged over 2000 and 10000 independent simulations of 900 s each. On that note, errors bars are shown in all figures and are based on a confidence level of 95%.

Results are analyzed in the rest of this section, and are based on two performance metrics: the Packet Delivery Ratio (PDR) relates to the user traffic between both traffic nodes. Then, the end-to-end delay is defined here from source to destination traffic node, for the same user traffic. We additionally examine the Cumulative Distribution Function (CDF) of this delay.

4.2 Simulation Results

We first took interest in how the considered schemes behave with an increasing number of nodes. We followed a similar approach as taken in our previous study [15] regarding the performance of VFP1 and RWP1 with respect to PDR and end-to-end delay, this time using those results as a comparison basis to

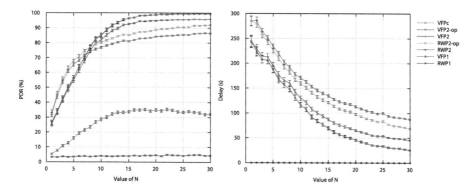

Fig. 5. PDR (left) and end-to-end delay (right) of the CBR packets received by destination node D, versus the number N of initial survey nodes.

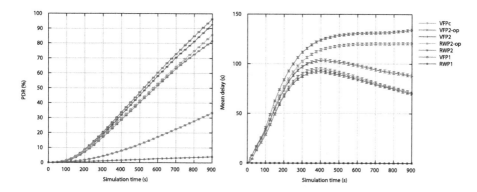

Fig. 6. PDR (left) and average end-to-end delay (right) of the CBR packets received by destination node D, versus simulation time. $N = 15$.

evaluate the other schemes. Figure 5 shows the PDR and end-to-end delay of the CBR transmissions between both traffic nodes for all considered schemes, with a varying initial number N of survey nodes in the network, such that $1 \leq N \leq 30$.

General performance outcomes. It can first be observed that both RWP1 and VFP1 exhibit low end-to-end delays (below 15 ms for all values of N) compared to the other schemes. However, VFP1, with a PDR reaching a maximum of around 35% for $16 \leq N \leq 18$, represents a significant improvement over RWP1 and its lower PDR, consistently below 5%. As we detailed in [15], the low performance of RWP1 can easily be explained by the low node density and the relatively high velocity of the survey nodes, in the range of [5, 10] m/s, which goes beyond the regular pedestrian-type speeds found in most MANET deployment scenarios. Likewise, the PDR results of VFP1 are understandably constrained by the unavoidable time needed for the mobile nodes to physically move and connect the traffic endpoints. We however verified in [15] that the performance of VFP1 is close to that of an ideal-theoretic mobility control scheme where node positions would always be known. Consequently, further design improvements of VFP1 based solely on packet-forwarding routing would offer limited perspectives, especially regarding PDR. In contrast with RWP1 and VFP1, the five epidemic-enabled strategies share a sharp improvement in terms of PDR, at the expense of significantly lengthened delays, as Fig. 5 illustrates. In any case, these schemes have an increasing PDR with N and, in this regard, will systematically outperform RWP1 and VFP1 for $N > 3$.

At this stage, a sharp distinction can also be made between the five epidemic-enabled schemes, on both the criteria of PDR and end-to-end delay:

- VFPc, VFP2 and VFP2-op systematically yield better PDR results than RWP2 and RWP2-op for $N \geq 9$: for instance, when $N = 15$, the former set outperforms the latter, in terms of PDR, by 7% to 15%. Figure 6 (left), which shows how PDR evolves with time for $N = 15$, confirms that the VFP-and-epidemic-based schemes consistently outmatch the RWP-and-epidemic-based strategies at all times. Further, while for $N = 30$, all PDRs are contained between 96% and 99.5%, VFPc, VFP2 and VFP2-op obviously converge faster and are already above 96% for $N \geq 15$. However, for $N \leq 7$, RWP2 and RWP2-op always outperform VFPc, VFP2 and VFP2-op regarding PDR. This is easily explained by the fact that low values for N do not allow the VFP-based controlled mobility protocol to successfully establish communication chains. Instead, the VFP-enabled strategies here forms incomplete chains which are not sufficiently long to create an end-to-end path between both user traffic endpoints, and which waste intermediate nodes which could otherwise explore the overall area and opportunistically transmit packets, hereby increasing PDR. Yet, for $N > 7$, using a VFP controlled mobility protocol starts making more sense than applying a simple RWP mobility scheme to the network nodes.
- With respect to delay, Fig. 5 (right) shows two general trends: first, VFPc, VFP2 and VFP2-op yield lower end-to-end delays for any value of N. In addition, as already observed for PDR, the optimized schemes (i.e. VFP2-op and RWP2-op) behave better than their counterpart with default values, which was expected by construction. Moreover, Fig. 6 (right) illustrates how the average end-to-end delays from CBR packets received since the start of the simulation evolves with time, for $N = 15$. VFPc, VFP2 and VFP2-op clearly exhibit a maximum at simulation time $t \approx 400$ s, which corresponds, for the considered scenario, to the statistical time at which the VFP-based communication chain is established, and packets can be transmitted along the multi-hop path formed by the relay nodes. Subsequent CBR packets are then likely to reach their destination endpoint with significantly lower delays, decreasing the average end-to-end delay accordingly. Instead, the average end-to-end delays of RWP2 and RWP2-op never decrease with time. At the end of the total simulation time, this delay reaches a steady point with RWP2-op while it still increases with RWP2.

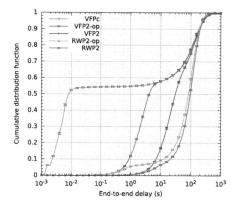

Fig. 7. Cumulative distribution function of the end-to-end delays.

Performance of VFPc. The specific case of the full-featured, dual routing stack, VFPc, is now singled out and analyzed. Although VFPc uses the epidemic stack with default parameter values, its PDR results are however almost identical to that of the optimized VFP2-op, as Fig. 5 (left) and Fig. 6 (left) illustrate. The same can be said in terms of mean delays, as shown by Fig. 5 (right). Figure 6 (right) shows that VFP2-op slightly outperforms VFPc during a part of the simulation, although the average delays of both schemes eventually match. As a result, both schemes exhibit a comparable performance in terms of PDR and mean delay, although VFPc has a more frugal behavior regarding overhead, thanks to the use of default epidemic parameter values which generates less control messages.

Furthermore, an in-depth study of the delay distributions reveals a solid argument, besides epidemic control message overhead mitigation, to consider the use of VFPc. Figure 7 displays the CDF of the end-to-end delays related to the CBR traffic for all epidemic schemes. The RWP-based schemes yield the longer delays: only 1% and 6% of the CBR packets are respectively received within 1 s when using the RWP2 and RWP2-op schemes. For the VFP-enabled schemes, the distributions significantly vary: while about 56% of CBR packets are received within 7 s when using VFPc or VFP2-op, only 8% of CBR packets are received in that time windows with VFP2. However, almost no packet is received within 100 ms for VFP2-op, versus more than 54% with VFPc. Even more significantly, a dual pattern can be observed from the CDF curve of VFPc: 52% of the user traffic is received synchronously, within 10 ms, through the VFP-enabled communication chains, while the rest is received asynchronously, with delays exceeding 1 s, via DTN-based opportunistic exchanges. This confirms the interest of enforcing controlled mobility principles: with VFPc, as much user traffic as possible is received with low delays when the VFP-based topology is fully formed, while the rest is conveyed via packet replication, with longer delays.

5   Conclusion

In this work, we presented VFPc, a distributed controlled mobility strategy relying on virtual forces, which enables a flock of network nodes to move cooperatively and form multi-hop communication links where needed. The use of VFPc is particularly justified in the context of disaster-relief communications and more generally, rapidly formed networks, which need to provide an efficient network coverage with a reduced set of network equipment. In that regard, we presented the architectural principles of VFPc and a dual routing framework that allows switching from packet-replication to packet forwarding (and reciprocally), based on whether a VFP communication chain is established or not. We then evaluated this strategy via a set of simulations which confirmed that the joint use of the VFP controlled mobility and a dual packet forwarding-replication routing stack yields the best performance in terms of packet delivery and delays, compared to other MANET- or DTN-based schemes. In addition, delay CDF results show that VFPc incurs two distinct communications phases during which the user traffic may be transmitted with very low delays, when VFPc communication chains are established, or with more elastic delays otherwise. An application aware of the times at which it enters synchronous or asynchronous modes may offer new perspectives in terms of user experience despite challenging network conditions. In the future, we plan to implement VFPc and its routing framework on a swarm of UAVs and further assess VFPc performance via experimentation.

References

1. Sanou, B.: ICT Facts and Figures 2015. International Telecommunication Union (ITU) Fact Sheet (2015)
2. Nelson, C.B., Steckler, B.D., Stamberger, J.A.: The evolution of hastily formed networks for disaster response: technologies, case studies, and future trends. In: IEEE Global Humanitarian Technology Conference, Seattle, USA, pp. 467–475 (2011)
3. Corson, S., Macker, J.: Mobile Ad hoc Networking (MANET): Routing Protocol Performance Issues and Evaluation Considerations. RFC 2501 (1999)
4. Reina, D.G., et al.: A survey on multihop ad hoc networks for disaster response scenarios. Int. J. Distrib. Sens. Netw. **2015** (2015)
5. Sassatelli, L., et al.: Reliable transport in delay-tolerant networks with opportunistic routing. IEEE Trans. Wireless Commun. **13**(10), 5546–5557 (2014)
6. Fall, K.: A delay-tolerant network architecture for challenged internets. In: Proceedings of ACM SIGCOMM 2003, Karlsruhe, Germany (2003)
7. Li, Y., Hui, P., Jin, D., Chen, S.: Delay-tolerant network protocol testing and evaluation. IEEE Com. Mag. **53**(1), 258 (2015)
8. Spyropoulos, T., Psounis, K., Raghavendra, C.S.: Spray and wait: an efficient routing scheme for intermittently connected mobile networks. In: Proceedings of the ACM Workshop on Delay-Tolerant Networking (2005)
9. Burgess, J., Gallagher, B., Jensen, D., Levine, B.N.: MaxProp: routing for vehicle-based disruption-tolerant networks. In: IEEE INFOCOM, Barcelona, Spain (2006)
10. Balasubramanian, A., Levine, B.N., Venkataramani, A.: DTN routing as a resource allocation problem. In: Proceedings of the ACM SIGCOMM (2007)

11. Clausen, T., Jacquet, P.: RFC3626, Optimized Link State Routing Protocol (OLSR). Experimental. http://www.ietf.org/rfc/rfc3626.txt
12. Alenazi, M.J.F., Cheng, Y., Zhang, D., Sterbenz, J.P.G.: Epidemic routing protocol implementation in ns-3. In: Workshop on ns-3, Barcelona, Spain (2015)
13. Spears, W.M., Spears, D.F., Hamann, J.C., Heil, R.: Distributed, physics-based control of swarms of vehicles. Auton. Robots **17**(2/3), 137–162 (2004)
14. Reynaud, L., Guérin Lassous, I.: Design of a force-based controlled mobility on aerial vehicles for pest management. Ad Hoc Net. J. **53**, 41–52 (2016). Elsevier
15. Reynaud, L., Guérin Lassous, I.: Physics-based swarm intelligence for disaster relief communications. In: International Conference on Ad Hoc Networks and Wireless, Lille, France (2016)
16. Zhao, W., Ammar, M., Zegura, E.: Controlling the mobility of multiple data transport ferries in a delay-tolerant network. In: IEEE INFOCOM, Miami, USA (2005)
17. Basagni, S., et al.: Controlled sink mobility for prolonging wireless sensor networks lifetime. Wireless Netw. J. **14**(6), 831–858 (2008)
18. Brambilla, M., Ferrante, E., Birattari, M., Dorigo, M.: Swarm robotics: a review from the swarm engineering perspective. Swarm Intell. **7**(1), 1–41 (2013)
19. Nouyan, S., Campo, A., Dorigo, M.: Path formation in a robot swarm: self-organized strategies to find your way home. Swarm Intell. **2**(1), 1–23 (2008)
20. Sperati, V., Trianni, V., Nolfi, S.: Self-organised path formation in a swarm of robots. Swarm Intell. **5**, 97–119 (2011)

Protocols

SDN Coordination for CCN and FC Content Dissemination in VANETs

Ridha Soua[1], Eirini Kalogeiton[2], Gaetano Manzo[2,3], Joao M. Duarte[2,4],
Maria Rita Palattella[1], Antonio Di Maio[1], Torsten Braun[2], Thomas Engel[1],
Leandro A. Villas[4], and Gianluca A. Rizzo[3(✉)]

[1] SnT, University of Luxembourg, Esch-sur-alzette, Luxembourg
{ridha.soua,maria-rita.palattella,antonio.dimaio,thomas.engel}@uni.lu
[2] University of Bern, Bern, Switzerland
{kalogeiton,duarte,braun}@inf.unibe.ch
[3] HES-SO, Delémont, Switzerland
{gaetano.manzo,gianluca.rizzo}@hevs.ch
[4] Institute of Computing, University of Campinas, Campinas, Brazil
leandro@ic.unicamp.br

Abstract. Content dissemination in Vehicular Ad-hoc Networks has a
myriad of applications, ranging from advertising and parking notifica-
tions, to traffic and emergency warnings. This heterogeneity requires
optimizing content storing, retrieval and forwarding among vehicles
to deliver data with short latency and without jeopardizing network
resources. In this paper, for a few reference scenarios, we illustrate how
approaches that combine Content Centric Networking (CCN) and Floating Content (FC) enable new and efficient solutions to this issue. Moreover, we describe how a network architecture based on Software Defined
Networking (SDN) can support both CCN and FC by coordinating distributed caching strategies, by optimizing the packet forwarding process
and the availability of floating data items. For each scenario analyzed,
we highlight the main research challenges open, and we describe a few
possible solutions.

Keywords: VANETs · Software defined networking · Content centric
networking · Floating content · Content caching and replication

1 Introduction

Vehicular Ad-hoc Networks (VANETs) [1] allow communications among vehicles
and between vehicles and fixed infrastructure, aiming to support a wide range of
services and applications to make travel experience pleasant, safe, and informed
[2]. Applications envisioned for VANETs vary from traffic conditions and accident warnings, to infotainment services such as live video streaming, live gaming,
etc. The main technical challenges in VANET communications are related to the
high dynamicity and volatility of the vehicular environment. Therefore, mechanisms for online adaptation of the network configuration to the wireless medium,

© ICST Institute for Computer Sciences, Social Informatics and Telecommunications Engineering 2017
Y. Zhou and T. Kunz (Eds.): ADHOCNETS 2016, LNICST 184, pp. 221–233, 2017.
DOI: 10.1007/978-3-319-51204-4_18

to highly varying inter-user distance and to node density, etc. are required [1]. Though a lot of work has been done in proposing mechanisms for efficient content dissemination, a lot of work remains to be done to reliably support infotainment applications with acceptable Quality of Service (QoS) and Quality of Experience (QoE).

This paper describes a possible approach to tackle this issue, based on combining three paradigms: Floating Content (FC), Software-Defined Networking (SDN), and Content-Centric Networking (CCN). *Content-Centric Networking* [3] allows messages to be exchanged throughout the network based on their content and not on the location of the hosts. In settings characterized by high node mobility and volatility, such as VANETs, this may greatly increase the chance of delivering the requested content in case of disrupted links and frequent changes of network topologies. However, mechanisms are required to adapt routing decisions to such changes and to achieve a good balance between optimizing content delivery likelihood and resource utilization.

In warning applications, data packets are usually small and can be disseminated between vehicles and between vehicles and infrastructure such as Road Site Units (RSUs) without significantly affecting the available communication bandwidth. For these types of services, the *push-based* communication approach, usually adopted in publish/subscribe applications, is well suitable. Therefore, *Floating Content*, an opportunistic communication scheme, which supports infrastructure-less distributed content sharing over a given geographic area, could be a good candidate for the implementation of these services. The basic formulation of FC is push-based and conceived to suit settings, in which a large fraction of nodes in a given area (the Anchor Zone, AZ) are interested in receiving small messages [4,5]. The idea behind FC is to store a given content object in a spatial region without any fixed infrastructure, making it available through opportunistic communications to all users traversing that region. Whenever a node possessing the content object is within the transmission range of some other nodes not having it, the content object is replicated. When a node with the content object moves out of its spatio-temporal limits, it deletes the content object. The content object may be available on a set of nodes and moves over time within the AZ, even after the node that has originally generated the content has left the AZ. Thus, the content object 'floats' within the AZ. Besides the advantages that both FC and CCN may bring in storing and disseminating content in VANETs, their performance could be improved by the coordination of a centralized entity that has global knowledge of the nodes, mobility, their interest in a given content, as well as other information. It can properly set FC and CCN parameters (e.g., AZ size, time to replicate content, strategy for content caching) and select which approach is more suitable for a given use case [6].

Initially designed for wired networks, *Software Defined Networking* (SDN) has been recognized as an attractive and promising approach for wireless and mobile networks too [7]. These networks can benefit from the flexibility, programmability and centralized control view offered by SDN, under different aspects, such as wireless resource optimization (i.e., channel allocation, interference avoidance), packet routing and forwarding in multi-hop multi-path scenarios as well as efficient mobility and network heterogeneity management. The pioneering SDN-based architecture for VANETs was proposed by Ku et al. [8], and afterwards enhanced [9,10] by adding cloud and fog computing components. In a scenario with a variety of vehicular services with very diverse communication requirements, SDN holds the potential to improve the management of content delivery [11]. A first Type-Based Content Distribution (TBCD) method was proposed in [12] to improve content caching and forwarding in a SDN-enabled VANET. TBCD adopts a push-and-pull approach for delivering content, based on the type of content, and on the number of users interested in it.

In line with that, we investigate how SDN could be beneficial in VANETs for improving CCN-based content caching and FC-based content replication within a given geographical area. The rest of the paper is organized as follows. Section 2 first describes how CCN and FC can be applied in vehicular network scenarios. Section 3 envisages the research areas where the use of a SDN controller may help improving the performance of CCN in vehicular networks. Section 4 illustrates how FC can benefit from some form of coordination among nodes, when implemented through a SDN controller. Finally, Sect. 5 concludes the paper.

2   Content Dissemination in VANETs

In this section, we discuss how to support content dissemination in highly mobile VANETs using CCN and FC. Some related works already proposed CCN modifications to better suit the paradigm to VANET features. Goals were to decrease packet forwarding load and dealing with difficulties in maintaining Forwarding Information Base (FIB) entries. In [13], an additional field was introduced in the Data message to help requesters in selecting the best content provider. To increase content availability, authors in [14] propose that each vehicle caches all received Data and forwards all Interest messages to all available interfaces. However, increasing the message size and forwarding multiple Interest messages for given content increase traffic load. Therefore, new approaches are needed.

When content objects are bound to a specific geographic region, and stored on all (or a subset of) nodes within the region itself, FC could play a key role for improving CCN. By replicating content, FC introduces redundancy, which can be useful when nodes cannot be continuously active and hence do not make content permanently available. Reasons for this could be high node failure rates, intermittent connectivity, load balancing, or energy management involving node duty cycling. Moreover, floating content can make searching and finding content more efficient, if the geographic area of the floating content can be derived from the content name. Normally, to retrieve content, requesters issue Interest messages including a content name describing the requested content. To ensure that the content bound to a geographic area is found, the content name could be mapped to that area. Consequently, an Interest message is forwarded to that geographic area to meet the requested content. Mapping can be supported by either having a geographic ID in the content name or by having an intermediate node, which translates the original content name into a geographic name. After the Interest has met a content source, the Data message including the content is sent back to the requester based on the stored information in the PIT.

2.1 Receiver and Source Mobility Support

In VANETs, both source and requester (receiver) nodes might become mobile. Although CCN has implicit support for *receiver mobility*, certain problems can occur in VANETs in case of very high receiver speeds [15]. If both the RTT of Interest/Data messages and the velocity of the mobile requester are rather low, then standard CCN based on FIB and PIT forwarding will work, because returned Data messages will meet valid PIT entries. However, in case of longer RTTs and fast mobile receivers, PIT entries might be invalid for returning Data messages and the content cannot be delivered.

Receiver mobility can be supported by FC. An Interest could be sent towards the source. In the Interest message an area could be indicated, specifying where the content shall be sent (e.g., the area that the receiver will visit soon). The Data is then sent to the indicated area, where it might float. The receiver can then pick it up when it arrives at the indicated area. However, this approach somehow contradicts the fundamentals of CCN, since Data messages are not sent along the reverse path (considering Interests) using the information stored in PITs.

Source mobility, not natively supported by CCN, makes content search more difficult. Source mobility either requires frequent content advertisement updates or frequent content searches possibly causing significant overhead by broadcasts. It can be supported efficiently by FC if content is assigned to a certain geographic area. If it is ensured that always some nodes storing the content are within the assigned area, then the content can be provided by at least one of the nodes independent on source mobility.

2.2 Agent-Based Content Retrieval (ABCR)

Another approach for requesting and receiving content is based on using agents. Requesters can delegate content retrieval to one or more agents [16] and retrieve from them the content later. A requester broadcasts an Interest to potential agents, which respond by Data messages describing their offer to retrieve content on behalf the requester. The requester then confirms the agent selection. Later, when the requester meets the agent again, it can retrieve the content using traditional CCN message exchange.

In a VANET, a user might like to retrieve a content object from a location that is not in range, and where the vehicle is not scheduled to go. By letting requester and agent know their respective typical or expected paths, the content source could select the right agent (the one with the right path) for sending back the content object. Here we assume that the content source knows the path of the requester and of the candidate agents for carrying the content and that the content source will give information about the path of the requester to the agent carrying the content. Uncertainty in space and time can make the rendezvous hard to implement. To face this issue, the agent carrying the content could implement an AZ, in that the desired content floats and that the requester will traverse with high likelihood.

Agents can also be used to come up with a solution for receiver mobility without the need to modify CCN. To address receiver mobility, a mobile requester intending to visit an indicated area can delegate content retrieval to an agent located in the indicated area. The agent requests the information from the content source and disseminates it using FC mechanisms in the indicated area. The mobile receiver then picks up the content from the agent or from another FC node when visiting the indicated area.

2.3   FC-Assisted, Geographic Content Centric (DTN) Routing

In settings where volatility makes it unfeasible or inefficient to maintain a path between content requester(s) and content source, FC may be used to support so-called geographic content centric routing. We assume that all nodes and content objects can be identified unequivocally, and that nodes know their location and future trajectory with some uncertainty. Such knowledge could be derived from past spatio-temporal patterns. By letting each node share through FC its own data on network state (e.g., node positions and trajectories, node content), FC can be used to build at each node a common, shared representation of the network and its evolution over time. Such representation could be used to implement various strategies for Delay Tolerant Network (DTN) routing.

As a complement, when nodes are sparse and/or their mobility does not allow to build a stable path for content dissemination, Interest packets could be made to float in an area around the originator node. The floating content would be used as a "fishing net" to improve the chance of finding content among passing nodes. As a result, each node within an area would then maintain and share a table (i.e., floating PIT), describing a list of requested content objects. For each requested content object, the table stores a list of requesters. When meeting a node with one or more of the requested content objects, those could be forwarded to the requesters using some form of DTN routing scheme, possibly using the floating network state.

3   SDN Support for CCN in VANETs

3.1   SDN for CCN Forwarding and Broadcasting

In large-scale networks the size of FIB tables on each CCN node can easily increase due to the large amount of exchanged content and the large number of content sources. Alternatively, the CCN routers that request content broadcast Interest packets to neighbor nodes in order to receive it. Subsequently, the possibility of congestion in paths through the network can be very high. There have been several attempts to encounter these problems; some of them try to identify a path between a requester and a content source and store this path for future references [17,18]. Others propose to forward Interests based on the distance between sender and receiver [19]. A similar approach has been adopted also in CCVN [13], where messages are broadcast in the entire VANET, but in case of collisions they are re-broadcast according to the distance from previous senders. To improve content retrieval, in [20] authors suggest to utilize the different interfaces of a CCN node to transmit Interests simultaneously through multiple interfaces. For the same aim, Udugama et al. [21] propose to split the Interest messages of the same request and send them through multiple paths at the same time.

A SDN controller, with its centralized view of the network, can support the design of more efficient CCN forwarding strategies. For instance, a controller can be responsible for selecting unicast paths between requester and content source in a given cluster of the network [22]. To integrate the SDN approach in CCN networks, Charpinel et al. [23] proposed a CCN controller that has a complete view of the network. It defines the forwarding strategy, and installs forwarding rules in the FIBs of CCN routers. In every CCN router there is a Cache Rules Table (CRT). The CCN controller sends cache rules (replacement policies) to the CCN switches, which are stored in the CRT table. The CCN router caches the content according to the rules that exist in the CRT table. The presence of a centralized controller could be further exploited to improve CCN performance. For instance, a SDN controller can determine the number of chunks to transmit through a face and send them through multiple faces simultaneously [24].

Our Approach. CCN content discovery mechanisms can be improved by integrating SDN logic. A SDN controller could setup a path between a requester and a content source and install forwarding rules in the FIB tables of the nodes [23]. Furthermore, flows could be set up similar to the Dynamic Unicast approach [17]. Then a SDN controller could guarantee the QoS for each flow and could transmit the message through the best flow. Moreover, as shown in Fig. 1(a), SDN could support the set up of multipath communication. When a vehicle sends or receives a message (Interest or Data), it could send (or receive) it through multiple (redundant) paths. A SDN controller could discover several paths that satisfy Interest messages of the same request. Then, the controller could send Interest messages through those paths in parallel, in order to retrieve Data much faster.

In a CCN-based VANET, several applications will generate requests with different names. The network must treat these requests in a different way. For instance, a safety application generating warning messages should forward these messages to the entire network. On the contrary, an entertainment application sending video requests should forward these messages through an appropriate face in order to find a content source. A SDN controller could install such forwarding rules to make sure that each message is treated differently, according to its name (Fig. 1(b)).

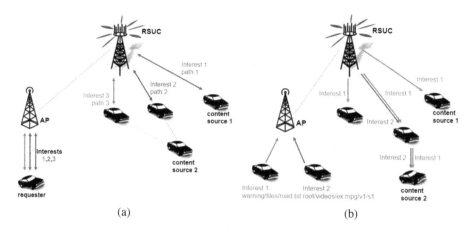

Fig. 1. SDN support for CCN forwarding in VANETs: (a) Forwarding of Interests along different paths, (b) Different forwarding strategy based on Interests names

3.2   SDN for CCN Caching

Caching techniques are used to avoid accessing the server storing the original copy for content requests. Specifically, the buffer memory in vehicles is considered as a cache space for replicas. Indeed, upon a cache hit at a Content Store, the Data message is sent back to the vehicle that sent the Interest. In case of a cache miss, the Interest message shall be forwarded according to a name-based routing strategy, if a similar request is not already pending. Caching enhances content discovery, retrieval and delivery in VANETs by providing multiple sources (caches) of the content. However, the explosion of infotainment applications with their ubiquitous replicas creates an increasing demand for the scarce spectrum in VANETs. Basically, caching is coupled with two fundamental questions: what content to cache and where to cache it.

Caching all content along the delivery paths as suggested in [25] may cause serious performance degradation. Even if the cache size is not a major concern in VANETs, it is not obvious if this cache can keep up the pace with the increasing scale of multimedia content distribution over VANETs. For these reasons, it was proposed in [26–28] to cache only popular content (i.e., content which has been requested a number of time equal or larger than a fixed Popularity Threshold, PT). Other works rely on the user interest for deciding what content to keep [29], i.e., nodes will cache only the content they are interested in.

Concerning where to cache, one trivial solution is to cache replicas in every vehicle [30]. The short-lived nature of links in VANETs, coupled with the time and space-relevant nature of content, accentuates the concern about the selection of relevant nodes that can cache specific content. Therefore, most new caching schemes attempt to reduce the nodes' caching redundancy by only selecting a subset of nodes in the delivery path. This subset of nodes has high probability to get a cache hit and hence nodes in this subset are called "central" nodes.

Centrality-based solutions for node selection were proposed in the context of static networks. However, their extension to highly mobile networks, such as VANETs, is a thorny problem as centrality is not trivial [31]. Recent works [32,33] focus on the social aspect of caching. Users in the same social space are likely to request the same content. Hence, the content is proactively pushed to the cache of users' proximate neighbors. The definition of these proximate neighbors is challenging. While in [32] proximate nodes are $1 - hop$ neighbors, authors of [33] define these nodes as the nodes having a specific ratio of common content items with the user.

Our Approach. Given the unique features of VANETs, SDN can provide centralized cached content management. A SDN controller could instruct Content Providers (CP) to trigger on-demand caching when popular content has been identified. In the same way, the SDN controller could fix the value of the PT according to different metrics. Moreover, the SDN controller, being the owner of the caching logic, will determine which vehicles (Content Cachers) should store replicas of the popular content at each period of time. This centralized caching logic helps authorities to program and deploy the desired caching behavior.

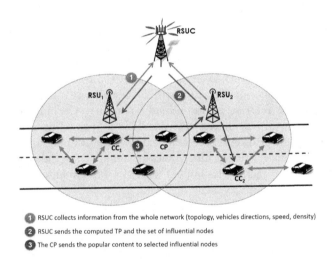

Fig. 2. SDN-based caching approach in VANETs.

Figure 2 illustrates an example of our proposed SDN-based caching approach in VANETs. Two RSUs are orchestrated by one RSU Controller (RSUC). Each RSU is covering a specific geographical area on the highway. The RSUC fixes the PT and the set of influential vehicles [34] in the zone. The set of influential nodes is time-variant since vehicles can enter and leave the network frequently. Therefore, the RSUC should periodically (or any time is needed) notify vehicles in the network about the updated set of influential nodes. Once the number of requests for the content produced by the CP, reaches the PT, the RSUC will trigger its caching spreading strategy. The popular content will not be sent to the immediate 1-hop neighbors of CP like in [32], but will be sent to the influential vehicles (CC_1 and CC_2). This mechanism avoids overloading the network with replicas of the same content. The CP is aware of the set of influential nodes and send them accordingly the replicas. While CC_1 is a 1-hop neighbor of the CP, CC_2 is out of its transmission range. Therefore, the content is sent first to RSU_2 and then RSU_2 forwards the popular content to CC_2. Vehicles interested in the content of CP can ask CC_1 or CC_2 to retrieve it. Thus, vehicles far away from the CP (the node that generated the original content) can get a replicas by asking the nearest CC.

4   SDN Support for FC in VANETs

In FC, information dissemination is geographically limited by the AZ. However, there is no rule on how a user defines the geographic origin, radius and expiration time of the AZ. Simulation results [35] show that the ratio between AZ radius and communication range has an impact on information availability. Besides, authors of [4], provide the criticality condition under which a population of mobile nodes can support the floating content. This condition is related to node density, transmission range, and AZ. Hence, defining the AZ size and shape is a complex task due to vehicles' high mobility and most importantly the lack of centralized network administration.

Furthermore, the main goal of FC is to ensure that each node inside the AZ gets the content and replicates it through opportunistic message exchanges. This exchange can involve a huge number of vehicles (especially in urban areas) and a large amount of information. A majority of studies [4,5,35] assume that when a node with a piece of content in the AZ comes within the transmission range of some other node that does not have it, the piece of content is replicated. This replication takes into account neither the popularity of the content nor the size of set of nodes carrying the content. Unfortunately, more information is not always better, at least not in VANETs, when a massive number of messages increase both spectrum congestion and management challenges.

Our Approach. The centralized intelligence and configuration flexibility that SDN offers allows vehicles to flexibly define the AZ size and shape to cope with content deletion. In challenging environments with highly varying density, SDN can ensure better persistence and availability of floating content by accurately tuning the AZ size. Once an AZ has been defined, the RSUC could reshape the existing AZ or trigger the activation of a new AZ based on its global knowledge. This on-demand creation or extension of an AZ saves network resources. Figure 3(a) shows a use case where an AZ is created upon demand. An emergency vehicle enters the zone covered by RSU_1 where there is no pre-established AZ. The emergency vehicle asks RSU_1 to define an AZ via V2I communication. Then the request is forwarded by RSU_1 to the RSUC, which sends back the parameters of the AZ (shape, size, lifetime) based on the content lifetime,

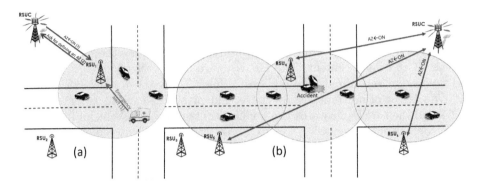

Fig. 3. (a) SDN-based FC activation, (b) AZ reshaping in an accident intersection scenario

nodes density, and mobility. Since communication in FC is infrastructure-less, this temporary on-demand AZ creation was not possible without using SDN. Moreover, SDN could be involved in reshaping of an existing AZ. One interesting use case is information dissemination in an intersection, which is a highly localized area as depicted in Fig. 3(b). To ensure that the content is floating outside the AZ defined by the intersection, one possible solution would be to reshape the AZ and make it bigger. However, as argued in [5], the bigger the AZ size is, the lower is the success probability (i.e. probability for a node entering in AZ to get the content), if node transmission ranges and density are fixed. Hence, the RSUC will define a new AZ's shape based on its global knowledge of the geographical area, node's space distribution, RSUs' traffic load, and the desired size of the floating zone.

Content replication could also be supported by SDN. The RSUC can define, through information collected from RSUs, the floating content's priority. Subsequently, a node can decide to receive and/or ignore content based on a priority rule that can take into account the content lifetime, number of nodes replicating the content, and the intermittent nature of the connections. Thus, the SDN controller implemented in RSUC could figure out the content popularity and decide with which frequency the content is replicated. The RSUC can also define a Range of Interest (ROI) inside the AZ. Indeed, the content is replicated mainly inside the ROI (resp. outside in other applications). For instance, if there are not so many copies of a content object inside the ROI, the SDN controller could allow the exchange of content even outside (resp. inside) the ROI's coverage range. Consequently, the probability of loosing the content is very small.

5   Conclusions

In this work, we have identified CCN and FC as two suitable enablers for improving content storing, dissemination, and forwarding in future VANETs. Both should be adapted to the high dynamic and volatile network environment. To this aim,

SDN could provide support for CCN and FC, e.g., to select, according to node mobility, the influential nodes where the content should be cached, or to activate the AZ where the content should float. Future work will investigate the envisioned integration of CCN, FC and SDN in VANETs in more depth.

Acknowledgment. This work was undertaken under the CONTACT project, CORE/SWISS/15/IS/10487418, funded by the National Research Fund Luxembourg (FNR) and the Swiss National Science Foundation (SNSF) project No. 164205. The authors also like to thank the São Paulo Research Foundation (FAPESP) for the financial support by grant 2016/09254-3.

References

1. Al-Sultan, S., Al-Doori, M.M., Al-Bayatti, A.H., Zedan, H.: A comprehensive survey on vehicular ad hoc network. J. Netw. Comput. Appl. **37**, 380–392 (2014)
2. Da Cunha, F.D., Boukerche, A., Villas, L., Viana, A.C., Loureiro, A.A.: Data communication in VANETs: a survey, challenges and applications. Ph.D. thesis, INRIA Saclay; INRIA (2014)
3. Jacobson, V., Smetters, D.K., Thornton, J.D., Plass, M.F., Briggs, N.H., Braynard, R.L.: Networking named content. In: Proceedings of the 5th International Conference on Emerging Networking Experiments and Technologies, pp. 1–12. ACM (2009)
4. Hyytiä, E., Virtamo, J., Lassila, P., Kangasharju, J., Ott, J.: When does content float? Characterizing availability of anchored information in opportunistic content sharing. In: INFOCOM, 2011 Proceedings, pp. 3137–3145. IEEE (2011)
5. Ali, S., Rizzo, G., Mancuso, V., Marsan, M.A.: Persistence and availability of floating content in a campus environment. In: 2015 IEEE Conference on Computer Communications (INFOCOM), pp. 2326–2334 (2015)
6. Rizzo, G., Palattella, M.R., Braun, T., Engel, T.: Content and context aware strategies for QoS support in VANETs. In: 2016 IEEE 30th International Conference on Advanced Information Networking and Applications (AINA), pp. 717–723 (2016)
7. Yang, M., Li, Y., Jin, D., Zeng, L., Xin, W., Vasilakos, A.V.: Software-defined and virtualized future mobile and wireless networks: a survey. Mob. Netw. Appl. **20**(1), 4–18 (2014)
8. Ian, K., You, L., Gerla, M., Gomes, R., Ongaro, F., Cerqueira, E.: Towards software-defined VANET: architecture and service. In: Conference on Annual Mediterranean Ad Hoc Networking Workshop MEDHOCNET (2014)
9. Salahuddin, M.A., Al-Fuqaha, A., Guizani, M.: Software-defined networking for rsu clouds in support of the internet of vehicles. IEEE Internet Things J. **2**(2), 133–144 (2015)
10. Truong, N.B., Lee, G.M., Ghamri-Doudane, Y.: Software defined networking-based vehicular adhoc network with fog computing. In: 2015 IFIP/IEEE Conference on Integrated Network Management (IM) (2015)
11. Broadbent, M., King, D., Baildon, S., Georgalas, N., Race, N.: Opencache: a software-defined content caching platform. In: 2015 1st IEEE Conference on Network Softwarization (NetSoft), pp. 1–5. IEEE (2015)
12. Cao, Y., Guo, J., Yue, W.: SDN enabled content distribution in vehicular networks. In: 2014 Fourth International Conference on Innovative Computing Technology (INTECH), pp. 164–169. IEEE (2014)

13. Amadeo, M., Campolo, C., Molinaro, A.: Enhancing content-centric networking for vehicular environments. Comput. Netw. **57**(16), 3222–3234 (2013)
14. Grassi, G., Pesavento, D., Pau, G., Vuyyuru, R., Wakikawa, R., Zhang, L.: VANET via named data networking. In: 2014 IEEE Conference on Computer Communications Workshops (INFOCOM WKSHPS), pp. 410–415. IEEE (2014)
15. Anastasiades, C., Braun, T., Siris, V.A.: Information-centric networking in mobile and opportunistic networks. In: Ganchev, I., Curado, M., Kassler, A. (eds.) Wireless Networking for Moving Objects. LNCS, vol. 8611, pp. 14–30. Springer International Publishing, Cham (2014). doi:10.1007/978-3-319-10834-6_2
16. Anastasiades, C., Schmid, T., Weber, J., Braun, T.: Information-centric content retrieval for delay-tolerant networks. Comput. Netw. (2016)
17. Anastasiades, C., Weber, J., Braun, T.: Dynamic unicast: information-centric multi-hop routing for mobile ad-hoc networks. Comput. Netw. (2016)
18. Baccelli, E., Mehlis, C., Hahm, O., Schmidt, T., Wählisch, M.: Information centric networking in the IoT: experiments with NDN in the wild. In: 1st ACM Conference on Information-Centric Networking (ICN-2014). ACM (2014)
19. Grassi, G., Pesavento, D., Pau, G., Zhang, L., Fdida, S.: Navigo: interest forwarding by geolocations in vehicular named data networking. In: 2015 IEEE 16th International Symposium on a World of Wireless, Mobile and Multimedia Networks (WoWMoM), pp. 1–10. IEEE (2015)
20. Yi, C., Afanasyev, A., Wang, L., Zhang, B., Zhang, L.: Adaptive forwarding in named data networking. ACM SIGCOMM Comput. Commun. Rev. **42**(3), 62–67 (2012)
21. Udugama, A., Zhang, X., Kuladinithi, K., Goerg, C.: An on-demand multi-path interest forwarding strategy for content retrievals in CCN. In: 2014 IEEE Network Operations and Management Symposium (NOMS), pp. 1–6. IEEE (2014)
22. Son, J., Kim, D., Kang, H.S., Hong, C.S.: Forwarding strategy on SDN-based content centric network for efficient content delivery. In: 2016 International Conference on Information Networking (ICOIN), pp. 220–225. IEEE (2016)
23. Charpinel, S., Santos, Celso Alberto Saibel Vieira, A.B., Villaca, R., Martinello, M.: SDCCN: a novel software defined content-centric networking approach. In: 2016 IEEE 30th International Conference on Advanced Information Networking and Applications (AINA), pp. 87–94. IEEE (2016)
24. Lee, D.H., Thar, K., Kim, D., Hong, C.S.: Efficient parallel multi-path interest forwarding for mobile user in CCN. In: 2016 International Conference on Information Networking (ICOIN), pp. 390–394. IEEE (2016)
25. Yu, Y.T., Gerla, M.: Information-centric VANETs: a study of content routing design alternatives. In: 2016 International Conference on Computing, Networking and Communications (ICNC), pp. 1–5 (2016)
26. Bernardini, C., Silverston, T., Festor, O.: Mpc: popularity-based caching strategy for content centric networks. In: 2013 IEEE International Conference on Communications (ICC), pp. 3619–3623 (2013)
27. Lee, S.B., Wong, S.H.Y., Lee, K.W., Lu, S.: Content management in a mobile ad hoc network: beyond opportunistic strategy. In: INFOCOM, 2011 Proceedings IEEE, pp. 266–270, April 2011
28. Xu, Y., Ma, S., Li, Y., Chen, F., Ci, S.: P-CLS: a popularity-driven caching location and searching scheme in content centric networking. In: 2015 IEEE 34th International Performance Computing and Communications Conference (IPCCC), pp. 1–8, December 2015

29. Iqbal, J., Giaccone, P.: Interest-based cooperative caching in multi-hop wireless networks. In: 2013 IEEE Globecom Workshops (GC Wkshps), pp. 617–622, December 2013

30. Wang, L., Afanasyev, A., Kuntz, R., Vuyyuru, R., Wakikawa, R., Zhang, L.: Rapid traffic information dissemination using named data. In: Proceedings of the 1st ACM Workshop on Emerging Name-Oriented Mobile Networking Design - Architecture, Algorithms, and Applications, NoM 2012, pp. 7–12, New York, NY, USA. ACM (2012)

31. Gallos, L., Havlin, S., Kitsak, M., Liljeros, F., Makse, H., Muchnik, L., Stanley, H.: Identification of influential spreaders in complex networks. Nat. Phys. **6**(11), 888–893 (2010)

32. Bernardini, C., Silverston, T., Festor, O.: Socially-aware caching strategy for content centric networking. In: Networking Conference, 2014 IFIP, pp. 1–9, June 2014

33. Moualla, G., Frangoudis, P.A., Hadjadj-Aoul, Y., Ait-Chellouche, S.: A bloom-filter-based socially aware scheme for content replication in mobile ad hoc networks. In: 2016 13th IEEE Annual Consumer Communications Networking Conference (CCNC), pp. 359–365, January 2016

34. Katsaros, D., Basaras, P.: Detecting influential nodes in complex networks with range probabilistic control centrality. In: Schuppen, J.H., Villa, T. (eds.) Coordination Control of Distributed Systems. LNCIS, vol. 456, pp. 265–272. Springer International Publishing, Cham (2015). doi:10.1007/978-3-319-10407-2_32

35. Kangasharju, J., Ott, J., Karkulahti, O.: Floating content: information availability in urban environments. In: 2010 8th IEEE International Conference on Pervasive Computing and Communications Workshops (PERCOM Workshops), pp. 804–808, March 2010

A Bandwidth Adaptation Scheme for Cloud Radio Access Networks

Yuh-Shyan Chen, Fang-Yu Liao, and Yi-Kuang Kan[(✉)]

National Taipei University, Taipei, Taiwan, R.O.C.
yschen@mail.ntpu.edu.tw, fangyu1313@gmail.com, david3343559@gmail.com
http://www.csie.ntpu.edu.tw/~yschen/

Abstract. With the advent of Cloud Radio Access Networks (C-RAN) where base band units (BBU) have the ability to dynamically support the bandwidth allocation and centrally manage the bearers transmitted from the remote radio heads (RRH), bandwidth adaptation (BA) mechanism emerges as a promising solution to provide the required resources for C-RAN to provide resources to the bearers during congestion. Existing studies use BA mechanism on congestion management only for LTE system but rarely for C-RAN. To achieve this goal, this paper takes the advantage of dual connectivity with the ability of bearer split to cooperatively provide resources by the serving RRHs, at the same time the mobility robustness and the throughput performance can be improved. The main contribution of this work is to propose an improved BA mechanism for C-RAN with dual connectivity in a centralized concept. More specifically, this work designs a "downgrading index" includes two additional centralized contribution attribute to decide the proportional resource contribution of the bearers which are transmitted by the chosen RRHs and are grouped to assist the serving RRHs. Finally, simulation results illustrate the proposed BA mechanism for C-RAN with dual connectivity significantly reduces the probabilities of the handoff bearer dropping and the bearer blocking.

Keywords: C-RAN · Dual connectivity · Bandwidth adaptation · Call admission control · Handoff

1 Introduction

The fifth generation (5G) cellular wireless networks are expected to overcome the amount of data traffic which is being increasing dramatically in recent years [1]. As a consequence, the mobile operators need to spend more for building, operating and upgrading the traditional Decentralized Radio Access Networks (D-RAN) while the revenue does not balance with the cost. To find feasible solutions emerging in the future, China Mobile has been developing and deploying Cloud Radio Access Networks (C-RAN) which is believed to be an answer to the challenges [2]. Moreover, Mobile and wireless communications Enablers for

© ICST Institute for Computer Sciences, Social Informatics and Telecommunications Engineering 2017
Y. Zhou and T. Kunz (Eds.): ADHOCNETS 2016, LNICST 184, pp. 234–245, 2017.
DOI: 10.1007/978-3-319-51204-4_19

Twenty-twenty (2020) Information Society (METIS) proposed that the 5G architecture trends to embrace C-RAN to have a scalable and centralized control [3]. With centralized processing of C-RAN by separating the base band units (BBU) and remote radio head (RRH), the architecture has the advantage of upgrading and expanding the network capacity. In this respect, operators can have a large saving in cost. However, the bearer would be blocked if there is no idle bandwidth to be allocated by the BBU in C-RAN. To overcome the problems mentioned above, this work proposes an improved BA mechanism with centralized concept in C-RAN. The improvement downgrades resources not only from the requested RRH but from the managed RRHs. To find suitable RRHs to assist the requested RRHs, the BBU discovers the candidate group, G_c and the assistance group, G_a. The detailed definition of G_c and G_a is described in Sect. 3. Moreover, to decide how much resources each bearer from G_a shall release, according to the proportional resource contribution of downgrading index which includes four contribution attributes. This paper designs two contribution attributes in a centralized concept. Therefore, each bearer from G_a only needs to partially release resources to provide enough resources and the resources from the serving RRHs can be decreased.

A GBR bearer is associated with a bearer priority, denoted as i, and a GBR parameter denoted as R_{gi} and a Maximum Bit Rate (MBR) parameter denoted as R_{mi}. To propose an improved BA mechanism includes the above-mention two main QoS parameters, R_{gi} and R_{mi}, this paper also takes dual connectivity which is in 3GPP LTE release 12 [4], is considered in the future of 5G architecture, into account. A term "dual connectivity" refers to the operation with the ability of bearer split where a given UE consumes radio resources provided by one master eNB (MeNB) and one secondary eNB (SeNB). Adopt dual connectivity to C-RAN architecture, the serving RRH with larger transmission coverage can substitute for the role of MeNB, where RRH_m is denoted as the master RRH. Similarly, the other serving RRH with smaller transmission coverage can substitute for the role of SeNB, where RRH_s is denoted as the secondary RRH. Hence, the mobility robustness and the per-user throughput can be improved in C-RAN. Furthermore, a UE consumes radio resources provided by RRH_m and RRH_s causes RRH_m and RRH_s can cooperatively provide resources to the UE. In sum, the objective of this paper is not only proposing an improved BA mechanism that is suitable in C-RAN but taking the advantage of dual connectivity to deal with the resource management in congestion.

The remainder of this paper is organized as follows. In Sect. 2, related work and motivation are described. Section 3 describes the system model, problem formulation and basic idea of the proposed scheme. Section 4 describes the proposed improved BA mechanism for C-RAN with dual connectivity. Simulation results are presented Sect. 5. Section 6 concludes this paper.

2   Related Works

Regarding other existing BA mechanisms by taking the advanced QoS requirements into account, some results are discussed. Khabazian *et al.* [5] proposed a fairness-based preemption algorithm which takes into consideration the bearer priority as well as the amount of the QoS over-provisioning of the bearers. According to the two matrix to have the downgrading index, the contribution of each bearer is computed to release resources in a fairness way. Moreover, the downgrading index can be fine-tuned to have a optimize performance gain by the operator. More recently, Khabazian $et\ al.$ [6] proposed a multi-objective and distributed BA mechanism which takes three bearer attributes into account namely bearer priority, bearer QoS over-provisioning and bearer communication channel quality. With fine-tuning exponents, the performance measurement can be reached a tradeoff. All the related works are proposed the designed BA mechanisms focus on D-RAN architecture, which are not designed for C-RAN.

3   Preliminaries

Considered the downlink C-RAN architecture of handover scenario, the system model is illustrated as shown in Fig. 1(a), where all the RRHs and the BBUs are separated from the BSs and then the RRHs are connected to the BBUs in centralized BBU pool through the fronthaul. In the proposed protocol, only the resource management of resource blocks usage in bandwidth is considered. Moreover, the bandwidth can be dynamically allocated by the BBUs according to the load of controlled RRHs. Therefore, the traffic congestion in some RRHs can be avoided due to the RRHs contribute the idle PRB resources to the busiest RRHs by the centralized BBUs during congestion.

On the other hand, due to the fact that dual connectivity is considered in C-RAN, the UE can consume radio resources provided by RRH_m and RRH_s. In the proposed protocol, RRH_m is managed by BBU_m, while $RRH^{k,r}$ that transmits the radio resources to the UE is also presented as RRH_s, where $RRH^{k,r}$ is the rth RRH managed by BBU_k. In Fig. 1(a), when the UE is moving to the transmission range of $RRH^{1,1}$, where RRH_m transmit resources to the UE plays the role of master RRH, and $RRH^{1,1}$ plays the role of secondary RRH is also presented as RRH_s. Both the RRH_m and RRH_s transmit radio resources to the UE.

To avoid experiencing strong interference from the macro and small RRHs when they are operated on the same frequency in the C-RAN architecture with dual connectivity [4], the bandwidth of channel is divided into two different parts. As shown in Fig. 1(b), the RRH_m with larger transmission range operates the lower frequency bands, $i.e.$, a GHz. Meanwhile, the other RRHs with smaller transmission range operate the higher frequency bands, $i.e.$, b GHz. Note that if the RRHs have transmission range overlap, then the bandeidth can be reassigned. Besides, Fig. 1(b) illustrates the downlink resource block. In 3GPP LTE release 12 [7], a physical resource block is defined as N_{sy} consecutive OFDM symbols in one downlink slot, T_s, and N_{sc} consecutive subcarriers in the frequency domain. Therefore, a PRB in the downlink consists of $N_{sy} \times N_{sc}$ resource elements during two downlink slot T_s.

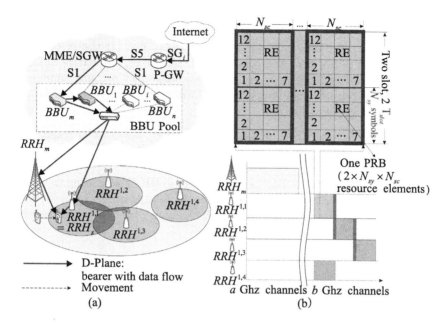

Fig. 1. System model of C-RAN with dual connectivity and channel operation.

3.1 Basic Idea

The basic idea of this work is by taking the advantage of C-RAN with dual connectivity to discover more resources from the chosen RRHs as assistants to assist RRH_m and RRH_s for establishing a new bearer. Moreover, this work designs a downgrading index that is suitable in the architecture in concept of centralized characteristic. When performing the centralized CAC scheme, the RRHs in G_c assist RRH_m and RRH_s to contribute more idle PRB resources. When performing the centralized BA algorithm to release enough resources, the RRHs in G_a assist to release more PRB resources for establishing new bearer. Based on this concept, the probability of a bearer request blocking and handoff bearer dropping can be significantly reduced by the assistance of the RRHs in G_c and G_a as well as the designed downgrading index to decide the proportional released resources of the active bearers.

4 A Bandwidth Adaptation Mechanism for Cloud Radio Access Networks

This section presents the improved BA mechanism for C-RAN with dual connectivity, three phases are described in the following.

4.1 Group Discovery Phase

The group discovery phase aims to find G_c and G_a first. If the resources are not enough when performing the centralized CAC phase, the RRHs in G_c can assist to release idle PRB resources. If the idle resources are not enough, the RRHs is G_a can release resources when performing the centralized BA phase. The procedure is given as follows.

S1. To discover G_c, the BBU which controls RRH_s selects the transmission range of the RRHs which are overlapped with RRH_s or the RRHs in G_c to form G_c.

S2. After discovering G_c, the BBU simply sorts the RRHs according to the QoS over-provisioning ratio $u^{k,r}$ of the RRHs in decreasing order. Let $u^{k,r}$ denote the QoS over-provisioning ratio of $RRH^{k,r}$, then $u^{k,r}$ can be expressed as follows.

$$u^{k,r} = \frac{\sum\limits_{i,j} \gamma_{i,j}^{k,r}}{\sum\limits_{k,r,i,j} \overline{b_{i,j}^{k,r}}},$$

$$\text{where } \gamma_{i,j}^{k,r} = \overline{b_{i,j}^{k,r}} - b_{i,j}^{k,r}(R_{gi}), \tag{1}$$

$$RRH^{k,r} \in G_c$$

where $\overline{b_{i,j}^{k,r}}$ denoted as the load contribution measured by the technology, $b_{i,j}^{k,r}(R_{gi})$ denoted as the resources the bearer $b_{i,j}^{k,r}$ requests to meet the guarantee bit rate R_{gi} with priority i. Therefore, the QoS over-provisioning resources of $b_{i,j}^{k,r}$ denoted as $\gamma_{i,j}^{k,r}$, where $\gamma_{i,j}^{k,r}$ is the resources $b_{i,j}^{k,r}(R_{gi})$ deducted from the allocated resource $\overline{b_{i,j}^{k,r}}$. The larger $u^{k,r}$ means the more bearers $b_{i,j}^{k,r}$ in $RRH^{k,r}$ take more resources that exceed their R_{gi}.

S3. After obtaining the QoS over-provisioning ratio order of each RRH, then G_a can be found. Select the $RRH^{k,r}$ in G_c with $u^{k,r}$ that higher than the threshold θ to form G_a. Note that the lower θ is, the more RRHs can be selected to form G_a, and the more bandwidth usage of RRHs are influenced to release resources. On the other hand, the higher θ is, the less RRHs are selected to form G_a, thus the more difficult to release enough resources.

An example of the group discovery phase is given in Fig. 2. The serving RRHs are RRH_m and $RRH^{1,1}$, where $RRH^{1,1}$ also represents RRH_s. To discover G_c, the transmission range of $RRH^{1,2}$ and $RRH^{1,3}$ are found overlapped with RRH_s. Therefore, $RRH^{1,2}$ and $RRH^{1,3}$ are added to G_c. Moreover, the transmission range of $RRH^{1,4}$ and $RRH^{1,5}$ are overlapped with that of the RRHs in G_c. Therefore, $RRH^{1,4}$ and $RRH^{1,5}$ are added to G_c. To discover G_a, according to the Eq. (1), the BBU sorts $u^{k,r}$ of $RRH^{k,r}$ in G_c and selects the $RRH^{k,r}$ with $u^{k,r}$ that higher than θ to form G_a. Assumes the order of QoS over-provisioning ratio is $u^{1,3} > u^{1,5} > u^{1,1} > u^{1,2} > u^{1,4}$, where $u^{1,3}$ and $u^{1,5}$ are higher than θ, which means $u^{1,3}$ and $u^{1,5}$ have more QoS over-provisioning resources than other RRHs to provide. Therefore, includes RRH_m and RRH_s ($RRH^{1,1}$), G_a is formed by $\{RRH_m, RRH^{1,1}, RRH^{1,3}, RRH^{1,5}\}$.

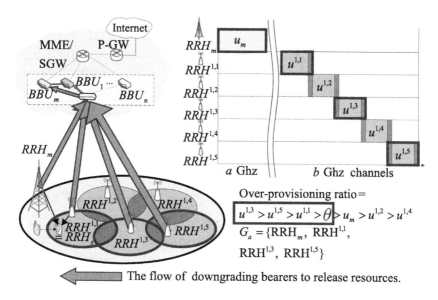

Fig. 2. Example of discovering the G_c and G_a.

4.2 Centralized CAC Phase

When an UE requests resources to establish a new bearer or maintain the rate of a handoff bearer, the CAC policy is performed to decide whether there are enough resources for the bearer admitting to establish. Here the centralized CAC phase aims to enhance the exiting CAC scheme [6] and propose a centralized CAC policy that is considered the centralized characteristic of C-RAN with dual connectivity. Let $n_i^{k,r}$ denote the amount of bearers with priority i that transmitted by $RRH^{k,r}$ and managed by BBU_k. A new bearer or a handoff bearer is the $(n_i^{k,r}+1)$-th. Let $b_{i,n_i^{k,r}+1}^{k,r}(R)$ denote a new bearer $(n_i^{k,r}+1)$-th requests resources to meet bit rate R and let $CQI^{k,r}$ denote the reported CQI value from the UE to $RRH^{k,r}$. Let $C^{k,r}$ denote the total capacity of $RRH^{k,r}$ in OFDM PRB. The procedure is given as follows.

S1. When a new bearer or a handoff bearer $b_{i,n_i^{k,r}+1}^{k,r}$ requests PRB resources $b_{i,n_i^{k,r}+1}^{k,r}(R)$ to meet bit rate R, RRH_m and RRH_s provide resources in a proportional way. The original request resources $b_{i,n_i^{k,r}+1}^{k,r}(R)$ divided to $b_{i,n_i^{k,r}+1}^{k,r}(R^{k,r})$ to request from the serving $RRH^{k,r}$ according to their $CQI^{k,r}$. To express the proportional resources each serving $RRH^{k,r}$ needs to provide, $b_{i,n_i^{k,r}+1}^{k,r}(R^{k,r})$ can be expressed as follows:

$$b_{i,j}^{k,r}(R^{k,r}) = b_{i,j}^{k,r}(R) \times \frac{CQI^{k,r}}{\sum\limits_{k,r,i,j} CQI^{k,r}},$$

$$\text{where}\quad RRH^{k,r} \in RRH_m \cup RRH_s, \tag{2}$$

the serving $RRH^{k,r}$ only need to provide the proportional contribution of $b_{i,j}^{k,r}(R^{k,r})$.

S2. To ensure that the total idle PRB resources in the serving $RRH^{k,r}$ is enough to provide their $b_{i,j}^{k,r}(R^{k,r})$, $Cp^{k,r}$ is denoted as the prediction remaining resources of $RRH^{k,r}$.

$$Cp^{k,r} = (C^{k,r} - \sum_{m=1}^{9} \sum_{n=1}^{n_i^{k,r}} \overline{b_{m,n}^{k,r}}) - b_{i,j}^{k,r}(R^{k,r}) \tag{3}$$

to obtain $Cp^{k,r}$, all the allocated resources to the bearers in $RRH^{k,r}$ need to be deducted from $C^{k,r}$, i.e., $(C^{k,r} - \sum_{m=1}^{9} \sum_{n=1}^{n_i^{k,r}} \overline{b_{m,n}^{k,r}})$, and then deduct the requested proportional resources $b_{i,j}^{k,r}(R^{k,r})$. Where the bearer priority is from 1 to 9. If one of the $Cp^{k,r}$ from the serving $RRH^{k,r}$ is negative, which means the idle PRB resources of $RRH^{k,r}$ is not enough to provide the proportional resource $b_{i,j}^{k,r}(R^{k,r})$, then the BBU checks the requested bit rate. If the requested bit rate $b_{i,j}^{k,r}(R)$ is larger than $b_{i,j}^{k,r}(R_{gi})$, which means the request bit rate is larger than the guarantee bit rate, then sets $b_{i,j}^{k,r}(R)$ to $b_{i,j}^{k,r}(R_{gi})$. Therefore, the procedure is back to S1 to compute $b_{i,j}^{k,r}(R^{k,r})$ again to obtain the new $b_{i,n_i^{k,r}+1}^{k,r}(R^{k,r})$ of serving $RRH^{k,r}$.

S3. If the requested bit rate is $b_{i,j}^{k,r}(R_{gi})$ and one of $Cp^{k,r}$ is still negative, then RRH_m and RRH_s need to assist each other. By summarizing all $Cp^{k,r}$, if the value of $\sum_{k,r} Cp^{k,r}$ larger or equal to zero which means one serving RRH whose $Cp^{k,r}$ is positive and can contribute the remaining resources to the other RRH whose $Cp^{k,r}$ is negative. Otherwise, the RRHs in G_c need to provide idle PRB resources to assist RRH_m and RRH_s. To obtain new request resources $b_{i,j}^{k,r}(R^{k,r})$ from each serving $RRH^{k,r}$, let $Cr^{k,r}$ denote the actual remaining resources, where $Cr^{k,r} = C^{k,r} - \sum_{m=1}^{9} \sum_{n=1}^{n_i^{k,r}} \overline{b_{m,n}^{k,r}}$, then Eq. (4) can re-assign the request resources in an assistance way.

$$b_{i,j}^{k,r}(R^{k,r}) = \begin{cases} b_{i,j}^{k,r}(R^{k,r}) + (Cp^{k,r} - \sum_{k,r} Cp^{k,r}), \\ \text{if } Cp^{k,r} > 0 \\ Cr^{k,r}, \text{if } Cp^{k,r} \leq 0 \text{ or } \sum_{k,r} Cp^{k,r} \leq 0 \end{cases} \tag{4}$$

therefore, the RRH whose $Cp^{k,r}$ is negative only need to provide the remaining resources $Cr^{k,r}$.

S4. If the steps from S1 to S3 cannot release enough resources by the RRHs in G_c and RRH_m with RRH_s, which can be expressed as follow.

$$S_{PRB}^{idle}(G_c) + \sum_{k,r} Cp^{k,r} < 0, \text{where } S_{PRB}^{idle} \in G_c \tag{5}$$

where S_{PRB}^{idle} denoted as the idle PRB resources, and $S_{PRB}^{idle}(G_c)$ denoted as the idle PRB resources from the RRHs in G_c. If $S_{PRB}^{idle}(G_c)$ cannot fill the insufficient resources of RRH_m and RRH_s, which means $S_{PRB}^{idle}(G_c)$ adds $\sum_{k,r} Cp^{k,r}$ still negative, then the proposed BA algorithm is performed to downgrade bearers to release request resources $b_{i,j}^{k,r}(R)$. Otherwise, the admission is failed then the bearer would be dropped.

4.3   Centralized BA Phase

The centralized BA phase aims to downgrade bearers from the RRHs in G_a to release enough resources. Before downgrading the bearers, the downgrading index is calculated to decide the proportional release resources. The downgrading index includes four attributes in this paper. In addition to take the two bearer attribute. After obtaining the value of downgrading index of the bearers, the centralized BA algorithm is performed to release resources in a fair way.

S1. To compute the needed resources of each bearers from RRHs in G_a to be released, the downgrading index $d_{i,j}^{k,r}$ need to be obtained, which includes four attributes. The first attribute is the bearer priority, which is denoted as i. The higher the bearer priority is, the lower the number is, and the less resources need to be released. The second attribute is the QoS over-provisioning resources, which is denoted as $\gamma_{i,j}^{k,r}$ in Eq. (1). At most the resources $\overline{b_{i,j}^{k,r}}$ can downgrade $\gamma_{i,j}^{k,r}$ to $b_{i,j}^{k,r}(R_{gi})$ in order to ensure each bearer can still meet its R_{gi}. The first two attributes are regarding the bearer itself, and the last two attributes are regarding the C-RAN architecture. Thus the attributes are in a hierarchical structure to decide the downgrading index.

S2. The last two attributes are in relation to RRH channel quality and BBU period popularity. The third attribute is the RRH channel quality. This paper takes the CQI as the channel quality from UE to the RRHs in G_a. The larger the CQI, the more resources the RRH needs to be released. It is because that with higher CQI, the RRH can provide less resource to meet the same bit rate compared to the RRH with lower CQI. Moreover, the attribute can have RRH_m and RRH_s provide more proportional of contribution. To denote $CQI^{k,r}$ of $RRH^{k,r}$ in G_a, the RRH channel quality $q_{k,r}$ can be written as follows.

$$q_{k,r} = \begin{cases} CQI^{k,r}, & \text{if } CQI^{k,r} > 1 \\ 1, & \text{otherwise.} \end{cases} \tag{6}$$

the UE may out of range from the RRHs in G_a, in order to avoid having the value of q_r be 0, the minimum value of q_r is set to 1. Finally, the fourth attribute is the period popularity of BBU which manages the RRHs in G_a. This paper takes the resource utilization to measure the popularity of a BBU in a period. If a BBU has larger period popularity, which predicts the next period the BBU may need to provide more resources to the coming bearers.

Let ξ_k denote the period popularity of BBU_k, then ξ_k can be written as follows.

$$\xi_k = \frac{\sum\limits_{i,j,r} \sum\limits_{t=T_s}^{T_e} \overline{b_{i,j}^{k,r}(t)}}{C(T_s - T_e)} \tag{7}$$

ξ_k is obtained from estimating the fraction of the total bandwidth capacity, which denoted as C, and the total resources allocated to the bearers which are established from BBU_k over an observation time period from T_s to T_e sub-frames. From the view of the prediction of the popularity, if BBU_k has higher ξ_k, then the bearers in BBU_k better not release more resources in order to save the resources to serve the future needed bearers. For the ease of the comparison of the resource utilization, ξ_k need to be normalized as follows.

$$\xi_{nor,k} = (1 - \frac{\xi_k}{\sum\limits_{k \in RRH^{k,r}} \xi_k}) \cdot \sum\limits_{k \in RRH^{k,r}} \xi_k \tag{8}$$

S3. After obtaining the four attributes to obtain the value of downgrading index for downgrading each bearers from G_a, the downgrading index is denoted as follows.

$$d_{i,j}^{k,r} = i^\alpha (\gamma_{i,j}^{k,r})^\beta q_{k,r}^\omega \xi_{nor,k}^\delta, \text{ where } \quad \alpha + \beta + \omega + \delta = 1 \tag{9}$$

the purpose of adding the four exponents including α, β, ω, and δ is that the downgrading index results in a different effect. With larger δ, the larger the BBU utilization need not to release more resources. Note that the sum of the four exponents is 1. To obtain the fraction of the resources each bearer need to be released from the RRHs in G_a, $d_{i,j}^{k,r}$ need to be normalized. The normalized downgrading index $\overline{d_{i,j}^{k,r}}$ can be expressed as follows.

$$\overline{d_{i,j}^{k,r}} = \frac{d_{i,j}^{k,r}}{\sum\limits_{k,r,i,j} d_{i,j}^{k,r}} \tag{10}$$

after obtaining the normalized downgrading index $\overline{d_{i,j}^{k,r}}$ of $b_{i,j}^{k,r}$ in G_a, the centralized BA algorithm is performed to release resources.

S4. With the obtained value of $\overline{d_{i,j}^{k,r}}$ of $b_{i,j}^{k,r}$, a centralized BA algorithm is presented to calculate the resources each bearer need to be released. This operation re-allocates the resources to all the bearers from G_a according to the fraction of $\overline{d_{i,j}^{k,r}}$ of $b_{i,j}^{k,r}$. After the operation, the CAC scheme can decide to admit the new bearer or not. In the centralized BA algorithm, this paper includes a centralized approach and applies the idea of the BA algorithm proposed by [6] because which can release resources in a proportional and fair way.

5   Simulation Results

This paper presents an improved BA mechanism for C-RAN with dual connectivity in centralized concept. To evaluate the improved BA mechanism (denoted as proposed scheme), the proposed scheme is simulated compared to the BA mechanism that proposed by Khabazian et al. [6] (denoted as distributed Khabazian scheme). However, the distributed Khabazian scheme is designed for D-RAN and is not under the assist of dual connectivity. In order to compare the two scheme in a fair way, this paper also compared the distributed Khabazian scheme with dual connectivity (denoted as distributed Khabazian scheme with DC) under D-RAN. These three protocols are simulated using the Network Simulator-2 (NS2) and the models and assumptions are based on 3GPP assumptions [4]. The main parameters and the classes of the bearers with their GBR and MBR are shown in Table 1. To ensure that the total value of $\alpha + \beta + \omega + \delta$ are equal to 1, the simulation aims to focus on the effect of the main exponents which are set to the same value (0.91) in the two protocol. To discuss the effect of the bearer request arrival rate and the number of RRHs of D-RAN and C-RAN, the performance metrics to be observed are:

- *Probability of bearer request blocking* (PBRB): The failure probability of establishing a bearer. If a bearer requests resources for establishing but there are not remaining resources for the bearer, then the bearer is not admitted by the CAC scheme.
- *Probability of handoff bearer dropping* (PHBD): The failure probability of maintaining a handoff bearer. If a handoff bearer requests resources from the target eNB or RRH but there are not remaining resources for the handoff bearer, then the bearer is not admitted by the CAC scheme.

5.1   Probability of Bearer Request Blocking (PBRB)

Figure 3(a) shows as the bearer request arrival rate increases, the PBRB increases because the more requests arrive causes congestion occurs and thus increases the PBRB. Observe that the PBRB of the proposed scheme is lower than the distributed Khabazian schem in each parameter of the exponential tune value, as shown in Fig. 3(a). Due to there are more resources can be discovered by the centralized characteristic and the assistance of the RRHs in G_a and the RRHs in G_c, therefore, the more bearers are not be dropped. Figure 3(b) shows the PBRB observed by tuning the umber of RRHs. The higher the number of RRHs is, the higher the PBRB will be. As the number of RRHs increases, the more idle PRB resources can be discovered and because the BBU finds other unused bandwidth to assist to establish the new bearer. The proposed scheme is better than the distributed Khabazian scheme because the distributed Khabazian scheme cannot find other RRHs to contribute their resources. On the other hand, the proposed protocol discovers more resources thus the PBRB reduces when the bearer request arrival rate and the number of RRHs is increased. Moreover,

Table 1. Simulation parameters.

Parameter	Value
Simulation time	1000 s
System bandwidth	Macro cell:10 MHz, Small cell:10 MHz
Carrier frequency	Macro cell:2.0 GHz, Small cell:5.0 GHz
	UE:2.0/5.0 GHz
Moving speed	3 km/h
Inter-site distance	500 m
Number of small cells	7 in macro cell sector
Number of UEs	Uniform distribution with max = 60
Bearer classes	GBR-bearer with priority 1
	GBR = 32 kb/s, MBR = 64 kb/s
	GBR-bearer with priority 2
	GBR = 128 kb/s, MBR = 256 kb/s
	GBR bearer with priority 3
	GBR = 10 kb/s, MBR = 64 kb/s

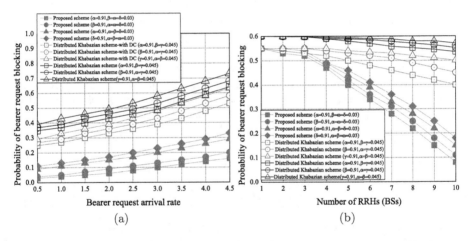

Fig. 3. (a) Probability of handoff bearer dropping vs. bearer request arrival rate. (b) Probability of handoff bearer dropping vs. number of RRHs (BSs).

the distributed Khabazian scheme with DC has higher performance than distributed Khabazian scheme for the reason that with DC, the UE can have more bandwidth to use.

6 Conclusions

An improved BA mechanism that is suitable in C-RAN with dual connectivity is proposed. The proposed improved BA mechanism takes C-RAN with dual connectivity architecture and designs G_a from G_c to assist to release resources. Moreover, the designed downgrading index adds two additional contribution attributes in a centralized concept. Finally, simulation results illustrate the proposed BA mechanism for C-RAN with dual connectivity significantly reduces the probabilities of handoff bearer dropping and bearer blocking.

Acknowledgement. This research was supported by the Ministry of Science and Technology of the ROC under Grants MOST-103-2221-E-305-005-MY3.

References

1. Demestichas, P., Georgakopoulos, A., Karvounas, D., Tsagkaris, K., Stavroulaki, V., Lu, J., Xiong, C., Yao, J.: 5G on the horizon: key challenges for the radio-access network. IEEE Veh. Technol. Mag. **8**(3), 47–53 (2013)
2. White Paper Version 2.5: C-RAN The Road Towards Green RAN. China Mobile Research Institute, June 2014
3. The METIS 2020 Project-Laying the foundation of 5G. https://www.metis2020.com/
4. 3GPP TR 36.842 V1.0.0, Evolved Universal Terrestrial Radio Access (E-UTRA); Study on Small Cell Enhancements for E-UTRA and E-UTRAN Higher Layer Aspects (Release 12). 3rd Generation Partnership Project (3GPP), November 2013
5. Khabazian, M., Kubbar, O., Hassanein, H.: A fairness-based preemption algorithm for LTE-Advanced. In: Proceedings of IEEE Global Telecommunications Conference (GLOBECOM 2012), Anaheim, America, pp. 5320–5325, December 2012
6. Khabazian, M., Kubbar, O., Hassanein, H.: An advanced bandwidth adaptation mechanism for LTE Systems. In: Proceedings of IEEE International Conference on Communications (ICC 2013), Budapest, Hungary, pp. 6189–6193, June 2013
7. 3GPP TS-36.211 v12.1.0, Technical Specification Group Radio Access Network; Evolved Universal Terrestrial Radio Access (E-UTRA); Physical channels and modulation. 3rd Generation Partnership Project (3GPP), March 2014

Cooperative On-the-Fly Decision Making in Mobility-Controlled Multi Ferry Delay Tolerant Networks

Mehdi Harounabadi$^{(\boxtimes)}$, Alina Rubina, and Andreas Mitschele-Thiel

Integrated Communication Systems Group, Ilmenau University of Technology,
Ilmenau, Germany
{mehdi.harounabadi,alina.rubina,mitsch}@tu-ilmenau.de

Abstract. This paper presents a cooperative On-the-fly Decision maker for Multi Ferry (ODMF) delay tolerant networks. ODMF chooses the next node to visit for mobility-controlled data ferries. In ODMF, each ferry keeps a history about its last visit time to nodes and applies it in its decision making. Moreover, a ferry shares and exchanges the history information with other ferries through an indirect signaling. While there is no direct communication among ferries in our assumptions, ferries employ nodes as relays for the indirect signaling of control information. The simulation results show that ODMF outperforms the TSP and on-the-fly approaches in terms of message latency. In scenarios with high number of ferries, it can be seen that the impact of indirect signaling on the performance is more notable. In ODMF, the travel time is reduced for messages that are in the buffer of ferries by the cooperative decision making of ferries. In addition, we study and discuss the impact of increasing number of ferries and increasing ferry speed on the performance and cost of a message ferry network. We show that a required performance can be achieved with less cost by increasing speed of ferries than increasing number of ferries.

Keywords: Delay tolerant networks · Mobility-controlled · Multi ferry · On-the-fly decsion maker

1 Introduction

In Delay Tolerant Networks (DTNs), nodes can be scattered over a vast area and far from the radio transmission range of each other. In such situations, a direct communication among nodes is not possible. A data ferry is a specific mobility-controlled wireless node that travels among disconnected nodes in a DTN and exchanges messages among them [1]. A message ferry network is a DTN where the communication among nodes is only possible employing a data ferry. The trajectory of a data ferry is called 'ferry path'. It impacts the performance of a message ferry network. In a network with an asymmetric traffic load of nodes where the traffic generation rate in nodes is different and variable in time,

© ICST Institute for Computer Sciences, Social Informatics and Telecommunications Engineering 2017
Y. Zhou and T. Kunz (Eds.): ADHOCNETS 2016, LNICST 184, pp. 246–257, 2017.
DOI: 10.1007/978-3-319-51204-4_20

visiting nodes by a ferry based on offline ferry path planning approaches such as the solution for the Traveling Salesman Problem (TSP) may degrade the performance. The TSP solution provides the shortest path to visit all nodes without considering the traffic load in nodes and the flow of traffic. Therefore, applying an on-the-fly decision maker in a ferry to dynamically choose the next node to visit seems to be an efficient solution for such networks. An on-the-fly algorithm adapts the trajectory of a ferry to the data traffic in a message ferry network. However, a single ferry approach provides a limited resource that is not sufficient for large or highly loaded networks. In such scenarios, adding more ferries boosts the performance [6].

In this paper, we propose a cooperative On-the-fly Decision maker for Multi Ferry networks (ODMF) where ferries share their observations using nodes as relays (indirect signaling of control information among ferries). The main goal of ODMF is making an on-the-fly decision in a ferry about the next node to visit. In ODMF, each ferry keeps the history of its last visit to nodes. Keeping the track of last visit time in ODMF helps to reduce too frequent or too seldom visits to nodes. While there is no direct communication among ferries, they share this history information through an indirect signaling via nodes to cooperate in a message ferry network.

The simulation results show that ODMF outperforms existing on-the-fly decision makers and the TSP solution as an offline ferry path planner in multi ferry networks in terms of message latency. Moreover, we show the importance of sharing history information among ferries and its impact on the performance. A message latency consist of a message waiting time in a node buffer after its generation and the message travel time in a ferry buffer (travel delay) till it is delivered at its destination. We study the impact of increasing number of ferries on the constituent components of a message latency applying ODMF and existing on-the-fly approaches. In addition, we investigate the impact of ferry speed in multi ferry scenarios. To the best of our knowledge, this is the first work that applies and evaluates a cooperative on-the-fly decision making in multi ferry networks.

The remainder of this paper is organized as follows: First, we will discuss existing work for single and multi ferry networks in Sect. 2. In Sect. 3, we describe our network model. Our on-the-fly decision maker algorithm is presented in Sect. 4. Section 5 introduces the indirect signaling among ferries via nodes for the cooperative decision making. In Sect. 6, we evaluate the performance of ODMF in multi ferry scenarios and study the impact of the number and speed of ferries on the performance and costs of the network.

2   Related Work

Message ferry path planning is modeled as a TSP problem in [1,2]. Authors find the shortest path for a ferry to visit all nodes. However, some visits may be unnecessary and a waste of resources. Moreover, the TSP solution do not consider the traffic load and the flow of traffic in a network. In [3], authors present an approach to control the mobility of a ferry. The problem is modeled as a Markov Decision Process (MDP) and solved by a heuristic algorithm to find a sequence of nodes to visit in a Round-Robin (RR) fashion. A self-Organized Messages Ferrying (SOMF) algorithm in [4,5] was proposed which employs the on-the-fly decision making for mobility of a ferry in single ferry scenarios. The on-the-fly decision maker selects a node to visit based on some weighted functions. The weight for each function is learned by a ferry for a specific network topology. Learning process takes a long time and becomes useless with variations in the network. It also applies a fairness function that distributes the number of visits among all nodes uniformly which may cause frequent and useless visits of a node.

As single ferry approaches cannot be applied in large and highly loaded networks due to the performance needs. To overcome limitation of a single ferry network, authors in [6] proposed different architectures to deploy a multi ferry network. However, they did not focus on the ferry path planning. In [7], authors modeled multi ferry path planning as a Partial Observable Markov Decision Process (POMDP). In their assumptions, nodes are located in clusters and ferries visit the cluster heads to exchange messages. From a ferry point of view, the clusters are stationary. However, cluster heads are mobile within a cluster. The contribution of the algorithm is to plan the ferries paths based on the mobility model of cluster heads. The mobility model of cluster heads is given as an input for the POMDP.

None of the existing approaches take the advantages of cooperation among ferries. There is a lack of an on-the-fly decision maker for multi ferry networks in the literature which a ferry can adapt its trajectory based on the traffic of the network and the decision of other ferries.

3   Network Model

3.1   Assumptions

In our work, the network is modeled as follows: wireless nodes (N) are of two types; regular nodes $(R \subset N)$ and ferry nodes $(F \subset N)$.

$$System = (R \subset N) \cup (F \subset N) | R \cap F = \emptyset \qquad (1)$$

From now on, we call regular nodes only 'nodes'. Nodes are assumed to be disconnected stationary wireless nodes. They can generate and consume (receive) messages. Message generation in nodes is variable in time with a random inter-message arrival time. Ferries are mobility-controlled wireless nodes that travel among nodes and transfer their messages. Location of nodes is known by ferries. A ferry itself does not generate or consume any messages. It only collects messages from nodes and delivers them to their destinations. Ferries travel always with a constant velocity. In our network, there is no obstacle to limit the direct movement of ferries between any pair of nodes. Moreover, we assume that there is no limit in the size of buffers in both ferries and nodes.

The scale of distances between nodes are considered much bigger than their radio transmission range.

$$d(i, j \in R) \gg tx_{range} \tag{2}$$

Therefore, we neglect the radio transmission range for both nodes and ferries and consider it zero ($tx_{range} = 0$). Moreover, the required time for a ferry to travel among nodes is much bigger than the required message transmission time (T_{tx}) between ferries and nodes. Thus, we neglect T_{tx}.

$$T_{travel}(i, j \in R) \gg T_{tx} \tag{3}$$

We assume that our network is a pure message ferry network and communication among nodes is only possible via ferries. Moreover, there is no direct communication among ferries. A ferry can only obtain an observation whenever the ferry visits a node. In our network, ferries plan on-the-fly for their trajectory. A ferry visits a node, obtains an observation and chooses the next node to visit. Therefore, there is no predefined path for ferries.

Next, we introduce steps for a ferry when it visits a node.

3.2   Steps of a Ferry Visit to a Node

In our multi ferry DTN, whenever a ferry visits a node, a set of functions is triggered and run sequentially in the ferry. The functions are:

1. Exchange of messages with the node
   (a) The ferry collects all messages from the node's buffer
   (b) The ferry delivers all messages for which the current node is the destination
2. Exchange the history information with the node (indirect signaling among ferries)
3. Decide about the next node to visit using ODMF
4. Traveling towards the next (decided) node

In the next section, we describe our on-the-fly decision maker in ferries that chooses the next node to visit.

4   On-the-Fly Decision Maker for Multi Ferry Networks

In this section, we propose a novel cooperative On-the-fly Decision maker for Multi Ferry (ODMF) networks. The main goal of ODMF is to make on-the-fly decisions in a ferry about the next node to visit. ODMF works only based on the local observations of a ferry and the history of nodes that a ferry saves in its memory. In our multi ferry network, there is no direct communication between ferries (no long range communication), but ferries share control information for better cooperation using an indirect signaling. Indirect signaling among ferries is done by employing nodes as relays. ODMF can be applied in single and multi ferry networks without any modifications in the algorithm.

In our network, the ODMF in a ferry decides about the next node to visit applying a *Score* function. The score function is calculated in a ferry for each node r and a node with the maximum $Score(r)$ value is selected as the next node to visit. The *Score* for each node r is calculated as follows:

$$Score(r) = \frac{fb_{norm}(r) + lvt_{norm}(r)}{d(c,r)} \tag{4}$$

where $fb_{norm}(r)$ is a function that returns a normalized value for a node r based on the number of waiting messages in the ferry buffer. A candidate node will have a bigger fb_{norm} value if it is the destination for more messages. It is calculated as follows:

$$fb_{norm}(r) = \frac{msg.count(r)}{msg.count(r_{max})} \tag{5}$$

$msg.count(r)$ is the number of messages for the node r in the ferry buffer and $msg.count(r_{max})$ is the number of messages for the node with the maximum number of messages in the ferry buffer.

The second function is based on the history of nodes in the memory of a ferry. Each ferry keeps the history of its last visit time to all nodes and applies it in its decision maker. $lvt_{norm}(r)$ returns a normalized value for the node r based on the last visit time of ferry to node r. The value for each node r is calculated as follows:

$$lvt_{norm}(r) = 1 - \frac{last.visit.time(r)}{current.time} \tag{6}$$

lvt_{norm} in the decision maker avoids any visit starvation in nodes. Visit starvation degrades the performance of a message ferry network because messages in a starved node must wait for a long time to be collected by a ferry. It also prevents frequent visits of a node in a short time window. Frequent visits of a node may waste the resource in a message ferry network, when the visit rate for a node is higher than its message generation rate.

$d(c,r)$ in *Score* function is the distance between the current node c and a possible next node r.

In the next section, we introduce the indirect signaling among ferries for a cooperative decision making. Ferries exchange their history of nodes ($last.visit.time$), that they save in their memory, using nodes as relays.

5   Cooperation of Ferries by Sharing History Information Through Indirect Signaling

As mentioned before, each ferry keeps the history of its last visit time to all nodes to apply it in its decision maker. The last visit time history of nodes in a ferry is updated when the ferry visits a node.

In our multi ferry network, each node acts as a relay for signaling among ferries. The signaling information is the last visit time history of nodes in ferries.

Similar to ferries, each node keeps this history in it's memory. The history information in a node does not refer to the history of any specific ferry. All ferries can update the history information in nodes to share their information with each other. Whenever a ferry visits a node, the ferry exchanges the history information with the node. Older history information is updated with more up-to-date information in both sides. Therefore, a ferry receives history information of other ferries through an indirect signaling via nodes. Moreover, the ferry shares its own history by updating it in the node, if the ferry has more up-to-date information.

As mentioned in Sect. 3.2, the history information exchange between a ferry and a node occurs before the decision making of a ferry about its next node to visit. Therefore, the ferry can apply more up-to-date information to its decision maker if it is available in the node. Indirect signaling among ferries can lead to better cooperation of ferries. For instance, a ferry can receive an information about a node that has been visited by other ferries recently. Thus, the ferry avoids visiting the node since it could likely be a waste of resource.

Algorithm 1 describes the update of last visit time history when a ferry visits a node. First, the ferry receives the history table that is in the node's memory. Then, the ferry updates its own history table and the history table of the node by comparing values in both tables for all nodes in the network. Finally, ferry sends the updated table to the node and the latter saves the updated history table in its memory.

Algorithm 1. Update of the last visit time (lvt) history in a ferry

1: $receive(lvt\_hist\_in\_node)$ ▷ $last.visit.time$ is the history in the ferry
2: **for** each node r **do**
3:      **if** $last.visit.time(r) < lvt\_hist\_in\_node(r)$ **then**
4:         $last.visit.time(r) \leftarrow lvt\_hist\_in\_node(r)$
5:      **else**
6:         $lvt\_hist\_in\_node(r) \leftarrow last.visit.time(r)$
7: $send(lvt\_hist\_in\_node)$

---

6   Simulation Study

In this section, we evaluate and study the performance of the proposed ODMF and existing approaches. In our comparisons, we evaluate two versions of ODMF. In the first version, indirect signaling of control information among ferries exists. We name this version 'ODMF'. In the second version, there is no signaling among ferries (no sharing of history information among ferries) but the decision function in a ferry is same as ODMF and we call it 'ODMF-NC' (non-cooperative). To do this, we extended the Python based single ferry simulator introduced in [4]. The main objectives of our simulations are as follows:

1. Comparison of the ODMF versions with existing on-the-fly decision makers and the TSP path planner as an offline approach in terms of message latency and the constituent components of a message latency
2. Study on the impact of increasing number of ferries on the average message latency in the network applying existing on-the-fly decision makers
3. Study on the impact of ferry speed on the performance and costs of single and multi ferry networks

In our simulations, nodes are placed randomly in each simulation run. The position of a node is restricted to a $1000 \times 1000 \, \text{m}^2$ area. Message generation in nodes is variable in time. It starts at $t = 0$ and runs for 1000 s. Then, the simulation is continued till delivery of all messages. We consider an asymmetric traffic load in our network. Nodes can be categorized in our model according to their message generation rates to very high rate (10% nodes), high rate (10% of nodes), normal rate (70% of nodes) and no message generation (10% of nodes) with mean inter-message arrival time of 1 s, 5 s, 10 s, ∞, respectively.

6.1 Comparison of ODMF with Existing Approaches

In the first simulation, we model the network with 20 nodes and 15 ferries and compare two versions of ODMF with existing on-the-fly decision makers and the TSP path planner as an offline approach. We run the simulation 10 times for each algorithm. In each run the topology of the network, i.e. the placement of nodes is different. Ferries start their travel from different nodes with a constant velocity of 10 m/s. The on-the-fly decision makers (other than ODMF versions) that we apply in our comparisons are 'fb' (ferry buffer) and 'SOMF' (Self-Organized Message Ferrying) [4,5]. fb is a basic on-the-fly decision maker that applies a ferry buffer function to choose a node to visit based on the number of messages in a ferry buffer for a destination. SOMF applies a visit count function in addition to the ferry buffer function and a distance function. Visit count function in SOMF distributes the number of ferry visits among nodes.

Figure 1 shows the end to end latency of messages in 5 different approaches. The end to end latency of a message refers to the time difference between a message generation in its source and the delivery of message at its destination.

In terms of the median of end to end latencies, TSP has the worst result as it is an offline path planner and does not consider the load of messages in nodes and the flow of traffic in the network. However, regarding to the dispersion of latency values and the maximum latency in the network fb is the worst approach. fb only considers messages in a ferry buffer for its decision. This causes long waiting of messages in nodes.

It can be observed that ODMF outperforms all existing approaches. Two lessons is learned from the ODMF results. First, applying the last visit time history leads to better end to end message latency in a network comparing with other metrics such as the visit count that is applied in SOMF. This can be seen by comparing ODMF-NC and SOMF which both approaches do not share any control information and their difference is in their decision functions. Second, sharing history

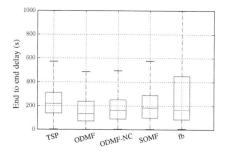

Fig. 1. The end to end latency of messages

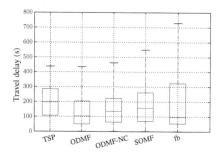

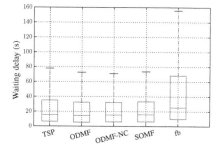

Fig. 2. Messages travel time (waiting time in a ferry buffer).

Fig. 3. Messages waiting time (in a node buffer).

information through indirect signaling in ODMF is an efficient solution for the cooperation of ferries in a multi ferry network. This can be seen by comparing ODMF and ODMF-NC. To have a deeper insight about the performance of all approaches, we divide the end to end message latency to its constituent elements. The end to end message latency consists of two parts:

1. Message waiting delay in a node buffer after its generation till it is collected by a ferry- $delay_{wait}$
2. Message waiting delay in a ferry buffer after the ferry collected it and before it is delivered at its destination. It is the time that a message travels in a ferry buffer- $delay_{travel}$.

Therefore the end to end delay of a message can be calculated as following:

$$Delay_{e2e} = delay_{wait} + delay_{travel} \tag{7}$$

Figures 2 and 3 illustrate the messages waiting delay '$delay_{wait}$' in nodes buffer and messages travel delay '$delay_{travel}$' in ferries buffer, respectively. TSP has the highest median of travel delays, as it does not consider the destination of messages in a ferry buffer. It always visits a predefined sequence of nodes in order to ensure that all the nodes are visited and the path length is minimal.

The best median value for $delay_{travel}$ occurs in fb while it chooses a node to visit only based on the waiting messages in a ferry buffer. However, it always chooses a node with maximum number of waiting messages in a ferry buffer and has no other metrics. This results in increasing the $delay_{travel}$ of messages for a destination that are in minority. ODMF has similar results to fb in terms of median value for $delay_{travel}$ but ODMF has the least maximum $delay_{travel}$ and better dispersion of values. In ODMF, cooperation of ferries by sharing the history information leads to better strategy in visiting nodes. Due to the indirect signaling of ferries, a ferry avoids visiting a node that has been visited by other ferries. It causes faster delivery of messages that are waiting in the buffer of a ferry. This is also the reason why ODMF is better than ODMF-NC. In ODMF-NC, no signaling among ferries occurs. Therefore, a node may be visited by different ferries in a short time window which impacts on the $delay_{travel}$ of messages in the buffer of ferries. In SOMF frequent visits of a node in a short time window occurs not only by several ferries, but also it may occur by one ferry while it tries to visit all nodes equally. The visit count function in SOMF may force a ferry to visit a node several times in a short time window that impacts the $delay_{travel}$ of messages in the ferry buffer.

Looking at waiting delay of messages in nodes '$delay_{wait}$', it can be seen that ODMF, ODMF-NC, SOMF and TSP have similar results. However, waiting delays in ODMF versions is slightly better due to the last visit time history in nodes. fb has the worst $delay_{wait}$ for messages while it only serves the messages in a ferry buffer. Messages in the buffer of nodes may wait for a long time to be visited by a ferry that applies fb. In fb, a ferry visits a node only if it has any message for the node or the ferry buffer is empty and in this case fb chooses a node randomly to visit.

Comparing constituent elements of end to end latency, we can conclude that the last visit time history of nodes and indirect signaling in ODMF impacts mostly on the travel delay of messages '$delay_{travel}$' while in multi ferry networks the waiting delay of messages '$delay_{wait}$' are not so different applying different strategies to visit nodes.

6.2   Study on the Impact of Increasing Number of Ferries on the Performance of Message Ferrying

In this section, we study the impact of increasing the number of ferries in a message ferry network applying ODMF and existing on-the-fly decision makers. We run the simulation 100 times and in each new run, the number of ferries is increased by one. Therefore, the number of ferries is increased from 1 to 100 and we measure the average latency of messages after each run. Moreover, we evaluate the impact of the number of ferries on the constituent elements of the average end to end latency in the network which are the average waiting delay ($delay_{wait}$) and the average travel delay ($delay_{travel}$) of messages. The network consist of 20 nodes and the placement of nodes are kept identical for all 100 runs to see only the impact of the number of ferries in the network. Ferries start their travel from one specific node (a depot) with a constant velocity of $10\,\mathrm{m/s}$ and

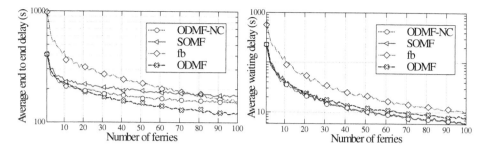

Fig. 4. Impact of the number of ferries on the average end to end latency of messages

Fig. 5. Impact of the number of ferries on the average waiting delay of messages (waiting in a node buffer).

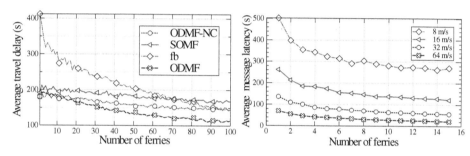

Fig. 6. Impact of the number of ferries on the average travel latency of messages (waiting in a ferry buffer).

Fig. 7. Increasing number of ferries with different speeds (ODMF).

keep it till delivery of all messages in the network. The traffic generation model in nodes is same as in Sect. 6.1 for all 100 runs.

Figure 4 demonstrates the average end to end latency of messages in a network employing 1 to 100 ferries. The average end to end latency decreases by increasing the number of ferries in the network. ODMF is always the best approach and shows better performance than ODMF-NC employing more ferries in the network. However, both versions of ODMF have similar performances having few number of ferries. The difference between versions of ODMF illustrates the importance of indirect signaling in a network with high number of ferries. Sharing the history information among ferries results in better cooperation and coordination of ferries when there are high number of ferries in the network.

SOMF and ODMF-NC show a saturation in decreasing the average message latency by adding 20 ferries or more in the network. On the other hand, the average end to end latency in fb is not saturated and always decreases. fb is a better decision maker than SOMF having more than 80 ferries. However, it is the worst approach with less number of ferries. To find the reasons for the behavior of different algorithms, we look at the constituents of a message latency. Figures 5 and 6 show the average ($delay_{wait}$) and the average ($delay_{travel}$), respectively.

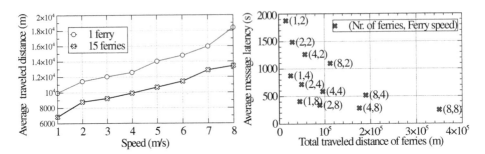

Fig. 8. The average traveled distance of a ferry (cost per ferry).

Fig. 9. Performance and total cost of different networks setups.

With more ferries in the network, the $delay_{wait}$ tends to zero in all approaches. Therefore, having 80 ferries or more, the $delay_{travel}$ has the main impact on the average end to end latency. In fb, increasing the number of ferries is more effective in $delay_{travel}$ comparing with SOMF and ODMF-NC because it only serves the waiting messages in a ferry buffer.

6.3   Study on the Impact of Ferry Speed

In this section, we study the impact of ferry speed on the performance and cost of message ferry networks. To do this, we increase the number of ferries from 1 to 15 for different speeds of ferries. The decision maker in ferries is ODMF and 20 nodes exist in the network. Traffic model in the network is same as in Sect. 6.1. Figure 7 shows the average latency of messages in the network for different number of ferries and different speeds. The average message latency is decreased by speeding up ferries. However, we can observe that a saturation in the network occurs by increasing the number of ferries. Saturation means that the increasing number of ferries does not have a tangible impact on the average end to end latency in the network. We can see that the saturation occurs earlier employing faster ferries.

Figure 8 illustrates the impact of ferry speed on the traveled distance of each ferry in the network to complete a message ferry mission. Increasing the ferry speed causes more traveled distance for each ferry. This is the cost of improvement in message latency by increasing the ferry speed.

Figure 9 shows the total traveled distance of all ferries and the average message latency for different setups in networks with 10 nodes. Each network setup is shown in a pair of (number of ferries, ferry speed). For instance, (4,8) is a network setup with 4 ferries that each ferry travels with the speed of 8 m/s. To achieve a required average message latency in a network, there are possibilities to increase the speed of ferry(ies) or number of ferries. For instance, to achieve the average message latency below 500 s, we can have both (1,8) or (2,4) setups. The best setups are those which have the least average message latency and the least total traveled distance of ferries (total cost) in a network. Therefore, setups closer to the lower left corner are desirable setups in a network. Looking at different setups, it can be inferred that adding up the number of ferries in a network imposes more total cost to a network than speeding up less number of ferries to achieve the required performance.

7   Conclusion

In this paper, we proposed a cooperative On-the-fly Decision maker for Multi Ferry networks (ODMF) where each ferry keeps a history of nodes and shares it with other ferries through an indirect signaling. The results show that ODMF outperforms existing approaches (offline and on-the-fly) in terms of message latency. Sharing the last visit time history through an indirect signaling in ODMF causes cooperative decisions in ferries and decreases the travel time of messages that are in the buffer of ferries. The impact of indirect signaling is more in multi ferry networks with high number of ferries. Moreover, we studied the impact of number and speed of ferries in message ferry networks. It can be seen from results that with increasing the number of ferries, the saturation in a network occurs earlier if we employ faster ferries. Increasing number of ferries after the saturation of a network does not improve the performance as before. Besides, increasing number of ferries imposes more cost to a network than increasing the speed of ferries to achieve a required performance. In addition to the cooperative decision making, ferries my cooperate in message forwarding by using nodes as relays to improve the performance of a message ferry network. This is our future work.

References

1. Zhao, W., Ammar, M., Zegura, E.: A message ferrying approach for data delivery in sparse mobile ad hoc networks. In: Proceedings of the 5th ACM International Symposium on Mobile Ad Hoc Networking and Computing (2004)
2. Heimfarth, T., de Araujo, J.P.: Using unmanned aerial vehicle to connect disjoint segments of wireless sensor network. In: 28th International Conference on Advanced Information Networking and Applications (AINA) (2014)
3. Mansy, A., Ammar, M., Zegura, E.: Deficit round-robin based message ferry routing. In: IEEE Global Telecommunications Conference (GLOBECOM) (2011)
4. Simon, T., Mitschele-Thiel, A.: A self-organized message ferrying algorithm. In: 14th International Symposium and Workshops on a World of Wireless, Mobile and Multimedia Networks (WoWMoM), pp. 1–6 (2013)
5. Simon, T., Mitschele-Thiel, A.: Next-hop decision-making in mobility-controlled message ferrying networks. In: First Workshop on Micro Aerial Vehicle Networks Systems, and Applications for Civilian Use (2015)
6. Zhang, Z., Fei, Z.: Route design for multiple ferries in delay tolerant networks. In: IEEE Wireless Communications and Networking Conference (WCNC) (2007)
7. He, T., Swami, A., Lee, K.-W., Dispatch-and-search: dynamic multi-ferry control in partitioned mobile networks. In: Twelfth ACM International Symposium on Mobile Ad Hoc Networking and Computing (2011)

Multipath Routing Optimization with Interference Consideration in Wireless Ad hoc Network

Junxiao He[1(✉)], Oliver Yang[1], Yifeng Zhou[2], and Omneya Issa[2]

[1] School of Electrical and Computer Engineering, University of Ottawa,
161 Louis Pasteur, Ottawa, ON, Canada
jhe052@uottawa.ca, yang@site.uottawa.ca
[2] Communication Research Centre, Ottawa, ON, Canada
yifeng.zhou@outlook.com, mneya.Issa@canada.ca

Abstract. This paper proposes a multipath routing optimization algorithm for allocating bandwidth resources to nodes that are subject to interference from flows in other parts of the network. The algorithm consists of three steps: path discovery, path selection and load distribution. In addition to delay, power and hop count, the routing metric also takes into account the interference of flows from other parts of the network during path selection and load distribution. An optimization model is formulated based on the flow cost and the bandwidth usage by the other flows. The AIMMS package is used to solve the optimization problem to obtain an optimal solution with the minimum total flow cost. Finally, we use computer simulations to assess the performance and effectiveness of the proposed routing technique.

Keywords: Optimization · Multipath routing · Interference · Wireless ad hoc network

1 Introduction

In contrast to the infrastructure wireless networks, where each user directly communicates with an access point or based station, the wireless ad hoc network does not rely on a fixed infrastructure for its operation [HoMo14]. When a node tries to send information to other nodes out of its transmission range, one or more intermediate nodes are needed. Many research works have been studied in this area, e.g., [KoAb06, LuLu00, TsMo06].

Routing protocol is one of the most important challenges in wireless ad hoc network. The goal is to find the appropriate paths from source to destination. Generally speaking, the routing protocols can be classified into two categories: Unipath Routing and Multipath Routing as shown in Fig. 1 [AaTy13, MuTs04, YiJi07].

There are already many works in routing protocols such as Proactive routing protocols and Reactive routing protocols. Proactive routing such as DSDV (Destination Sequenced Distance Vector) Routing [PeBh94] and WRP (Wireless Routing Protocol) [MuGa96] is also called Table-Driven routing because they keep track of routes for all

© ICST Institute for Computer Sciences, Social Informatics and Telecommunications Engineering 2017
Y. Zhou and T. Kunz (Eds.): ADHOCNETS 2016, LNICST 184, pp. 258–269, 2017.
DOI: 10.1007/978-3-319-51204-4_21

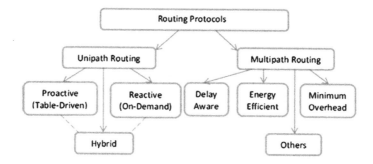

Fig. 1. Classification of routing protocols

destinations and store the route information in tables. When the application starts, a route can be immediately selected from the routing table. Reactive routing such as DSR (Dynamic Source Routing) [JoMa96] and AODV (Ad hoc On-demand Distance Vector) [PeRo99] is also known as On-Demand routing protocol because it does not need to maintain the routing information or routing activity if there is no transmission between two nodes. Routes are only computed when they are needed. These two classes of routing protocols are usually Unipath routing which do not consider the bandwidth limitation along the path. Since the bandwidth is usually limited in wireless ad hoc networks, routing along a single path may not provide enough bandwidth for transmission. This is why Multipath Routing is becoming more and more popular. This routing technique uses multiple alternative paths through a network, which can yield variety of benefits such as increasing fault tolerance, bandwidth aggregation, mini-mizing end-to-end delay, enhancing reliability of data transmission and improving security [BeGa84, Gall77, TsMo06]. The multiple paths computed might be over-lapped, edge-disjointed or node-disjointed with each other [Wiki15b].

There are three fundamental components when designing the multipath routing: Path Discovery, Path Selection and Load Distribution. These three have always been the most important issues in the multipath routing. However, many papers are just concerned with only one or two of these issues. For examples, the SMR (Split Mul-tipath Routing) protocol is an on-demand MSR (Multipath Source Routing) protocol that is concerned with the path discovery and path selection [LeGe00]. Some papers are mainly concerned with the path selection [MaDa01, MaDa06]. Based on the AOMDV (Ad hoc On-demand Multipath Distance Vector) [YeKr03], an NS-AOMDV (AOMDV based on the node state) protocol [ZhXu13] was proposed to choose the path with the largest path weight from the node state for data transmission. All of the above papers do not discuss how to allocate the bandwidth to different multiple paths because once several multiple paths are selected; an algorithm is required to distribute the load. Early papers usually assume an unlimited bandwidth, e.g. [GiEp02, Vand93]. A distributed routing and scheduling algorithm based on link metric is proposed [GiEp02] to decrease the consumption of limited resources such as the power. However, it does not take into account of limited bandwidth. An arbitrarily large bandwidth is also assumed in the derivation of the lower and upper bounds of a uniform capacity in a power-constrained wireless ad hoc network [ZhHo05].

There has been much work on optimization approaches in wireless ad hoc network. Since energy is a major concern, one obvious objective is to minimize the power consumption by formulating the routing problem as a LP (Linear Programming) optimization model [KaTa08]. It usually leads to other problems such as the need to improve the end-to-end delay while minimizing the power consumption during optimization [SiWo98]. There has been optimization work based on limited resources such as [KrPa06, LeCi98]. However, those papers do not consider the interference of transmissions from other parts of the network.

Based on the above short comings, we would like to study a multipath routing algorithm where bandwidth is limited, and to assign a Flow Cost as a function of several essential factors which can be used during the path selection step. In the third step of the multipath routing (load distribution). We would like to formulate an optimum bandwidth allocation algorithm for the multiple paths under the constraint of limited bandwidth available at each link. At the same time, we will also consider about the interference from the other network flows in the network during our optimization model.

In order to achieve our objectives, we would like to first provide a network model for multipath routing in a wireless ad hoc network where nodes with limited bandwidth can be shared by several network flows. For the path selection step of the multipath routing, we would formulate an algorithm that assigns to every routing path in the network a FC (Flow Cost) as a function of 4 different factors: interference in addition to end-to-end delay, power consumption and hop count. The interference can come from flows receiving influence from other parts of the network, but not just the physical interference from other nodes like the noise. Our algorithm is designed for more practical networks where bandwidth is limited so that congestion can arise due to the competition of limited resource. A CN (Crowded Node) is identified for each multiple routing path with the purpose to formulate a LP (Linear Programming) optimization model and to obtain the minimum cost in all CNs. The optimization results allow us to choose the best bandwidth allocation scheme for the CNs.

The contributions of our paper as the following: (1) Accounting for interference from other flows in the network in addition to the traditional power consumption, end-to-end delay and hop count. (2) Taking into account the bandwidth usage and the interaction of other network transmission in the CNs. As far as we know, there is no bandwidth allocation with this interaction carefully studied so far. (3) Solving the optimization model in AIMMS and using the optimization results to obtain the bandwidth allocation trend related with the flow cost.

The remainder of this paper is organized as follows: Sect. 2 presents the network model used in our research and the related assumptions. Section 3 explains the 3 steps of the multipath routing we propose/use in this paper. Section 4 provides the optimization procedure for use in the bandwidth allocation algorithm and discusses the optimization results. Section 5 summarizes our findings and future work. For the remainder of this paper, the following symbols and notations pertain.

C_{max}^k Maximum capacity of the CN in k^{th} routing path.

C_u^k Occupied bandwidth by other network flows in the CN of kth routing path.

D_S Data rate from source node S.

D_{vu} Link delay from node v to node u_o.
D_{max} Maximum delay of all links in the network.
H_k Number of hops of the kth routing path.
k   Row number of RPT.
l Number of links in the network.
M Set of nodes located in the routing path.
n Number of nodes in a network.
P_{vu} Total power consumption from node v to node u.
P_{max} Maximum power consumption.
R Distance between the source node S and destination node D.
R' Radius of the Half-Circle.
r Maximum coverage distance of a node.
U_k FC value of the kth routing path.
X_k Allocated network flow rate of the kth routing path.

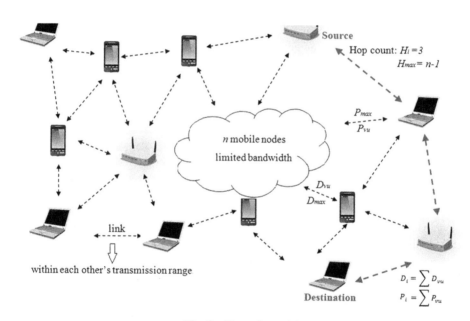

Fig. 2. Network model

2   Network Operation, Modeling and Assumptions

As shown in Fig. 2, we consider a network with n mobile nodes, each with a limited bandwidth. The nodes are equipped with Omni-directional antenna that has a maximum transmission power P_{max} and maximum interference I_{max}. A link between two nodes would exist if they are within the transmission range of each other. Along each link (v, u), let P_{vu} be the total power (such as the transmission power and the processing power) required to deliver data. Also let D_{vu} be the link delay consisting of the propagation delay and the processing delay; I_{vu} be the interference which can be anything that alters, modifies, or disrupts a message (e.g., noise) as it transmits the data from node v to node u. A routing path between a source and a destination consists of the concatenation of different links, and its length can be measured in hops H (called hop count). Another measure is the end-to-end delay which is the sum of all link delays along the routing path. Likewise we can associate a total power and total interference consumed along the path. The node with the minimum available bandwidth is identified as the CN (Crowded Node) which is most likely to have congestion in the presence of many network flows.

Unless otherwise specified, the following assumptions are pertained:

(a) There is no link breakage during the transmission: This allows us to focus on the channel conditions in this first stage of analysis without various complications. This will be relaxed in our future research.
(b) The number of network flows and their occupied bandwidths for the crowded node are known for the current transmission.
(c) The bandwidth of a node is limited. This is practical because the total bandwidth of its outgoing links is limited. This important assumption is different from many other papers which assume the bandwidth is big enough for the transmission of all network flows in the network.
(d) The end-to-end delay for each link is fixed until the next route discovery.

3   Multipath Routing Algorithm

This section provides the details of the 3 steps of the multipath routing: path discovery, path selection and load distribution. During our path selection step, we not only consider the delay, power and path length, but also take the interference into consider. In the load distribution, we create a LP (Linear Optimization) model to optimally distribute the limited bandwidth to different multiple paths.

3.1   Path Discovery

Before a data packet is sent from the source to its destination, an end-to-end route must be determined. During its routing discovery phase [MaDa01, WaZh01], a source would initially flood the network with RREQ (Route REQuest) packets. Each intermediate node receiving an RREQ will reply with an RREP (Route REPly) along the reverse path back to its source if a valid route to the destination is available; else the RREQ is rebroadcast. Duplicate copies of the RREQ packet received at any node are discarded. When the destination receives an RREQ, it also generates an RREP. The RREP is routed back to the source via the reverse path. As the RREP proceeds towards the source, a forward path to the destination is established.

One can see that after each routing discovery process, an intermediate node can acquire all the related information (such as its next-node number in the routing path and the end-to-end delay from the source) from the RREP packets it received, and therefore can detect all the routing paths through it to a given destination. These paths will be saved in the RPT (Routing Path Table) in the increasing order of the end-to-end delays initially.

3.2   Path Selection

Before selecting the appropriate paths, the first step we need to do is to sort all the paths in the RPT based on their FCs which consists of the following 4 parameters:

(a) P_{vu}: the total link power consisting of factors such as transmission power and processing power.
(b) H_k: the path length of a route saved in the kth row in RPT. Obviously, we have $H_k \leq n-1$ $(k = 1, 2, 3, ...)$.
(c) D_{vu}: link delay as introduced in the network model.
(d) I_{vu}: interference when transmitting packets from node v to node u.

We can now determine the FC of kth routing path in the RPT as the total contributions of all the above parameters from all links along the path. Let U_k be the Flow Cost of the kth path in the RPT and M be the set of nodes in the routing path such that $(v, u) \in M$ is the set of concatenated links to form the path. Then we have

$$U_k = \sum_{(v,u) \in M} \left(\frac{P_{vu}}{P_{max}} + \frac{D_{vu}}{D_{max}} + \frac{I_{vu}}{I_{max}} \right) + \frac{H_k}{n - 1} \tag{1}$$

Since the interference, power, delay and hop count take on different units and different magnitudes, we have normalized each parameter with their respective maximum value (i.e., P_{max}, D_{max}, I_{max} and $n-1$ respectively) so that their contributions become values between 0 and 1.

We can now sort/update all the paths in the RPT according to their FCs in their ascending order. The routing path with the smallest FC is saved in the first row of the RPT. Its path index number is 1. The routing path with the second smallest FC is saved in the second row with an index number 2, and so on and so forth. A smaller FC indicates a routing path with a combination of lower power consumption, lower interference, lower time delay and smaller hop count.

After the first step of updating the RPT, we can get new routing path information in the table which is in an ascending order of FC. Thus, our second step is to simplify the table by the procedure of node-disjoint scheme. We will compare all the routing paths to see if they share the same node. When two or more routing paths share one same node, we will delete the path with higher FC (larger row number in RPT) until all the remaining routing paths are node-disjoint. For example, we firstly obtain the node numbers in the first routing path (the one in the first row with lowest FC), then we compare it with the node numbers saved in the second row. If they have one or more same node numbers, we will delete the second row from the RPT. Next, compare the node numbers with the third row, fourth row, etc. After the first iteration, all the paths in RPT will be node-disjoint with the first row. So we begin our second iteration to compare the second row with the others with larger row numbers. Then third iteration to compare the third row with the others with larger row numbers, the fourth iteration, etc., until all the paths in RPT are node-disjoint.

After the above two steps, we can obtain a simplified Routing Path Table. Every path in the table is node-disjoint with the others and the paths are in an ascending number according to their FCs. Assume there are k rows in total in the table. We use U_1 as the FC value for the first routing path saved in the first row in RPT; U_2 as the FC value for the second routing path saved in the second row in RPT; etc. U_k as the FC value for the kth routing path saved in the last row in RPT. There are usually many routing paths in the RPT even after the node-disjoint selection scheme. If we consider all these paths, the efficiency of the algorithm will decrease. Therefore, we just consider the first 2 routing paths here (it's easy to expand the 2 routing paths to more).

3.3   Load Distribution

For each of the two multiple routing paths selected in Sect. 3.2, we need to distribute the limited bandwidth to different paths. For each path, we can find a Crowded Node with minimum available bandwidth. We attempt to optimize the bandwidth allocation of the limited bandwidth available at the two CNs by taking into account the usage of bandwidth by the other flows that can arise from anywhere in the network. In addition, we plan to use the utilization factor (packet arrival rate/transmission rate) combined with the FC in the optimization objective function to find the optimum bandwidth allocation scheme which can achieve the minimum flow cost. The results of the optimization would allow us to choose the best bandwidth allocation scheme among all paths between a source-destination pair.

We use U as the total flow cost and U_k as the flow cost of kth path. Because we just choose the first two paths in RPT, so k can be equal to 1 or 2. If we want to bring in more multiple paths, we just extend the values of k. We assume the occupied bandwidth for the CN in k^{th} path is C_u^k and its maximum capacity is C_{max}^k. The bandwidth will be allocated to the kth path is X_k. Then the optimization formulation is as the following.

$$\text{Minimize } U = \sum_{k=1,2} U_k * \frac{C_u^k + X_k}{C_{max}^k} * X_k \qquad (2)$$

Subject to:

$$C_u^k + X_k \leq C_{max}^k, \ k = 1, 2 \qquad (3)$$

$$\sum_k X_k \geq D_s, \ k = 1, 2 \qquad (4)$$

$$X_k \geq 0, \ k = 1, 2 \qquad (5)$$

Constraint (3) says that the sum of arrival rates from all network flows cannot exceed the maximum capacity (data rate) of a CN. Constraint (4) says that the sum of all outgoing path capacities supporting the flows from the same source should be greater than the source data rate. Constraint (5) is just a regular condition to ensure the non-negativity of a flow value.

4   Optimization Performance

We shall use the AIMMS-CPLEX (Advanced Integrated Multidimensional Modeling Software) solver to solve our optimization problem. We will take a 20 nodes network for example as shown in Fig. 3.

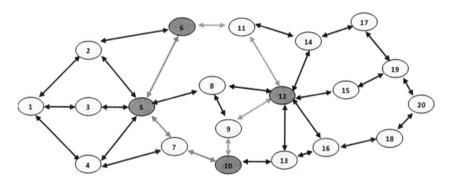

Fig. 3. A 20-nodes network example. (Color figure online)

Assume we have decided the two node-disjoint routing paths from source node 6 to destination node 10: the first one with smallest FC is *6-5-7-10* (red path); and the second path is *6-11-12-9-10* (green path). The two Crowded Nodes for these two multiple paths are Node 5 and Node 12. For node 5, we assume there are two other flows is using this node when we transmit the packets from node 6 to node 10, they are *1-3-5-7* and *4-5-8*. These two flows occupy $C_u^1 = 120$ Kbps. For node 12, we assume there are three other flows is using this node for transmission: *13-12-14, 9-12-15-19 and 8-12-16*. These three flows occupy $C_u^2 = 145$ Kbps . The maximum capacity for these two routing paths is 200 Kbps. The data arrival rate for current transmission (from source node 6) is $D_S = 60$ Kbps. Based on the above data information, we can create an optimization model in the AIMMS to solve the problem and obtain the optimal bandwidth allocation to the two multiple routing paths. We give some random values of Flow Cost for the first and second routing path. After running the model in AIMMS several times, we can get the different allocation results with different FC values as shown in Table 1 below.

Table 1. Allocation optimization results

The 1st routing path		The 2nd routing path		Total FC
FC	Allocated Bandwidth	FC	Allocated Bandwidth	
1.4	48.3 Kbps	1.8	11.7 Kbps	73
2.7	50.0 Kbps	3.6	10.0 Kbps	143
3.2	50.4 Kbps	4.3	9.6 Kbps	169
3.6	53.7 Kbps	5.2	6.3 Kbps	193
4.0	54.7 Kbps	5.9	5.3 Kbps	215
4.3	55.9 Kbps	6.5	4.1 Kbps	231
4.7	53.8 Kbps	6.8	6.2 Kbps	252
5.5	47.8 Kbps	7.0	12.2 Kbps	288
6.4	41.9 kbps	7.2	18.1 kbps	323

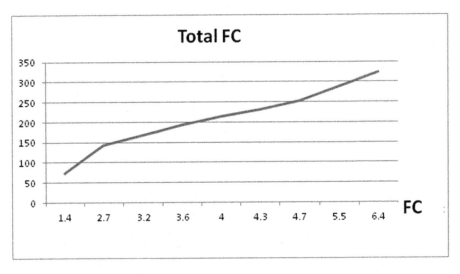

Fig. 4. Relationship between FC and total FC

From the table, it is obviously to see that the higher FC values for the two multiple paths, the higher Total FC as shown in Fig. 4.

The allocated bandwidth to a path is not an increasing function of FC as illustrated in Fig. 5 for the first multiple paths. A smaller FC does not guarantee to obtain more bandwidth because there is a FC for which a maximum bandwidth is obtained. So this relationship tells us that we need to find the more appropriate values for two multiple paths in order to optimally allocate the bandwidth according to our specific requirements.

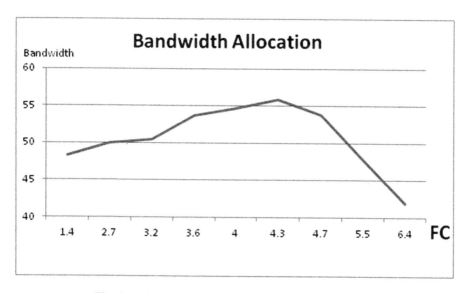

Fig. 5. Relationship between FC and allocated bandwidth

Table 2. Other random allocation schemes

	Bandwidth for 1st routing path	Bandwidth for 2nd routing path	Total FC
FC = 1.4 for the 1st routing path FC = 1.8 for the 2nd routing path	20 Kbps	40 Kbps	86
	15 Kbps	45 Kbps	91
	35 Kbps	25 Kbps	76
	30 Kbps	30 Kbps	79
	40 Kbps	20 Kbps	75
	10 Kbps	50 Kbps	97

In order to demonstrate the results from our optimization model is truly the optimum one, we will use the first group of FC values and results from Table 1 as an example. The FC values for the two routing paths are 1.4 and 1.8 respectively and the Total FC obtained is 73 based on the FC and its bandwidth allocation results: 48.3 Kbps for the first routing path and 11.7 Kbps for the second routing path. Therefore, we will give some other random bandwidth allocation schemes to see what their Total FC will be.

Table 2 shows that the Total FC from the other random allocation schemes are all bigger than our optimum Total FC value of 73. Therefore, we can demonstrate that the result we get from our optimization model is the minimum.

5 Conclusion

We have provided in this paper a multipath routing optimization for the allocation of limited bandwidth along multiple paths to a destination. The 3 steps of the multipath routing, path discovery, path selection and load distribution are discussed in details. We have integrated interference into the FC in both path selection and load distribution stages in addition to other three factors of end-to-end delay, power consumption, and hop count. An optimization model was created with consideration of interference by other flows to solve the problem and to obtain the minimum total flow cost. From the investigation of the relationship between the FC and Total FC, FC and Bandwidth allocated to one path, we can show that an optimum result is obtained.

The bandwidth allocation optimization methodology proposed in this paper has the benefit of increasing the reliability of the packet transmission and decreasing network congestion. Future work includes the simulations of our algorithms in a test-bed which will be created in Opnet and the applications of the beamforming directional antenna.

Acknowledgement. This work has been supported partially by CRC (Communication Research Center) and by an NSERC grant.

References

[AaTy13] Aarti, Tyagi, S.S.: Study of MANET: characteristics, challenges, application and security attacks. Int. J. Adv. Res. Comput. Sci. Softw. Eng. **3**(5), 252–257 (2013)

[BeGa84] Bertsekas, D., Gfni, E., Gallager, R.: Second derivative algorithms for minimum delay distributed routing in networks. IEEE Trans. Legacy Commun. **32**, 911–919 (1984)

[Gall77] Gallager, R.: A minimum delay routing algorithm using distributed computation. IEEE Trans. Legacy Commun. **25**, 73–85 (1977)

[GiEp02] Girici, T., Ephremides, A.: Joint routing and scheduling metrics for ad hoc wireless networks. In: Conference Record of the Thirty-Sixth Asilomar Conference on Signals, Systems and Computers, vol. 2 (2002)

[HoMo14] Hoebeke, J., Moerman, I., et al.: An overview of mobile ad hoc networks: applications and challenges. J. Commun. Netw. **3**(3), 60–66 (2014)

[JoMa96] Johnson, D.B., Maltz, D.A.: Dynamic source routing in ad hoc wireless networks. In: Imielinski, T., Korth, H.F. (eds.) Mobile Computing, vol. 353, pp. 153–181. Springer, US (1996)

[KaTa08] Kazemitabar, J., Tabatabaee, V., Jafarkhani, H.: Global optimal routing, scheduling and power control for multi-hop wireless networks with interference. In: Coping with Interference in Wireless Networks. pp. 59–77. Springer, Netherlands (2008)

[KoAb06] Kolar, V., Abu-Ghazaleh, N.B.: A multi-commodity flow approach for globally aware routing in multi-hop wireless networks. In: Proceedings of Fourth Annual IEEE International Conference on Pervasive Computing and Communications, pp. 306–317 (2006)

[LeGe00]   Lee, S.J., Gerla, M.: SMR: split mulitpath routing with maximally disjoint paths in ad hoc networks. Technical report. Computer Science Department, University of California, Los Angeles, August 2000

[LeCi98]   Leke, A., Cioffi, J.M.: Dynamic bandwidth optimization for wireline and wireless channels. In: IEEE Conference Record of the Thirty-Second Asilomar Conference on Signals, Systems & Computers, vol. 2, pp. 1753–1757 (1998)

[LuLu00]   Luo, H., Lu, S., Bharghavan, V.: A new model for packet scheduling in multihop wireless networks. In: Proceedings of ACM Mobicom Conference, pp. 76–86 (2000)

[MaDa01]   Marina, M.K., Das, S.R.: On-demand multipath distance vector routing in ad hoc networks. In: IEEE Ninth International Conference on Network Protocols, pp. 14–23 (2001)

[MaDa06]   Marina, M.K., Das, S.R.: Ad hoc On-demand multipath distance vector routing. Wirel. Commun. Mob. Comput. 6(7), 969–988 (2006)

[MuGa96]   Murthy, S., Garcia-Luna-Aceves, J.J.: An efficient routing protocol for wireless networks. App. J. Spec. Issue Routing Mob. Commun. Netw. 1(3), 183–197 (1996)

[MuTs04]   Mueller, S., Tsang, Rose, P., Ghosal, D.: Multipath routing in mobile ad hoc networks: issues and challenges. In: Calzarossa, M.C., Gelenbe, E. (eds.) MASCOTS 2003. LNCS, vol. 2965, pp. 209–234. Springer Berlin Heidelberg, Berlin, Heidelberg (2004). doi:10.1007/978-3-540-24663-3_10

[PeBh94]   Perkins, C.E., Bhagwat, P.: Highly dynamic destination-sequenced distance vector routing for mobile computers. Comp. Comm. Rev. 24(4), 234–244 (1994)

[PeRo99]   Perkins, C.E., Royer, E.M.: Ad-hoc On-demand distance vector routing. In: Proceedings of the 2nd IEEE Workshop on Mobile Computing Systems and Applications (1999)

[TsMo06]   Tsai, J., Moors, T.: A review of multipath routing protocols: from wireless ad hoc to mesh networks. In: Early Career Researcher Workshop on Wireless Multihop Networking, vol. 30 (2006)

[YiJi07]   Yi, J.: Summary of routing protocols in mobile ad hoc networks. Polytechnic School of University of Nantes (2007)

[YeKr03]   Ye, Z., Krishnamurthy, S.V., Tripathi, S.K.: A framework for reliable routing in mobile ad hoc networks. In: IEEE INFORCOM (2003)

[Vand93]   Vandendorpe, L.: Multitone system in an unlimited bandwidth multipath Rician fading environment. In: Seventh IEE European Conference on Mobile and Personal Communications (1993)

[WaZh01]   Wang, L., Zhang, L., Shu, Y., Dong, M., Yang, O.: Adaptive multipath source routing in ad hoc networks. In: IEEE International Conference on Communications, Helsinki, vol. 3, pp. 867–871, June 2001

[Wiki15b]   07 July 2015. https://en.wikipedia.org/wiki/Multipath_routing

[ZhHo05]   Zhang, H., Hou, J.C.: Capacity of wireless ad-hoc networks under ultra wide band with power constraint. In: 24th Annual Joint Conference of the IEEE Computer and Communications Societies, vol. 1 (2005)

[ZhXu13]   Zhou, J., Xu, H., Qin, Z., Peng, Y., Lei, C.: Ad hoc On-demand multipath distance vector routing protocol based on node state. Commun. Netw. 5(03), 408–413 (2013)

An Accurate Passive RFID Indoor Localization System Based on Sense-a-Tag and Zoning Algorithm

Majed Rostamian$^{(\boxtimes)}$, Jing Wang, and Miodrag Bolić

School of Electrical Engineering and Computer Science,
University of Ottawa, Ottawa, Canada
m_rostamian@ieee.org, {jwang226,mbolic}@uottawa.ca

Abstract. Localization and tracking of objects (e.g. objects or people) in indoor environment will facilitate many location dependent or context-aware applications. Localization of passive ultra high frequency (UHF) radio-frequency identification (RFID) tags attached to objects or people is of special interest because of the low cost of the tags and backscatter communication that is power efficient. An augmented RFID system for localization based on a new tag called Sense-a-Tag (ST) that communicates with the RFID reader as a passive tags and can detect and record communication of other passive tags in its proximity was introduced several years ago. In ST-based localization system, a large set of passive landmark tags are placed at the known locations. The system localizes ST based on the aggregation of binary detection measurements according to localization algorithm, such as weighted centroid localization (WCL). However, the aforementioned method is easily affected by the outlier detection of distant landmark tags by ST. To improve localization accuracy, this paper propose to iteratively refine the interrogation area of the reader so that it includes only the most relevant landmark tags. The performance of the proposed method is demonstrated by extensive computer simulation and realistic experiments.

Keywords: Internet of Things · Ultra High Frequency (UHF) Radio Frequency Identification (RFID) · Weighted centroid · Zoning algorithm

1 Introduction

In recent years, the concept of the Internet of Things (IoT) has been gaining popularity. The basic premise of the concept is that "things" are interconnected and have unique identifiers. One of the potential technologies for the IoT is radio frequency identification (RFID). However, coarse-grain location of a tag obtained by current RFID systems is not suitable for the context-awareness of identifiable objects in IoT. The objective of this paper is to analyse and improve the functionality of a novel semi-passive tag called "Sense-a-Tag (ST)" introduced in [1]. The ST can be used for accurate indoor localization based on

© ICST Institute for Computer Sciences, Social Informatics and Telecommunications Engineering 2017
Y. Zhou and T. Kunz (Eds.): ADHOCNETS 2016, LNICST 184, pp. 270–281, 2017.
DOI: 10.1007/978-3-319-51204-4_22

proximity detection. The ST is based on the idea of tag to tag communication that has been introduced earlier [2]. ST is able to capture both UHF RFID reader and tag signals and it is fully compatible with EPCGlobal Class 1 Generation 2 RFID protocol (we will refer to it as Gen2 protocol). The ST can also act as a Gen2 tag and send the information to the reader. The ST can overcome existing limitations of RFID systems including cost and localization accuracy. The existing systems for proximity detection and localization are mainly based on active RFID and run on specialized platforms. Since ST can be added as a new hardware to any current Gen2 UHF RFID system without any modification in hardware or firmware of off-the-shelf tags and readers, it can be considered as a realistic solution for proximity detection and localization applications.

The ST-based localization system requires that RFID readers are installed and a network of passive landmark tags is placed. The landmark tags are placed at known locations. Since the ST can detect communication of the RFID reader and the landmark tags in its proximity and send detected landmark tags' IDs to the reader. The host software estimates the position of the ST based on the position of the detected landmark tags. In [3], we investigated the "weighted-centroid" method of localization and activity tracking system based on proximity detection using ST and RFID passive tags. In [3], the RFID reader reads all landmark tags in each query round. Due to RF propagation characteristics, as well as relative orientation among the reader, ST and landmark tag's antennas, it is possible that sometimes the ST detects some distant landmark tags while missing to detect nearby landmark tags. This effect causes comparatively large estimation error. In this paper, we propose a method where we iteratively refine the interrogation area of the reader so that it includes only relevant landmark tags in order to improve the ST localization accuracy. A combination of weighted centroid localization method and the method of iteratively reducing the zone that the ST detects is called the zoning algorithm in this paper.

We have calibrated our newly developed UHF RFID simulation framework PASS based on the data obtained from the field experiment [4,5]. Using this simulator we could simulate behaviour of all components of the system including the reader, tags and ST in a given environment. The zoning algorithm is evaluated in both PASS simulator and real office environment.

The paper is organized as follows. Section 2 introduces the background and related work. In Sect. 3, based on the brief description about ST-based localization system, we present combined localization method that includes the zoning algorithm and Weighted Centroid Localization(WCL). Section 4 describes the PASS simulator. The simulation and experimental results are shown in Sect. 5. We conclude the paper with future discussion in Sect. 6.

2   State of the Art

Generally, we can divide any RF-based indoor positioning systems to two main categories: range-based and range-free. In range-based localization, methods are dependent on accurate measurement of the RF signal that we can get RFID tags or readers. Previous designs based on this methodology such as Cricket [6], Radar [7], SpinLoc [8], all required information extracted from wireless propagation such as RSSI, angle of arrival, relative velocity measurement or fine-grained point-to-point distance information. However, the propagation parameters of wireless signal is vulnerable to the non-line-of-sight situations, severe multipath fading effects and dynamic temporal changes of indoor environment. Therefore, the accuracy and robustness of range-based localization methods are relatively low.

The main idea of range free localization depends on discriminating the presence of reference nodes within the target's proximity [9, 10]. The most encouraging reason for using range-free localization systems is that they are less vulnerable to the effects of complex indoor environment. Centroid method is an example of a range free localization algorithm. Centroid method is based on finding the position of the unknown object based on the avarage of the coordinates of the coordinates of positions of the landmark nodes. LANDMARC [11] and WCL [12] improved centroid design by assigning weights to the different reference nodes. In weighted WCL the node, which is closer to the target node, would be assigned larger weight in the location estimation algorithm.

Later, several methods have been introduced to improve proximity based localization. APIT [10] segmented the area into a large number of triangular regions with different sets of landmark nodes. The target node receives the messages from those landmark nodes that have common coverage with its location. The overlapping area from all received data can be used to estimate the location of a target node.

P. Bahl et al. [7] offered a localization method with two phases based on RSSI and proximity respectively to increase the localization accuracy. They proposed concepts named Environment Independent Positioning (EIP) and Environment Dependent Positioning (EDP). In the initiation phase, the EIP provides a coarse location. Based on this information, a model determination module produces updated and appropriate signal propagation parameters. In the second phase, EDP generates refined position according to the acquired environmental parameters.

Vire is also a classical algorithm for indoor positioning which based on Landmarc solution [13]. Vire algorithm needs RFID readers that can get precise RSSI of each reference tags or detected tags. Vire present the concept of virtual tags. The advantage of Vire is the fact that it can rely on a smaller number of reference tags. The disadvantage is the need to calculate the position of virtual tags and the need to obtain signal strength information. Hence, the algorithm represents a trade-off between real-time performance and accuracy.

CY Cheng [14] introduced a localization algorithm using clustering on signal and coordination patterns. With this algorithm such technique increased the accuracy of LANDMARC and Vire. The localization algorithm combines the advantages of two models, applying the concepts of virtual tags and two-step clustering analysis. Augmented RFID Receiver (ARR) introduced in [15] uses a method called synchronous detection to be able to overcome frequency offset challenges associated with intercepting tag signals using a non-envelope detection scheme. This system captures the data by an off-the-shelf UHF RFID reader IC and then sends and processes the data using an FPGA hardware and soft processor core. Since ARR has high power consumption and it communicates with the host through Ethernet, it cannot be used for portable applications.

A. Atalye et al. in [1] introduced a device called Sense-a-tag that improves localization in UHF-RFID systems. A localization system based on ST and WCL algorithm has been investigated in [3]. The localization accuracy is directly dependent on the density of deployed landmark tags, speed of the movement of the ST and the distance between the ST and landmark tag on one side and the reader antennas on the other side. In these papers, the localization accuracy is studies for several different deployment scenarios. The accuracy achieved using WCL method when the landmark tags are 1 m apart in the direction of y-axis and 0.5 m apart in the direction of x-axis is about 32 cm.

3   Structure of the System

3.1   ST Structure

The ST is a semipassive device and can operate as a standard passive RFID tag that communicates with RFID reader via backscattering communication. However, it has an important additional functionality: it can detect the backscattered response from the tags to the reader.

The ST waits for the particular enquiry commands of the reader for landmark tags. Subsequently, if the ST captures the tag's response and the reader's corresponding acknowledgement, it is considered that tag is detected by ST. The proximity detection information about the detected tag is stored in the memory of the ST. When the reader singulates the ST, the ST transfers the information about the detected tag(s) to the reader.

Figure 1 presents the main operations of software executed on the host computer. The software is used to control the reader and to read tag information from the reader. However, in the figure we show only the part of the software that is needed to process ST. In order to incorporate STs into existing RFID systems, the readers treat the STs as standard tags. Since the readers are not aware of the STs and they treat them as ordinary tags, the host has to specify the procedure of reading the tags and the STs. Also, the host understands the STs operations and it controls the reader using standardized commands. Hence, the host is responsible for intelligent control of the system. The first task of the host is to make sure that the reader first reads tags and then STs. Thus, the reader has to be able to read tags rather than STs in the first reading cycle. During this cycle, the STs listen to the tags responses and store the detected tags information. In the second round, the reader reads only STs.

After initialization, the host initiates the query cycle. The tags' ID numbers and ST information are obtained by the reader and transferred to the host. Next, the host has to analyse the ST information and to relate the tags with the particular STs.

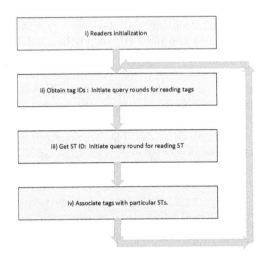

Fig. 1. Steps implemented by the host for obtaining information from STs.

3.2 Zoning and WCL

The method used for localization is *"Weighted Centroid"*. In the centroid method, the target calculates its position at the center of the detected reference tags, as shown in Fig. 2(a). To increase the accuracy of the localization, *weights* are introduced into this system. Weights are used to calculate the distance of the target from each respective landmark tag, as shown in Fig. 2(b). Depending on the system, weights can be a function of RSSI value or other factors that can reflect the distance from each tag. In our case, number of reads by the ST was used for weight factors.

The algorithm which can be performed on each unknown location of ST is shown in Eq. (1). In this formula, $P_i(x, y)$ indicates the position of unknown ST i. The known position of tag j is given by $B_j(x, y)$. The number of tags which are within the communication range of the unknown ST is indicated by m where w_{ij} describes the weight value for landmark tag j used by the ST i.

$$P_i(x, y) = \frac{\sum_{j=1}^{m} w_{ij} \cdot B_j(x, y)}{\sum_{j=1}^{m} w_{ij}} \tag{1}$$

where w_{ij} is the number of times the tag is read by the ST in each report.

There are a couple of limitations in the centroid method. The localization accuracy is dependent on the density of the landmark tags. On the other hand, the power strength of the backscattered signal from passive tags varies in time. That might cause ST to detect some landmark tags that are farther than some other landmark tags that are not detected. This will cause significant localization error.

Since ST can just detect the tags that complete communication with the reader, if we confine reader's interrogation zone to the landmark tags in proximity

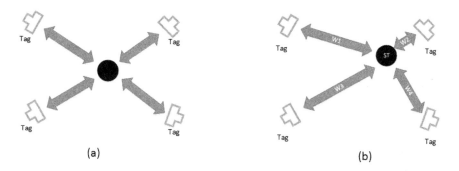

Fig. 2. (a) Centroid localization method. (b) Weighted-centroid localization method

of the ST, then the error will be smaller. In Gen 2 systems the reader can send a command to select particular tags that will be read in the next query round. During system setup, it is possible to assign the ID numbers to the landmark tags so that the only landmark tags in particular zones can be selected. The designer of the system has full freedom on how to assign the ID numbers and what kinds of zones to select. In this paper we assign the ID numbers so that we can divide the whole area into symmetric rectangular zones.

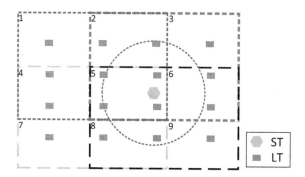

Fig. 3. Virtual zones with landmark tags (Color figure online)

Figure 3 shows an example configuration that consists of 16 landmark tags and one ST that needs to be located. One reader that interrogates the landmark tags and ST, is not presented in the figure. The 4 zones have been shown using four different color rectangles containing 9 tags each. For example, zone A with brown color covers landmark tags placed in rectangles 2, 3, 5 and 6. Zone B with dark green color covers landmark tags placed in 5, 6, 8 and 9. To avoid error in zone boundaries, there should be some overlapping area between the zones. In our example we have rectangles 5 and 6 as an overlap area between zones A and B.

First, the reader interrogates the whole area. There are 8 different landmark tags (LTs) in the detection range of the ST. As we mentioned before, there is not a linear relationship between the distance and number of detections by the ST. This will cause that in some situations, the landmark tags in rectangles 2 or 9 are detected more times by the ST compared to tags in the rectangles 5 or 8 and the estimated location error will be higher accordingly. In the second round, the reader just interrogates two zones (just the landmark tags in rectangles 2, 3, 5, 6, 8 and 9) by adding a mask in the query. In the third round, the reader will just interrogate one zone (landmark tags in 5, 6, 8 and 9). In this round we can see that the landmark tags in 2 will be filtered from our equations.

4   Simulator Core

The augmented RFID system is modelled in our newly developed Proximity-detection-based augmented RFID system simulator (PASS) framework. PASS is a time-domain system-level simulator, which is based on position aware RFID system (PARIS) simulation framework. The development environment is MAT-LAB 2012a on Windows 8.1 (64bits). From PARIS simulation framework, PASS inherits the behaviour model of a NXP UCODE G2XM based passive UHF RFID tag and wireless propagation channel, as well as a core framework to perform simple control and logging. While PARIS mainly focuses on ranging rather than the system behavior, we completed the functionality of generic reader and tag so that the simulator behaves according to ISO 18000-6C protocol [4,5]. The sensatag model is designed to emulate the specific sensatag device. Following the modular and hierarchical design pattern, the levels of PASS hierarchy are separated, and different models for reader, tag, sensatag, and wireless channel are also strictly divided.

5   Simulation and Experimental Results

We present the results of the simulation and the corresponding experiment in the office environment. To characterize parameters of the simulator, noise level, reader's transmitting power and other parameters are obtained based on the measurement data from the field experiment.

5.1   Simulation Results

To evaluate the idea of zoning-WCL based localization, first we establish the experimental scenario in the simulator. The area $4 \, \text{m} \times 2 \, \text{m}$ was covered with passive UHF RFID landmark tags. Landmark tags are placed in the simulator 0.5 m from one another in the direction of X axis in 3 different rows that are at the distance of 1 m from one another. A virtual robot that was tagged with an ST was moving along a trajectory defined in the simulator. The simulator ran 3 different scenarios: (a) pure WCL, (b) WCL + half-division zoning, (c) WCL + quarter division.

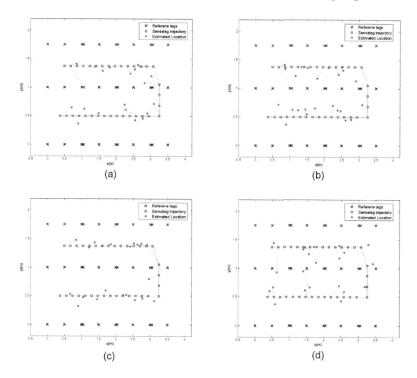

Fig. 4. Localizing the ST in landmarked area. (a) WCL method. (b) WCL + dividing zone to half in X direction. (c) WCL + dividing zone to half in Y direction. (d) WCL + dividing zone to quarters in X direction. (Color figure online)

Figure 4 shows the simulation results for the described scenarios. The blue path is the path that virtual robot moves on. The red dots are the location estimation of the ST at each time instance. Figure 4(a) illustrates the localization based on WCL. The error in Y direction is expected to be larger than the error in X direction due to more asymmetric deployment of landmarks in the direction of Y axis. In Fig. 4(b) the landmark tags are divided into two zones in the direction of X axis. That means we have two step localization procedure where in the first step the reader reads all landmark tags and the second one it interrogates the landmark tags in the half area where the ST is located. We do not see significant improvement in location estimation because the zones are still big enough to put unnecessary landmark tags in the detection area of the ST.

The same scenario was repeated in Y axis only and the results are shown in Fig. 4(c). In this figure we can see the significant decrease in localization error in the direction of Y axis error compare to the regular WCL. In Fig. 4(d) the zone divided to 4 subzones in X axis and the localization procedure now includes three steps. The root mean square error (RMSE) for zoning in X direction is shown in Fig. 5. Based on the graph, we can observe an evident improvement in X direction by dividing the zone to smaller sections. The error in Y direction increased because of involvement of less tags in Y direction. Note that if we divide the zone in Y direction as well,this error would be much smaller like the scenario that happened for X direction.

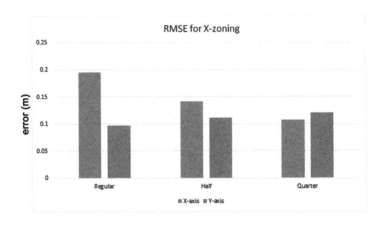

Fig. 5. RMSE for different X-axis zoning scenarios

Note that there is a non-linear relation between dividing zones and accuracy. We need also to take into consideration additional processing time required when more divisions are performed. Therefore, there is a trade-off between localization accuracy and processing time.

5.2   Real Experiments with ST

The same experimental setup has been done in a laboratory to test the assumptions in a real environment. ST was mounted on a robot that followed at a constant speed a black line path in a similar manner as in the simulator. A C# based GUI was developed to get the results from the reader, run localization algorithms and presents location of the ST on the GUI in real-time.

Stationary ST. First we placed the robot at a random location and stored 50 samples at that coordinate. We wanted to see if the zoning affects the localization accuracy. Figure 7 shows the localization results with and without quarter zoning for the random location. Real location of the ST is (175 cm, 50 cm). Each red dot is representation of 1 location estimation. Obviously the results after zoning are more precise compare to regular WCL. The average error in X direction (vertical) decreased from 33.5 cm to 18.8 cm. Figure 6 illustrates the average error and standard deviation for 50 samples on vertical direction with and without quarter zoning.

The same experiment has been run for 10 more points and the mean value of the error for all samples is bellow 21.6 cm compared to 38.1 cm for the regular WCL method.

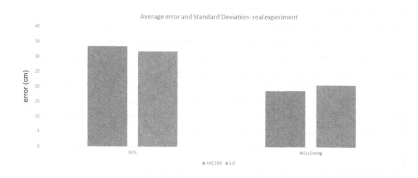

Fig. 6. Average error and standard deviation on X axis for 50 samples (real experiment).

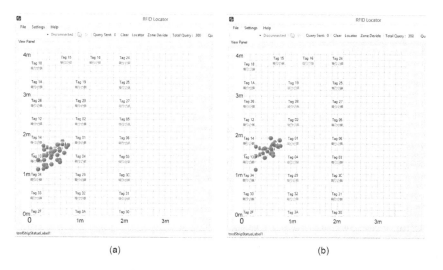

Fig. 7. GUI for real-time localization. Fifty location estimates of the ST are shown for the following cases: (a) Regular WCL, (b) WCL + quarter zoning

Mobile ST. To confirm simulation results, we repeated the same scenario of Fig. 4 but using real ST and RFID system. The ST is placed on a robot that follows the black line and move between the landmark tags. Figure 8(a) shows the location estimations during the robot movements. Considering the Y direction for example (horizontal axis), the red dots that are representation of location estimations, are mostly close to the center. That is because the farthest set of landmark tags on the right or on the left affect the estimation and we have larger error in Y direction. Figure 8(b) illustrates the same scenario but localization method is based on half zoning. We can clearly observe that the red dots converge to the path and are on both sides of the black line.

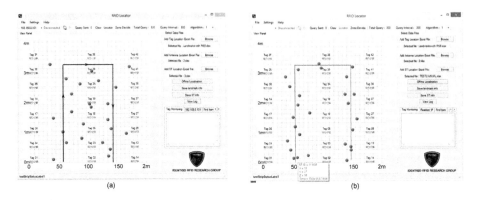

(a)   (b)

Fig. 8. Real-time localization of the mobile ST when location estimation is performed using: (a) Regular WCL. (b) WCL + Half zoning (Y direction) (Color figure online)

6 Conclusion

In this paper it is demonstrated that it is possible to improve the localization accuracy of ST-based augmented RFID localization system with the combination of WCL and zoning algorithm. Based on the zoning algorithm, the outlier detection of ST towards distant reference tags is dropped, correspondingly the estimated location of target from WCL is more accurate. The performance of the proposed method is evaluated in the simulation and field experiment.

Future work includes two directions. The first direction is to improve the three steps iterative zoning procedure. Applying the Kalman filter, the prediction of ST's future position will be available [16]. Based on the prediction, the zoning algorithm would decide which zone to interrogate, which would reduce time for running localization procedure significantly. The second direction is to formulate the flexible zoning scheme. Although the current zoning algorithm is more robust to outlier detection, its potential to counter the localization heterogeneity is not developed. The heterogeneity in proximity detection-based localization solution originates from the non-uniform distribution of reference nodes within target's proximity. In flexible zoning the division of landmark tags in zones is adjustable according to the estimated position of the ST. Furthermore, the zone could be refined iteratively into smaller size. The zoning algorithm would become "zooming" algorithm. However, this method would bring about implementation challenges. Since the RFID reader interrogates a group of tags by issuing command with a specific mask, then the assignment of EPC numbers would be much more complex.

References

1. Athalye, A., Savic, V., Bolic, M., Djuric, P.M.: A radio frequency identification system for accurate indoor localization. In: 2011 IEEE International Conference on Acoustics, Speech and Signal Processing (ICASSP), pp. 1777–1780. IEEE, May 2011
2. Maltse, P.A., Winter, S., Nikitin, P., Kodukula, V., Erosheva, E.: U.S. Patent No. 8,199,689, Washington, DC: U.S. Patent and Trademark Oce. (2012)
3. Bolic, M., Rostamian, M., Djuric, P.M.: Proximity detection with RFID: a step toward the internet of things. IEEE Pervasive Comput. **2**, 70–76 (2015)
4. Wang, J., Bolic, M.: Exploiting dual-antenna diversity for phase cancellation in augmented RFID system. In: 2014 International Conference on Smart Communications in Network Technologies (SaCoNeT), pp. 1–6. IEEE, June 2014
5. Wang, J., Bolic, M.: Reducing the phase cancellation effect in augmetned RFID system. Int. J. Parallel Emerg. Distrib. Syst. **30**, 494–514 (2015). doi:10.1080/17445760.2015.1035718
6. Priyantha, N.B., Chakraborty, A., Balakrishnan, H.: The cricket location-support system. In: Proceedings of the 6th Annual International Conference on Mobile Computing and Networking, pp. 32–43. ACM, August 2000
7. Bahl, P., Padmanabhan, V.N.: RADAR: an in-building RF-based user location and tracking system. In: Proceedings of INFOCOM 2000 Nineteenth Annual Joint Conference of the IEEE Computer and Communications Societies, vol. 2, pp. 775–784. IEEE (2000)
8. Chang, H.L., Tian, J.B., Lai, T.T., Chu, H.H., Huang, P.: Spinning beacons for precise indoor localization. In: Proceedings of the 6th ACM Conference on Embedded Network Sensor Systems, pp. 127–140. ACM, November 2008
9. Bulusu, N., Heidemann, J., Estrin, D.: GPS-less low-cost outdoor localization for very small devices. Pers. Commun. IEEE **7**(5), 28–34 (2000)
10. He, T., Huang, C., Blum, B.M., Stankovic, J.A., Abdelzaher, T.: Range-free localization schemes for large scale sensor networks. In: Proceedings of the 9th Annual International Conference on Mobile Computing and Networking, pp. 81–95. ACM, September 2003
11. Ni, L.M., Liu, Y., Lau, Y.C., Patil, A.P.: LANDMARC: indoor location sensing using active RFID. Wirel. Netw. **10**(6), 701–710 (2004)
12. Schuhmann, S., Herrmann, K., Rothermel, K., Blumenthal, J., Timmermann, D.: Improved weighted centroid localization in smart ubiquitous environments. In: Sandnes, F.E., Zhang, Y., Chunming, R., Yang, L.T., Ma, J. (eds.) UIC 2008. LNCS, vol. 5061, pp. 20–34. Springer, Heidelberg (2008)
13. Zhao, Y., Liu, Y., Ni, L.M.: VIRE: active RFID-based localization using virtual reference elimination. In: International Conference on Parallel Processing, ICPP 2007, pp. 56–56. IEEE, September 2007
14. Cheng, C.Y.: Indoor localization algorithm using clustering on signal and coordination pattern. Ann. Oper. Res. **216**(1), 83–99 (2014)
15. Borisenko, A., Bolic, M., Rostamian, M.: Intercepting UHF RFID signals through synchronous detection. EURASIP J. Wirel. Commun. Netw. **2013**(1), 1–10 (2013)
16. Kalman, R.E.: A new approach to linear filtering and prediction problems. J. Fluids Eng. **82**(1), 35–45 (1960)

Workshop on Practical ad hoc Network Security and Vulnerability

Communication Links Vulnerability Model for Cyber Security Mitigation

Eman Hammad$^{(\boxtimes)}$, Abdallah Farraj, and Deepa Kundur

Department of Electrical and Computer Engineering,
University of Toronto, Toronto, Canada
{ehammad,abdallah,dkundur}@ece.utoronto.ca

Abstract. We consider the problem of defining a metric to capture communication links vulnerability that is a function of threat models of concern. The model is based on the Confidentiality-Integrity-Availability (C-I-A) framework and combines communication links parametric models with dynamical historical models. The proposed model arrives at a vulnerability matrix to describe the cyber component of a cyber-physical system. The vulnerability matrix is used for flexible adaptive constrained routing implemented on Software Defined Networks (SDNs) as a mitigation approach for threats of concern.

Keywords: Smart grid · Cyber security · Metric · Vulnerability · Constrained shortest path · Quality of Service · Routing · Software defined networks

1 Introduction

Cyber security is perceived as a challenge on different levels of cyber-physical systems such as the smart grid. This is, in part, due to the fact that current standard communication protocols were not designed with a cyber-security perspective. Complex interconnected cyber-physical systems offer a multitude of opportunities and challenges within the cyber security context. Vulnerabilities resulting from the cyber enablement of the physical system may not be fully uncovered or understood, and this unveils many challenges into how to reenforce the system against the unknown. A distinguishing characteristic of the cyber-physical systems (e.g. smart grid) is that cyber-security approaches cannot be considered without studying their impact on real-time operations of the physical system.

One of the coupled interactions in the smart grid is between communication network infrastructure and cyber-enabled control; in this context developing a functional cyber-security assessment framework that can be used for flexible cyber-physical mitigation approaches is still lacking. Information Technology (IT) based security did not prove to be efficient due to its focus on the cyber plane of the system. This motivated the emergence of several impact based cyber-security frameworks; of which the C-I-A (Confidentiality-Integrity-Availability)

© ICST Institute for Computer Sciences, Social Informatics and Telecommunications Engineering 2017
Y. Zhou and T. Kunz (Eds.): ADHOCNETS 2016, LNICST 184, pp. 285–296, 2017.
DOI: 10.1007/978-3-319-51204-4_23

framework has prevailed as an adequate tool for high level cyber-security impact analysis in cyber-physical systems.

Given a certain cyber infrastructure for a cyber-physical system such as the smart grid, system engineers and operator should be able to asses the existing system according to various cyber-security threat models. This establishes a baseline for system security planning and testing: highlighting weaknesses in the system, and providing guidance and input to different mitigation schemes. Previously, authors have developed a communication link vulnerability mitigation framework that satisfies Quality of Service (QoS) constraints of the underlying power system. The mitigation framework is enabled through the utilization of Software Defined Networks (SDN), as a flexible adaptive communication infrastructure and control platform. The problem was formulated as a constrained shortest path routing problem, that optimizes for the least vulnerable route with satisfactory QoS (delay). A model for the vulnerability of a communication link was roughly outlined. In this work we further develop the proposed communication link vulnerability. The proposed vulnerability model is directly assessed in a dynamic constrained QoS routing setting.

Ad hoc networks are considered a promising solution for networking on the distribution level [7], specifically for smart-metering systems and applications. Moreover, previous works have considered optimizing ad hoc network management using SDN [5,8]. This suggests that the proposed vulnerability metric could be extended to ad hoc networks, with proper parameters pertinent to the network specifics and operation.

The main contributions of this work include the following:

1. propose and develop a vulnerability metric model for communication links in cyber-physical systems,
2. employ the proposed vulnerability metric via an SDN based adaptive QoS routing.

2   System Model

Let N denote the number of nodes in the power system; for this discussion let N refer to number of buses in the power grid. Then, we can assume a communication network connecting the N buses in a topology that parallels that of the electrical grid.

Consider a graph representation of the corresponding communication network. The weighted undirected graph model $G(V, E, w)$ describes an N-node and M-link network, where the node set $V = \{v_1, \ldots, v_N\}$ and the edge set $E = \{e_{ij}, i, j = 1, 2, \ldots, M\}$ denote the buses and communication links, respectively. The weight w on the edge between two nodes is defined as the cost of the corresponding communication link. Then, the adjacency matrix A can be defined as

$$A_{i,j} = \begin{cases} w_{ij} & i \neq j, \text{ for } (i, j) \in E \\ 0 & \text{otherwise.} \end{cases} \tag{1}$$

Consider next the routing problem of communicating data between a source node s and destination node t in the graph G. The shortest path route between the pair can be found using various algorithms. Due to its simplicity and optimality, Dijkstra-based routing algorithm has long been the most used algorithm to arrive at the shortest path.

The SDN framework allows us to obtain a dynamic delay cost matrix A_d sampled from the network at pre-defined intervals. Similarly, provided that a vulnerability cost metric is defined, then a corresponding vulnerability cost matrix A_v can be evaluated for the network. Hence, the problem of QoS routing while mitigating link vulnerabilities is then formulated as a constraint shortest path (MCSP) problem.

Within the smart grid, cyber-enabled control systems require information delivery between relevant nodes with certain delay requirements; as an example, the IEC 61850 GOOSE messaging specifies the message delay constrains for performance class P2/3 to be within 3 ms [2]. Accordingly, if we define the delay cost matrix A_d, then the problem of finding paths that satisfy the delay constraints can be formulated as a constraint shortest path (CSP) problem. While the CSP is NP-hard, many algorithms have been developed to find a feasible or a set of feasible solutions [16].

In the context of smart grid systems, networked sensory and control impose many constraints on data communications; nevertheless, in this paper we are more focused on the development of a tractable vulnerability metric for communication links. The proposed vulnerability metric will be utilized for various threat models (Fig. 1).

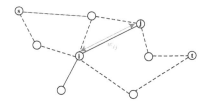

Fig. 1. Communication network graph

2.1 Vulnerability Metric

The increased cyber coupling of the smart grid through more cyber-enabled sensory and control increases the vulnerability surface of the smart grid. The Confidentiality-Integrity-Availability (C-I-A) framework provides a neat classification of vulnerabilities based to their impact with respect to information. In this work, we consider a vulnerability threat model directly related to the C-I-A framework, where threats and attacks can be classified using the aforementioned framework. Developing a metric that captures the elements affecting the vulnerability level of a certain communication link will enable us to develop a corresponding network response. It is intuitive that a link vulnerability is not a binary characteristic; i.e., a simple label of a link as vulnerable or not-vulnerable is not very informative for system operation.

The communication link vulnerability metric proposed in this work satisfies on the following "security metric" properties [14]: a security metric should be quantitative, objective, employs a formal model, not boolean (0, 1), and reflects time dependance [14]. Further, it would be advantageous if the metric self-reflects the associated risk level to better provide an insight.

We next try to formalize our definition of vulnerability from a cyber-security perspective; based on several definitions from information technology and computing, to disaster management; the vulnerability of a system or group of systems is defined as a weakness in that system that hinders its ability to withstand threats [3]. Extending this definition to communication networks in cyber-physical systems, leads us to consider the attributes and installed mechanisms to arrive at a measure relating how vulnerable communication links are to threat models of concern. In a smart grid's communication network infrastructure, few of the communication link attributes can be combined to describe and quantify a link vulnerability metric. These attributes can be grouped into categories as shown in Fig. 2, and as follows

1. **Dynamic**; attributes in this category dynamically vary over time and in response to events.
   - History L_H^{ij}; a link that was previously targeted by an attacker is more likely to be targeted again by a passive/active adversary.
2. **Parametric**, attributes in this category tend to be static and are characteristic of the link. The parameters are scored and ranked in a hierarchical fashion, in the first layer the threat model is linked to the parameters through an impact analysis based on the C-I-A impact framework, and in the second level each of the link parameter sets is internally ordered based on relative vulnerability preference to establish set weights.

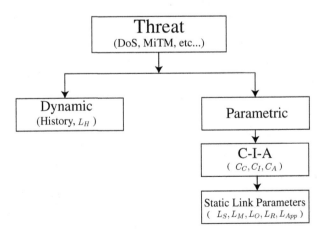

Fig. 2. Link vulnerability model

i. C-I-A
   - C_C: cost on confidentiality;
   - C_I: cost on integrity;
   - C_A: cost on availability.
ii. Ordered static link parameters, and these could include the following among others.
   - L_S^{ij}: security installed measures: a link with strong encryption is typically less vulnerable,
   - L_M^{ij}: physical-channel modality: a wireless link is usually more vulnerable than a fiber optic due to technology,
   - L_O^{ij}: ownership: a self-owned and operated channel is typically less vulnerable than a shared and/or leased channel,
   - L_R^{ij}: redundancy: a link with installed redundancy mechanisms is less vulnerable compared to a single link,
   - L_A^{ij}: application: a link dedicated for command/control traffic is more vulnerable to being targeted.

It is important to acknowledge the existence of interdependencies between the various parameters, yet we argue that virtually decoupling the interdependencies as a list is informative for our problem of interest. Link history L_H is modeled by a Markov model as is described in Sect. 2.2. Moreover, link parameters $\{L_S, L_M, L_O, L_R, L_A, \ldots\}$ are subsets with embedded monotonically increasing ordering where a higher order is related to a more vulnerable state. The subsets are constructed by system engineers and operators where an exhaustive enumeration of the related parameter possible values is established with the proper ordering. Let the ordered subset related to a generic parameter L_x be denoted $\mathbb{X}$, then a plausible mapping of the set entries to a normalized weight is defined by

$$X_w = \frac{1}{|\mathbb{X}|} \times \{X_1, X_2, \ldots\} \tag{2}$$

Lets consider link security as an example: the set $\mathbb{S}$ corresponding to L_S includes a ranking of the different installed and configured security measures. For simplicity let a subset of security measures include three different encryption mechanisms $\{Encr_1, Encr_2, Encr_3\}$, where corresponding L_S will be assigned based on the strength of the encryption. Let $\rhd$ denote a security comparison operator where left hand is a stronger encryption than the right hand operator, then if $Encr_1 \rhd Encr_2 \rhd Encr_3$, a possible assignment could be $L_S^{ij} \in \{0, 0.33, 0.66\}$.

A similar approach can be used to arrive at the communication link parameters subsets and their corresponding weights. We next consider how to combine these parameters into a single representative vulnerability metric (L_V^{ij}). The vulnerability metric should reflect the attributes that make a link more probable to be targeted by an attempted adversary action, as well as being affected by that targeting. We base our vulnerability metric definition on the C-I-A framework threat model. Consider the threat set Γ, where the set could include threats such as denial of service (DoS), false data injection (FDI), among others.

And consider a general link parameter ordered subset $\mathbb{X} = \{X_1, X_2, X_3\}$. A threat Γ_m can be decomposed using the C-I-A model into a three part binary indicator based on the threat cost/impact, as is shown below

$$\mathbf{\Gamma_m} = \mathbb{1}_m^{C-I-A} = [\mathbb{1}_m(C_A) \quad \mathbb{1}_m(C_I) \quad \mathbb{1}_m(C_C)] \tag{3}$$

Further, as shown in Fig. 2, a link vulnerability can be modeled as a function of dynamic link history L_H and parametric vulnerability L_P of the link.

$$L_V^{ij}(\Gamma_m) = L_H^{ij}(\Gamma_m) + L_P^{ij}(\Gamma_m) \tag{4}$$

where L_P, the parametric link vulnerability is further developed below.

This indicator is then embedded in the scoring of the ordered parameters subsets $\mathbb{X}$ to arrive at the communication link parametric vulnerability $L_P^{ij}(\Gamma_m)$.

To facilitate the threat model embedded scoring of link parameters, ordered parameter subsets are further clustered into sub-subsets according to their correlation with the C-I-A decomposition of any threat. I.e consider the following clustering of the ordered subset $\tilde{\mathbb{X}} = \mathbb{X}_C \cup \mathbb{X}_I \cup \mathbb{X}_A$. This is extended to link parameter subsets $\tilde{\mathbb{P}} = \{\tilde{\mathbb{S}}, \tilde{\mathbb{M}}, \tilde{\mathbb{O}}, \tilde{\mathbb{R}}, \tilde{\mathbb{A}}, \ldots\}$ This leads us to the following model for the parametric link vulnerability (the superscript ij have been removed for clarity, wherever L is presented it is related to communication link ij

$$L_P(\Gamma_m) = L_S + L_M + L_O + L_R + L_A + \ldots$$
$$L_P(\Gamma_m) = \mathbf{\Gamma_m} \cdot \tilde{\mathbb{P}} \tag{5}$$

$$L_P(\Gamma_m) = \frac{1}{|\mathbb{P}|} \sum_{n=1}^{|\mathbb{P}|} \left([\mathbb{1}_m(C_A) \quad \mathbb{1}_m(C_I) \quad \mathbb{1}_m(C_C)] \cdot \begin{bmatrix} \mathbb{S}_C & \mathbb{M}_C & \mathbb{O}_C & \ldots \\ \mathbb{S}_I & \mathbb{M}_I & \mathbb{O}_I & \ldots \\ \mathbb{S}_A & \mathbb{M}_A & \mathbb{O}_A & \ldots \end{bmatrix} \right) \tag{6}$$

To further illustrate this proposal, if the threat model is focused on DoS attacks, then the attack/threat decomposition vector will explicitly model the DoS as $\Gamma_m = [1 \quad 0 \quad 0]$, in terms of direct cost on availability. Similarly, the corresponding link parametric sub-subsets will identify the relevant components from each parameter subset that will be part of the vulnerability metric.

Finally, necessary normalization is performed when combining the different metrics, and the vulnerability cost matrix for the whole communication network is constructed such that $A_v^{ij} \in [0-100]$. We next discuss the dynamic link history vulnerability.

2.2 Link History

Most of the communication link attributes considered above are static and do not change with time, unless advanced functionalities are installed such as service-adaptive cryptography levels. However, vulnerability history of a link is affected by events and status of the network; thus a link history L_H is best modeled by a dynamical model. It is important to note that the goal of the proposed dynamical model is to provide a tool to quantify the probability of a communication link in the system being targeted based on detected events history. However, we do not intend to provide an intrusion detection/prediction capability. For our purposes it suffices to consider a Markov model, which is simply a system with one step history. On the other hand, a longer history of the link would be beneficial to tune the Markov model parameters (i.e., transition probabilities) [1,17].

Let the event of "targeting" the communication link between nodes i and j be a stochastic event that happens with a probability P_A^{ij}. Further, let the link status be termed as $S_{ij} \in \{G, T\}$, where G denotes a good link and T denotes a targeted link, and is modeled by a Markov Model. A finite-state Markov chain process is described by its transition matrix P where the $P(l, m)$ element is defined as the probability of state $X^{k+1} = m$ given that the previous state is $X^k = l$. This is commonly known as the Markov property where the next state of the system depends only on the current state, and is described as $P(l, m) = \mathbb{P}(X^{k+1} = m | X^k = l)$.

The transition matrix P for the 2-state model is mathematically described by the following probabilities

$$\mathbb{P}(L_{ij}^{k+1} = G \mid L_{ij}^k = T) = P_D$$
$$\mathbb{P}(L_{ij}^{k+1} = T \mid L_{ij}^k = G) = P_A$$
$$\mathbb{P}(L_{ij}^{k+1} = G \mid L_{ij}^k = G) = 1 - P_A$$
$$\mathbb{P}(L_{ij}^{k+1} = T \mid L_{ij}^k = T) = 1 - P_D \tag{7}$$

and $P = \begin{bmatrix} 1 - P_A & P_A \\ P_D & 1 - P_D \end{bmatrix}$.

Given that the Markov chain described above is time-homogeneous, then we consider the stationary equilibrium/distribution of the chain for important insights such as the probability of being in a certain state. Specifically, we can obtain the probability that the communication link will be in state T (i.e., targeted by an attacker) L_H^{ij}. This is then combined with the parametric attributes of the channel according to Eq. (4) to arrive at the vulnerability metric of the link, L_V^{ij}.

2.3   Sustainable Security

Above referenced communication link parameters are mostly characteristic of the communication network, its usage and any ancillary additions to it. From a cyber-security perspective, system engineers and administrators are responsible for tracking the various network components and their corresponding configurations. Based on this perspective, the proposed vulnerability model assumes that system administrators should be able to log communication links respective parameters and flag each parameters sub-subcomponent relationship with the C-I-A framework. Further, it assumes a fundamental understanding of the scope and limits of each installed and configured component, their relative ordering with respective to vulnerability, and the planned/unplanned interactions between the cyber-physical components and the cyber-cyber components.

The usefulness of the proposed metric relies on a sustainable cyber-security environment [13,15]; which can be described by two characteristics: (1) the establishment of a cyber-security eco-system where validation and frequent updates are regulated to ensure up-do-date match between envisioned and actual system state. This is necessary for any algorithms or network defined functionality such as the one proposed by the authors via adaptive routing. (2) existence of defined policies and system procedures for active recovery and mitigation feedback, where system engineers continuously adapt by applying necessary measures to ensure a minimum future risk.

Further, the proposed model can be used as a tool for cyber-security assessment of the communication network in use. As it will pinpoint the most vulnerable links based on system configuration and dynamic history of cyber events. This assessment is helpful to (1) sketch a system update-upgrade plan (2) develop a cyber-security monitoring procedure/application check points, (3) revise response/recovery procedures. An Autonomous cyber-security system is a future vision that will require tremendous intelligence and adaptability, and is probably a threat to itself.

3   Software Defined Networks and Adaptive Constrained Routing

Provided that we can obtain an updated communication network vulnerability matrix A_v that is regularly updated, then we can adaptively route information based on a set criteria. The previous is valid if we have a communication network paradigm/architecture that is able to: (1) have a dynamically updated global network state, (2) be programmed with additional intelligence to control network traffic, (3) be managed and configured with low complexity. Software Defined Networks (SDN) is a very promising network architecture that is capable of supporting and enabling our adaptive routing. Moreover, it allows a more complicated processing to optimize vulnerable link avoidance to minimize both delay (of extreme importance in smart grid) and information leakage through vulnerable links.

3.1   Software Defined Networks

Software defined networking offers the potential to change the traditional way networks operate. Current communication networks are typically built from a large number of network devices, with many complex protocols implemented on them. Operators in traditional communication networks are responsible for configuring policies to respond to a wide range of network events and applications. Consequently, network management and performance tuning is quite challenging and error-prone [1,10].

The main characteristic of SDN, is the separation of control and data planes, where the network is decomposed to an SDN controller and various SDN data forwarding switches. This architecture enables revolutionary approaches to network

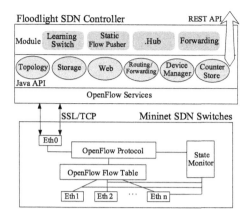

Fig. 3. SDN architecture

management, including adaptive networks that can dynamically be configured and programmed to respond to changes in the network. Network layer applications can acquire detailed traffic statistics from network devices to construct an up-to-date network view. One common standard for the implementation of software defined networks is OpenFlow [11]. The OpenFlow standard defines a communication protocol between network switches forming the data plane and one or multiple controllers forming the control plane.

In this work the SDN system setup is built using free open source tools. We use Floodlight v1.0 [12] as the SDN controller and Mininet 2.2.0 [9] for the SDN switches. Floodlight is an Apache-licensed, Java-based OpenFlow SDN Controller. Mininet can create a realistic virtual network. The SDN controller can communicate with the switches via the OpenFlow protocol through the abstraction layer present at the forwarding hardware.

The architecture of an SDN network is illustrated in Fig. 3 and is comprised of Floodlight controller and Mininet switches. An OpenFlow controller typically manages a number of switches, and every switch maintains one or more flow tables that determine how packets belonging to a flow will be processed and forwarded. Communication between a controller and a switch happens via the OpenFlow protocol, which defines a set of messages that can be exchanged between these entities over a secure channel. The state monitor module can be used to collect switch state and transmit it to the controller.

3.2 Constrained Shortest Path Problem and LARAC Algorithm

Given a network $G(V, E)$, assume every link $L_{u,v} \in E$ has two weights $c_{uv} > 0$ and $d_{uv} > 0$ (denoting, cost and delay). For source and destination nodes (s, t) and maximum delay $T_{max} > 0$, let $\mathbf{P}_{st}$ denote the set of paths from s to t. Further, for any path p define

$$c(p) = \sum_{(u,v) \in p} c_{uv}, \; d(p) = \sum_{(u,v) \in p} d_{uv}. \tag{8}$$

CSP problem seeks to arrive at the shortest path between s and t nodes with a certain link cost c. However, when the path is constrained by more than one constraint, the problem is termed an MCP problem. Given that there are multiple paths between s and t, a modified MCP problem, often called the multiconstrained optimal path (MCOP) problem, is defined where the goal is to retrieve the shortest path among a set of feasible paths.

A feasible path $s \to t$ is defined as path p_{st} that satisfies $d(p_{st}) \leq T_{max}$; let $P_{st}(T_{max})$ be the set of all feasible paths from s to t. Then, the CSP problem can be formulated as an integer linear program (ILP) with a set of zero-one decision variables [4,6,16]. The CSP NP-hard problem have many algorithmic approaches that successfully arrived at feasible solutions. The Lagrangian Relaxation Based Aggregated Cost (LARAC) algorithm developed in [4] solves the integer relaxation of the CSP problem efficiently.

3.3   Vulnerable-Link Adaptive Avoidance (VLAA) via SDN

We adopt the CSP formulation to capture the problem of best-effort avoiding vulnerable links while maintaining a QoS constraint (specifically, a delay constraint). We propose a Vulnerable-Link Adaptive Avoidance (VLAA) algorithm that uses previously-defined vulnerability metric in addition to communication delay in order to arrive at a set of feasible paths between source node s and destination node t. Link delays are observed through SDN switches at each update interval, and if changes are observed, the OpenFlow Floodlight controller is updated. Similarly, link vulnerability costs are observed and the controller cost matrix is updated when changes are sensed.

The VLAA algorithm is implemented in two parts; a controller function which is implemented in Floodlight using Java, and a switch function implemented in Mininet using Python. The flowchart of the VLAA algorithm that is implemented in the controller side is shown in Fig. 4(a). Here, the algorithms perform the following main tasks [1]:

- Listening to messages from switches and calculating link delay value of each link, and then constructing the link-delay cost matrix.
- Calculating the link-vulnerability cost matrix according to the proposed metric; this matrix can be modified and calibrated by network engineers.
- Running a topology-update thread, and checking the link cost matrix updates regularly; if a change is detected, the controller recalculates the routing paths.
- Calculating the routing paths based on the link cost metrics of interest, and updating the flow table of each switch by advertise a PACKET OUT message to switches.

The main function of the VLAA algorithm in the switches' side is to collect the values of link delays for the directly-connected switches as shown in Fig. 4(b). This is achieved by periodically testing the link between that switch and all connected switches with higher ID. Link delay testing is done periodically and the average value is then compared with the last known value. If the new delay is

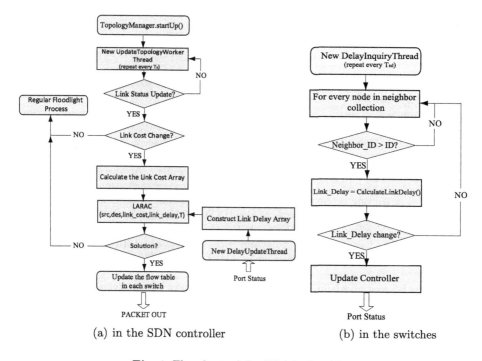

(a) in the SDN controller   (b) in the switches

Fig. 4. Flowchart of the VLAA algorithm

significantly different from the previous value, the switch updates the controller accordingly.

The mechanism in which an SDN switch exchanges link-delay updates with the Floodlight controller is implemented using Port Status messages. OpenFlow standards (v1.0–v1.4) expect the switch to send Port Status messages to the controller as port configuration state changes. These events include change in port status (for example, if it was brought down directly by a user) or a change in port status as specified by 802.1D standard.

4   Conclusions

In this work, we proposed a communication link vulnerability metric that is suitable for cyber security study and mitigation. The proposed metric relies on different parametric and dynamic characteristics of the communication network, and is most useful in adaptive communication networks such as SDNs. Future work would evaluate the proposed metric performance for different attack and threat models within the mitigation framework via QoS routing implementation in SDN.

References

1. Hammad, E., Zhao, J., Farraj, A., Kundur, D.: Mitigating link insecurities in smart grids via QoS multi-constraint routing. In: IEEE ICC Workshops: Workshop on Integrating Communications, Control, and Computing Technologies for Smart Grid (ICT4SG) (2016)
2. Hohlbaum, F., Braendle, M., Alvarez, F.: Cyber security practical considerations for implementing iec 62351, ABB Technical Report (2010)
3. International Federation of Red Cross and Red Crescent Societies: What is vulnerability (2016). http://www.ifrc.org/en/what-we-do/disaster-management/about-disasters/what-is-a-disaster/what-is-vulnerability/. Accessed 14 June 2016
4. Jüttner, A., Szviatovski, B., Mécs, I., Rajkó, Z.: Lagrange relaxation based method for the QoS routing problem. In: Twentieth Annual Joint Conference of the IEEE Computer and Communications Societies (INFOCOM), vol. 2, pp. 859–868 (2001)
5. Ku, I., Lu, Y., Gerla, M., Gomes, R.L., Ongaro, F., Cerqueira, E.: Towards software-defined vanet: architecture and services. In: 2014 13th Annual Mediterranean Ad Hoc Networking Workshop (MED-HOC-NET), pp. 103–110. IEEE (2014)
6. Kuipers, F., Van Mieghem, P., Korkmaz, T., Krunz, M.: An overview of constraint-based path selection algorithms for QoS routing. IEEE Commun. Mag. **40**(12), 50–55 (2002)
7. Liotta, A., Geelen, D., van Kempen, G., van Hoogstraten, F.: A survey on networks for smart-metering systems. Int. J. Pervasive Comput. Commun. **8**(1), 23–52 (2012)
8. Mendonca, M., Obraczka, K., Turletti, T.: The case for software-defined networking in heterogeneous networked environments. In: Proceedings of the 2012 ACM Conference on CoNEXT Student Workshop, pp. 59–60. ACM (2012)
9. Mininet: Mininet (2015). http://mininet.org/. Accessed 9 June 2015
10. Nunes, B., Mendonca, M., Nguyen, X.-N., Obraczka, K., Turletti, T.: A survey of software-defined networking: past, present, and future of programmable networks. IEEE Commun. Surv. Tutorials **16**(3), 1617–1634 (2014)
11. Open Networking Foundation: Software-defined networking: the new norm for networks. ONF White Paper (2012)
12. Project Floodlight: Project Floodlight (2015). http://www.projectfloodlight.org/floodlight/. Accessed 9 June 2015
13. Stamp, J., Campbell, P., DePoy, J., Dillinger, J., Young, W.: Sustainable Security for Infrastructure Scada. Sandia National Laboratories, Albuquerque (2003). www.sandia.gov/scada/documents/SustainableSecurity.pdf
14. Wang, A.J.A.: Information security models and metrics. In: Proceedings of the 43rd Annual Southeast Regional Conference, vol. 2, pp. 178–184. ACM (2005)
15. White, G.B.: The community cyber security maturity model. In: 2011 IEEE International Conference on Technologies for Homeland Security (HST), pp. 173–178. IEEE (2011)
16. Xiao, Y., Thulasiraman, K., Xue, G., Jüttner, A.: The constrained shortest path problem: algorithmic approaches and an algebraic study with generalization. AKCE Int. J. Graphs Comb. **2**(2), 63–86 (2005)
17. Zhao, J., Hammad, E., Farraj, A., Kundur, D.: Network-aware QoS routing for smart grids using software defined networks. In: Leon-Garcia, A., et al. (eds.) Smart City 360, pp. 384–394. Springer, Heidelberg (2016)

Entropy-Based Recommendation Trust Model for Machine to Machine Communications

Saneeha Ahmed$^{(\boxtimes)}$ and Kemal Tepe

University of Windsor, Windsor, ON, Canada
{ahmed13m,ktepe}@uwindsor.ca

Abstract. In a vast data collection and processing applications of machine to machine communications, identifying malicious information and nodes is important, if the collected information is to be utilized in any decision making algorithm. In this process, nodes can learn behaviors of their peers in the form of recommendation from other nodes. These recommendations can be altered due to various motives such as bad-mouthing honest nodes or ballot stuffing malicious nodes. A receiving node can identify an incorrect recommendation by computing similarity between its own opinion and received recommendations. However, if the ratio of false recommendations is low, the similarity score will be insufficient to detect malicious misbehavior. Therefore in this paper, an entropy-based recommendation trust model is proposed. In this model, a receiving node computes the conditional entropy using consistency and similarity of received recommendations with respect to its own opinions. The computed entropy indicates the trustworthiness of the sender. The proposed model clearly distinguishes malicious nodes from honest nodes by iteratively updating trust values with each message. The performance of the model is validated by a high true positive rate and a false positive rate of zero.

Keywords: Recommendation trust · Similarity · Entropy · Consistency · Connected vehicles

1 Introduction

Machine to Machine (M2M) communications is offered as a solution to collect vast amount of information from sensor and control actuators. In this setting, information is collected distributively, and decisions can be made at nodes by using the collected information. This information sharing enables nodes to benefit from their neighbors' experiences and to learn important information faster, such as identifying emergency events in connected vehicles, determining the quality of products in e-commerce, and making friends in social networks. In verifying or identifying false and malicious activity using the disseminated information from node, recommendation about other nodes can play an important role. However, recommendation schemes can be manipulated to badmouth honest nodes

© ICST Institute for Computer Sciences, Social Informatics and Telecommunications Engineering 2017
Y. Zhou and T. Kunz (Eds.): ADHOCNETS 2016, LNICST 184, pp. 297–305, 2017.
DOI: 10.1007/978-3-319-51204-4_24

and ballot stuff malicious nodes in order to have a stronger influence in the network. By ballot stuffing and badmouthing, a malicious node can manipulate a novice receiver to exclude any important information coming from honest nodes or accept wrong information from malicious nodes. That is why trustworthiness of recommendations must be estimated before accepting them.

Recommendation Trust (RT) is often considered as feedback credibility [1] and its is defined as a measure of trustworthiness of a node's recommendation about another node. RT is estimated on the basis of similarity between an evaluator's own opinion and received recommendations. In case a node sends a false recommendation only for a few nodes, i.e. selective misbehavior, or switches between malicious and honest behaviors in an on-off manner, it will attain a very high similarity score. The evaluator may observe the pattern of recommendations over a period of time and determine how consistent is the sender in its recommendations. Consistency of information can allow a new evaluator to determine whether to trust a recommender when it provides information about new nodes.

In this paper, an entropy-based RT model is proposed which utilizes the consistency of information as well as its similarity with evaluator's own opinion. The evaluating node calculates the average entropy based on following two factors: (1) Jaccard similarity score [2] of the recommendations, (2) ratio of consistent information over the total number of recommendations. Similarity is defined as the fraction of recommendations of a node that are same as the opinion of the evaluator. Consistency is defined as the fraction of recommendations from a sender that does not change in consecutive messages. A lower entropy indicates higher trustworthiness of the source since the information sent by this source has less uncertainty. In order to predict the time dependent behavior of the source, the recent trust is derived from the observed entropy and previous trust value. With this proposed trust model, evaluator can identify the malicious nodes even if there is a selective misbehavior or an on-off attack.

Remainder of this paper is organized as follows: Sect. 2 provides a literature review, Sect. 3 provides details of the trust model Sect. 4 provides simulation results and Sect. 5 provides a conclusion and future works.

2   Literature Review

Recommendation systems have gained significant attention in M2M communications [3,4]. In these systems, nodes provide feedback about behaviors of other nodes in order to make communication secure and reliable. However, nodes can manipulate their recommendations in order to have a stronger influence in the network by supporting malicious nodes or by badmouthing honest nodes [5–7]. If the credibility of recommendations is not considered, an attacker can easily defame a target node by creating multiple fake identities to generate false recommendations [8]. In order to use recommendations in establishing trust for a seller in an e-commerce site or a user in a social network site, a personalized similarity metric is proposed in [9,10]. In these studies, similarity is a measure of difference of satisfactions between two users over a set of common items. In [1], a user's recommendation is assigned as a weight equal to the similarity and the weighted recommendations provide the trust in the recommended user. Schemes in [1,9,10] are resilient against on-off attacks and effectively prevent unsuccessful transactions. The similarity measure has been adapted for a mobile ad hoc network in [11] which uses a fuzzy collaborative filter to restrain malicious recommendations from effecting the trust computation. The proposed collaborative filter uses the similarity between the nearest most similar neighbors to compute trust in a node. While the proposed trust mechanism improves the network throughput and packet drop ratio, it is not studied how reliable is the trust estimation. Moreover, the system is only tested under the honest majority scenario.

The infrequent badmouthing attacks are successful in defaming honest nodes or credible items. That is why a scheme to prevent those attacks is proposed in [12], where a drastic change in product rating with respect to time indicates such attacks. In [12], Dempster-Shaffer Theory is used in a system to identify malicious users and recover reputations of products or users. The system is tested under honest majority assumption and provided good receiver operation characteristics. However, if malicious users form a majority, the system may fail. In [12], recommendation values are real numbers which are naturally different from each other. This difference leads to oscillations while computing trust. These oscillations are undesirable in M2M communications and can be masked out by using binary recommendations similar to the one used in [13] for an e-commerce system. However, the scheme proposed in [13] only filters most deviant recommendations and ignores their sources. Hence, malicious nodes can survive and may corrupt a larger portion of the network to have a stronger influence on the network. That is why, in this paper RT is established in order to determine whether a node's recommendations can be trusted and included in obtaining the true nature of the neighbors. The proposed entropy-based RT model is explained in the following section.

3 Entropy-Based Recommendation Trust Model

In order to establish RT and use it to eliminate malicious recommenders, it is assumed that in a given network some of the nodes have opinions about each other and they update their opinions based on their experiences. These nodes communicate their opinions in the form of recommendation to their neighbors in order to assist them to make their own judgments about other nodes. In this scenario, malicious recommenders modify values of their opinions, where they report a binary "1" for a malicious node and a binary "0" for an honest node in order to misguide the evaluating node. Let us consider that an evaluator node E, having a set of opinions $\mathbf{G}$, receives recommendations $\mathbf{R}^1, \mathbf{R}^2, ..., \mathbf{R}^d$ with cardinality A, from sources $s_1, s_2, ..., s_d$, as shown in Fig. 1. In our work, the cardinality A was fixed for all cases. The opinions as well as recommendations comprise of binary values such that a binary "0" indicates that a negative recommendation and a binary "1" indicates a positive recommendation. The node E

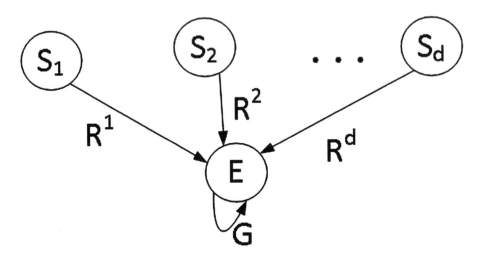

Fig. 1. Opinion of E and recommendations sent by neighbors

computes a Jaccard Similarity score [2] between its own opinion and the received recommendation from any node s_k as follows.

$$Sim(E, s_k) = \frac{|\mathbf{G} \cap \mathbf{R^k}|}{|\mathbf{G} \cup \mathbf{R^k}|}. \tag{1}$$

The similarity score computed in (1) is basically $P[G \cap R^k]$ where the universal set is assumed to be $\{G \cup R^k\}$. Hence, assuming that probability $p_{s_k} = Sim(E, s_k)$, then the entropy resulting from the similarity score is given by,

$$H_{s_k} = -p_{s_k} * log_A(p_{s_k}). \tag{2}$$

If the recommendation from the source s_k is similar to the evaluator E, the entropy will be low. However, if the similarity score is around 0.5, the entropy will be the maximum since the source is unreliable. In order to avoid detection, a malicious node may maintain a high similarity by giving incorrect feedback for a few nodes. Moreover such a node may only send incorrect feedbacks when they are most beneficial for the attacker, i.e. in a probabilistic manner. Hence the consistency of recommendations from this sender must be observed in subsequent messages in order to identify its true nature. Every time the node s_k sends a recommendation, E observes $\mathbf{R^k}$ for any inconsistencies in $\mathbf{R^k} = \{r_1, r_2, ..., r_A\}$. For simplicity, let us assume that symbols $r_1, r_2, ..., r_A$ always assume a binary value. Now let a given symbol r_i^k assume a value a_i in m_i out of N messages from sender s_k. Then the probability $p_{r_i}^k$ of r_i taking the value a_i in the next message is given by

$$p_{r_i}^k = P[r_i^k = a_i] = \frac{m_i}{N}. \tag{3}$$

The entropy of input symbols of this source is given by

$$H_{R_k} = \sum_{i=1}^{A} -p_{r_i} * log_A(p_{r_i}). \tag{4}$$

The trust of s_k for the message N is the average entropy of its similarity as well as entropy of input symbols and is given by

$$T_k(N) = \theta \cdot (1 - (H_{s_k} + H_{R_k})) + (1 - \theta) \cdot T_k(N-1), \tag{5}$$

where θ is a weight parameter that recognizes the importance of recent entropy and previous trust. For the first set of recommendations from the sender s_k, the trust is only measured by the similarity, that is, $T_k(0) = H_{s_k}$. The value of trust can be normalized however in this work, if $T_k(N)$ is undefined or below zero, it is assigned a value of zero. If $T_k(N)$ is greater than one, it is assigned a value of one.

The performance of the system is measure in terms of true and false positive rates which are defined as follows:

True Positive Rate (TPR). TPR determines how correctly are the malicious nodes identified and it is given by

$$TPR = \frac{P_{M|M}}{P_{M|M} + P_{H|M}},$$

where $P_{M|M}$ is the probability of detecting a malicious node as malicious and $P_{H|M}$ is the probability of detecting a malicious node as honest.

False Positive Rate (FPR). FPR determines the error produced by misclassifying an honest node as malicious and it is given by

$$FPR = \frac{P_{M|H}}{P_{H|H} + P_{M|H}},$$

where $P_{H|H}$ is the probability of correctly identifying the honest nodes as honest and $P_{M|H}$ is the probability of misclassifying the honest nodes as malicious. Two parameters, TPR and FPR, will be used to study the performance of the proposed scheme in later sections.

4   Simulation Results

In order to test the trust model, a network of connected vehicles is assumed where cars travel together for some distance. In this scenario, one vehicle may encounter the same neighbors over and over and forms opinions about them. These opinions may change with time, however, for honest vehicles the changes will be infrequent. Nodes will report their opinions in the form of recommendations, in scheduled broadcast messages. Some of these senders act maliciously and modify their recommendations about q of their neighbors. Malicious nodes send messages containing modified recommendations with a probability p. Honest nodes may misjudge some of the other nodes and amount of error thus introduced is assigned an error probability of 0.04. The error introduced by the honest nodes is treated as noise and does not affect the performance of the overall system.

It is assumed that the malicious nodes do not collude to provide incorrect recommendations about the same target nodes. However, it can be shown that even if the nodes collude, the consistency of the information and the similarity between evaluator and recommender can be used to identify malicious nodes and result will not be severely affected. Another assumption is that all malicious nodes misbehave with the same probability. This assumption would affect the true positive rate, since the threshold value for trust to classify nodes as malicious, will be affected.

Performance of the model will be tested in three aspects that include trust evolution with number of messages, true positive rates and false positive rate. First it will be studied how the trust of honest and malicious nodes evolve with the number of messages. For these experiments, malicious nodes constitute 50% of the nodes. These nodes send 100 messages in which they send their recommendations about 100 other nodes therefore the cardinality A is 100. The trust values of honest and malicious nodes are updated with each message, by averaging the previous trust, and combined entropy of similarity and consistency. For the simulations, equal weight is applied to the previous trust, and the combination of similarity and consistency, which means that $\theta = 0.5$. In the first message all nodes send correct information in order to attain a high initial trust value. First the evolution of trust is studied in Fig. 2. In Fig. 2 the honest nodes, despite of introducing noise in the information, attains a maximum trust value of over 0.999. However the malicious nodes send false recommendations about 10 out of 100 neighbors, i.e. $q = 10$, with different probabilities. In this scenario if p is low, the malicious node attains a relatively higher average trust value of about 0.6, although the trust of these nodes decreases gradually with the number of

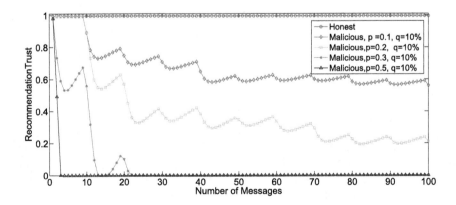

Fig. 2. Recommendation trust evolution

messages. It can be observed that at a low p, trust values of malicious nodes may increase with more correct messages, nevertheless the initial trust of the nodes can not be recovered completely. However if p is larger, the trust value is reduced at a much faster rate. The trust values of malicious nodes decline sharply with the first few messages and does not increase easily with correct messages. Thus the model identifies selective misbehavior and assigns low trust values to the malicious nodes even if the probability of incorrect information, p, is low.

Let us further examine the impact of number of false recommendations on trust evolution. For this experiment, a relatively larger value of q is considered in Fig. 3. In Fig. 3, the value of $q = 30\%$, which means that out of 100 nodes, recommendations for 30 nodes have been modified. With this setting, even if the malicious nodes send a very small number of messages with modified recommendations, hence p is small, the proposed scheme reduces their trust values drastically. Hence it is learned that if a large number of recommendations is modified then detecting malicious nodes becomes easier.

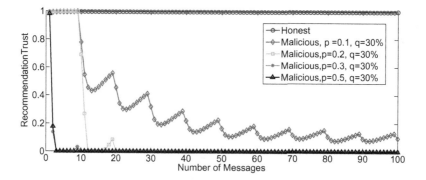

Fig. 3. Impact of increasing number of altered recommendations

In order to study the true positive rate, 50% of the nodes are configured as malicious and send false recommendations with different probabilities. The malicious nodes show selective misbehavior, such that they only ballot stuff or bad mouth q of its neighbors with a probability p and give honest recommendations about the others.

Figure 4 demonstrates that when p and q both are low, then the TPR is low, as the recommendations remain consistent most of the time and recommendations for only few neighbors are modified by the malicious nodes. These malicious nodes maintain a high similarity most of the time and show extremely infrequent misbehaviors. That is why, it is difficult to identify malicious nodes when p and q are low. However if the malicious activity becomes more obvious, either by increasing p or q or both, the true positive rate increases significantly. On the other hand, the trust of honest nodes increases with each message and therefore they are never misclassified. Hence the proposed trust model does not produce any false positives and the false positive rate is zero.

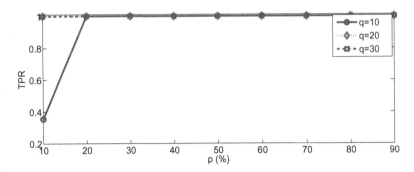

Fig. 4. TPR for a recommendation vector of length 100 bits where q bits are maliciously modified with a probability p

5   Conclusion

The essence of machine to machine communication is to identify reliable information even when some of the participants are malicious. In this study the malicious behavior has been modeled as false feedback about peers. These malicious peers are identified using the proposed entropy based trust model and their recommendations are excluded by the system. On the other hand the trust of honest nodes is increased with each interaction and their recommendations are used to predict the behavior of other nodes. How the system will make use of such recommendations is studied in our other research. The proposed trust model is tested only for binary recommendations. The scheme can be extended for non-binary recommendations which are used in most e-commerce systems.

References

1. Das, A., Islam, M.M.: SecuredTrust: a dynamic trust computation model for secured communication in multiagent systems. IEEE Trans. Dependable Secure Comput. **9**(2), 261–274 (2012)
2. Niwattanakul, S., Singthongchai, J., Naenudorn, E., Wanapu, S.: Using of Jaccard coefficient for keywords similarity. In: Proceedings of the International MultiConference of Engineers and Computer Scientists, vol. 1, p. 6 (2013)
3. Luo, J., Liu, X., Fan, M.: A trust model based on fuzzy recommendation for mobile ad-hoc networks. Comput. Netw. **53**(14), 2396–2407 (2009)
4. Kim, S., Kwon, J.: Recommendation technique using social network in internet of things environment. J. Korea Soc. Digital Ind. Inf. Manag. **1**(1), 47–57 (2015)
5. Ullah, Z., Islam, M.H., Khan, A.A., Sarwar, S.: Filtering dishonest trust recommendations in trust management systems in mobile ad hoc networks. Int. J. Commun. Netw. Inf. Secur. **8**(1), 18 (2016)
6. Khedim, F., Labraoui, N., Lehsaini, M.: Dishonest recommendation attacks in wireless sensor networks: a survey. In: 12th International Symposium on Programming and Systems (ISPS 2015), pp. 1–10. IEEE (2015)

7. Mousa, H., Mokhtar, S.B., Hasan, O., Younes, O., Hadhoud, M., Brunie, L.: Trust management and reputation systems in mobile participatory sensing applications: a survey. Comput. Netw. **90**, 49–73 (2015)
8. Yang, Y., Feng, Q., Sun, Y.L., Dai, Y.: Reptrap: a novel attack on feedback-based reputation systems. In: Proceedings of the 4th International Conference on Security and Privacy in Communication Networks, p. 8. ACM (2008)
9. Xiong, L., Liu, L.: Peertrust: supporting reputation-based trust for peer-to-peer electronic communities. IEEE Trans. Knowl. Data Eng. **16**(7), 843–857 (2004)
10. Srivatsa, M., Xiong, L., Liu, L.: Trustguard: countering vulnerabilities in reputation management for decentralized overlay networks. In: Proceedings of the 14th International Conference on World Wide Web, pp. 422–431. ACM (2005)
11. Luo, J., Liu, X., Zhang, Y., Ye, D., Xu, Z.: Fuzzy trust recommendation based on collaborative filtering for mobile ad-hoc networks. In: LCN, pp. 305–311. Citeseer (2008)
12. Liu, Y., Sun, Y., Liu, S., Kot, A.C.: Securing online reputation systems through trust modeling and temporal analysis. IEEE Trans. Inf. Forensics Secur. **8**(6), 936–948 (2013)
13. Whitby, A., Jøsang, A., Indulska, J.: Filtering out unfair ratings in bayesian reputation systems. In: Proceedings of 7th International Workshop on Trust in Agent Societies, vol. 6, pp. 106–117 (2004)

Secure Data Sharing for Vehicular Ad-hoc Networks Using Cloud Computing

Mehdi Sookhak[1], F. Richard Yu[1($\boxtimes$)], and Helen Tang[2]

[1] Department of System and Computer Engineering,
Carleton University, Ottawa, ON, Canada
`mehdi.sookhak@carleton.ca, richard.yu@carleton.ca`
[2] Defence Research and Development Canada, Ottawa, ON, Canada
`Helen.Tang@drdc-rddc.gc.ca`

Abstract. During the last decade, researchers and developers have been attracted to Vehicular Ad-hoc Network (VANET) research area due to its significant applications, including efficient traffic management, road safety, and entertainment. Several resources such as communication, on-board unit, storage, computing, and endless battery are embedded in the vehicles, which are used for enhancing Intelligent Transportation Systems (ITSs). One of the crucial challenges for VANETs is to securely share an important information among vehicles. In some cases, the data owner is also not available and unable to control the data sharing process, i.e., sharing data with a new user or revoking the existing user. In this paper, we present a new method to address the data sharing problem and delegate the management of data to a Trusted Third Party (TPA) based on bilinear pairing technique. To achieve this goal, we use a cloud computing, as the mainstream platform of utility computing paradigm, to store the huge amount of data and perform the re-encryption process securely.

Keywords: Vehicular ad-hoc networks · Cloud computing · Data access control · Bilinear pairing technique · Proxy re-encryption

1 Introduction

Recent improvements in hardware, software, and communication technologies lead to many improvements and developments in the networking area. One of the main networking technologies that has attracted the researchers' and industries consideration over the last decades is VANETs [1]. VANET is a self-organized network composed of mobile nodes connected with wireless links where the vehicles act as nodes [2]. Vehicular network is formed between moving vehicles equipped with wireless interfaces that could be homogeneous or heterogeneous technologies. These networks are considered as one of the real-life applications of the ad-hoc network, which enable communications among nearby vehicles as well as between vehicles and nearby fixed equipment (roadside equipment). Vehicles can communicate to infrastructure in a Vehicle-to-Infrastructure (V2I) design

© ICST Institute for Computer Sciences, Social Informatics and Telecommunications Engineering 2017
Y. Zhou and T. Kunz (Eds.): ADHOCNETS 2016, LNICST 184, pp. 306–315, 2017.
DOI: 10.1007/978-3-319-51204-4_25

where Road Side Unit (RSU) functions as an interface between On Board Unit (OBU) and main core network. Vehicles can also directly communicate to each other in Vehicle-to-Vehicle (V2V) design [3].

The main aim of VANET is to provide safety for drivers and passengers by developing novel applications and solutions. The idea of vehicular network has been expanded into ITS and Intelligent Vehicular Network as promising solutions to transportation and traffic-related problems in modern cities by creating safe, secure, and healthy environment [4]. However, the growth of vehicular network and its applications and services requires scalable infrastructure, computing capacity, and storage [5].

One of the main challenges of vehicular networks is to securely share the critical information among vehicles. To address this problem, the data owner who wants to share the data with other vehicles, can outsource the data to the remote servers of cloud computing [6,8]. However, delegating the management of such an important information to the untrusted cloud server is not reasonable, due to the confidentiality and integrity of outsourced files [13,14]. One way to solve this problem is using the proxy re-encryption technique that allows the data owner to encrypt the data before outsourcing to the remote servers, and delegate the management of encrypted data to the cloud server without requiring to decrypt them. Although there are several methods to support proxy re-encryption in cloud computing, the data owner must generate a re-encryption key for allowing a new user to access the data and decrypting it. Therefore, the existing methods are inapplicable when the data owner is unavailable. Moreover, sending the owner's private key to the third part to perform the management is not a usable solution.

This paper presents a new method for securely sharing data in the vehicular networks by using cloud computing and the bilinear pairing technique. After encrypting the data, the data owner, who wants to share the data, transfers it to the cloud computing, which is responsible for re-encrypting the ciphertext for the users. The data owner also delegates the key management of the encrypted data to the TPA while preserving the privacy of data by blinding the private key. As a result, when a new user requests to access the outsourced data, the TPA is able to generate the re-encryption key, which allows the user to decrypt it by the user's private key.

The rest of the paper is organized as follows: Sect. 2 presents a background on vehicular networks, and vehicular cloud computing. Section 3 explains the preliminaries and makes an overview on the bilinear pairing technique. Section 4 describes the architecture and system operation of the proposed method. The related works as well as the advantages and disadvantage of the existing methods are stated in Sect. 5. Finally, we conclude the paper in Sect. 6.

2 Background

This section explains the concept of cloud computing, mobile computing and vehicular networks, respectively, since these are the cornerstone of vehicle cloud computing concept.

2.1 Vehicular Ad-hoc Networks

In this era, by growing the new intelligent technologies as a remarkable contributor of transportation systems, the employing of ITS concept has significantly brought attention of governments and academia in this area [9]. Meanwhile, the use of VANET as a subset of Mobile Ad-hoc Network (MANET) is the significant wireless technology proposed exclusively for vehicular environment. The employment of this technology in ITS is the concept which is significantly demonstrated to enhance the road safety, efficiency, and services through real-time V2V and V2I communications [10].

In ITSs, each vehicle plays the role of sender, receiver, and router to broadcast data to the vehicular network, which then utilizes the data to ensure safe and free-flow of traffic [11]. To take place the communication between RSUs and vehicles, vehicles must be equipped with OBU that enables short-range wireless ad-hoc networks to be formed. In addition, vehicles must be equipped with a hardware, which permits detailed position information, such as Global Positioning System (GPS) or a Differential Global Positioning System (DGPS) receiver. On the other side, fixed RSUs that are linked to the backbone network must be mounted, to facilitate communication. Communication configurations in VANET contain V2V, V2I, and routing-based communications. They rely on very accurate information regarding the surrounding environment, which requires the employment of accurate positioning systems and well communication protocols for transferring information [12].

2.2 Vehicular Cloud Computing

Cloud computing is a comparatively new trend in the field of Information Technology (IT) that decreases computing, storage and other functions from traditional desktop and portable computer devices since all the functions can be virtualized in cloud computing platform [15,17]. Cloud computing provides ubiquitous, applicable, and on-demand network access to the vast shared computing resources, such as all networks, servers, storages, applications, and services. Consequently, end users only need some simple I/O devices to enjoy powerful processing ability and convenient service in cloud computing platform [18]. One of the main applications of cloud computing is in vehicular networks, as vehicular cloud computing.

Vehicular cloud computing can be divided into two categories: (1) Vehicular Computing, and (2) Vehicular using Cloud. In the first type of VCC, each vehicle can play a role as a datacenter, while in the VuC, the vehicles will be connected to the cloud for outsourcing data and augmenting the computation resources [6]. In the following, we briefly explain these two concepts:

1. Vehicular Computing (VC): The cloud computing paradigm enables the utilization of excess computing power in a way that vehicles are treated as underutilized computational resources, which can be used for providing public services. In this scenario, the parked vehicles can be counted as a huge idle

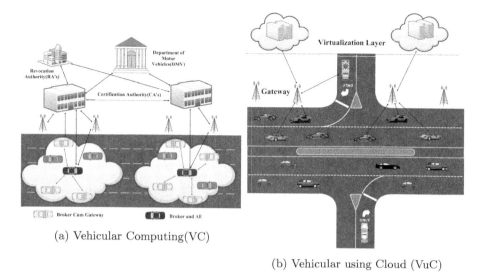

(a) Vehicular Computing(VC)

(b) Vehicular using Cloud (VuC)

Fig. 1. Vehicular cloud computing architectures

resource that is merely wasted. For instance, many people park their vehicles in the parking airports while traveling. In addition, some vehicles are stuck in congested traffic. These characters of such vehicles make them an ideal nominee for nodes in a cloud computing network [5,6]. Figure 1(a) shows three main components of VC, such as VANET infrastructure, gateways and brokers.

2. Vehicular using Cloud (VuC): It has been emerged as a new concept to efficiently solve the drivers' problem by using the cloud services instead of sharing their own resources. In VuC, vehicles utilize VANET infrastructure to connect to conventional clouds and use the real-time services, for example monitoring the real-time traffic information and infotainment. VANET infrastructure, gateways, and virtualization layer are three main components of VuC. RSUs act as gateways between the vehicles and clouds. They are also responsible to provide the virtualization layer. To connect the gateways to clouds, high speed wired communication (e.g. optical fiber) can be used, while wireless communication (e.g. V2V and V2I) is used to connect the vehicles to gateways [19,20]. It is important to mention that our proposed method is based on VuC. Figure 1b shows the general architecture of VuC.

By taking advantage of VCC, the problem of municipal traffic management centers, which is the lack of adequate computational resources, will be removed. This is because the vehicles assist local consultants to resolve traffic incidents in a timely fashion. The chief concentration of the VCC is to provide on demand solutions for unpredictable incidents in a proactive fashion. VCCs present a unified incorporation and reorganized management of on board facilities. Moreover, they adapt dynamically based on the system environments and application requirements. A federation of VCCs presents a decision support system and becomes the temporary infrastructure replacement in case of natural disaster that abolishes standing infrastructure. The Federal Communication Commission (FCC) allocated Dedicated Short-Range Communication (DSRC) for supporting the vehicular networks. Furthermore, road infrastructures such as cameras, access points, and inductive loop detectors are supportive for VCC.

3   Preliminaries and Definition

This section briefly reviews the cryptographic background about the bilinear map and the required security assumptions.

3.1   Bilinear Maps

In 2001, some researchers [21–23] introduced a special type of encryption method, which is called proxy re-encryption, on the basis of bilinear maps. Let G_1, G_2 be cyclic groups with prime order p; $g_1, g_2 \in G_1$ be the generators of the group G_1; and $a, b \in Z_p$ that indicates a, b are randomly selected from a finite set Z.

Function $e : G_1 \times G_1 \rightarrow G_2$ is a bilinear map with the following properties: (1) Bilinearity: for all $a, b \in Z_p$, it can be seen that $e\left(g_1{}^a, g_1{}^b\right) = e(g_1, g_1)^{ab}$, and (2) Non-degeneracy: If $g_1, g_2 \in G_1$ have the capability to generate G_1, then $e(g_1, g_1)$ can generate G_2.

3.2   Complexity Assumptions

Most of the cryptosystems that have designed on the basis of bilinear map properties, rely on the Decisional Bilinear Diffie-Hellman (DBDH) assumption. This assumption indicates that for any $g_1 \in G_1$, $a, b, c \in Z_p$, and $Q \in G_2$, it is hard to distinguish $e(g_1, g_1)^{abc}$ from the random given that $(g_1, g_1{}^a, g_1{}^b, g_1{}^c, Q)$.

4   Proposed Method

In this section, we propose our method on the basis of proxy re-encryption technique for providing a secure mechanism to share data in vehicular-based cloud computing.

4.1   Architecture of the Proposed Method

Figure 2 shows the architecture of the proposed method, which consists of four important components, as follows: (1) Data Owner (DO): who encrypts and outsources the data to the cloud server, and delegates the re-encryption process to the cloud service provider; (2) TPA: who is responsible for adding or revoking users based on the received information from DO; (3) Cloud Service Provider (CSP): who stores the revived data from DO, checks the access control of the files, and re-encrypts data for new users; and (4) User: who asks the CSP for accessing an encrypted file.

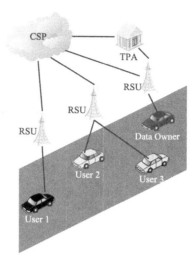

Fig. 2. The architecture of the proposed method for secure data sharing in VCC

4.2 Definition

- $KeyGen(1^k) \rightarrow (pk, sk, pp)$. This algorithm generates the public and secret key for the DO (pk_d, sk_d) and users (pk_d, sk_d) as well as some public parameters by using a security parameter 1^k.
- $KeyDelegation(sk_d) \rightarrow (Aux)$. This algorithm uses the secret key of the data owner to generate the auxiliary key that can be used by TPA to generate the re-encryption key for the users.
- $ReKeyGen(pk_u, sk_t, Aux) \rightarrow (rekey)$. The output of this algorithm is a new key that can be used by the CSP to re-encrypt the outsourced ciphertext.
- $Enc(F, pk_d, sk_d) \rightarrow (C)$. This algorithm decrypts the DO's file by using the public and secret key of the DO.
- $ReEnc(C, pk_u, rekey) \rightarrow (C')$. It is responsible to re-encrypts the outsourced ciphertext based on the users' public key and the generated key by the TPA.
- $Dec(C', sk_u)$ to M. The user can use this function to decrypt the re-encrypted outsourced file (C') using her secret key.

Table 1 shows the notation of the proposed method for secure data sharing in vehicular-based cloud computing.

4.3 System Operation

The designed data sharing method for VCC consists of the following phases:

(1) *Setup.* Our method operates over two groups G_1, G_2 of order p with the bilinear map properties $e : G_1 \times G_1 \rightarrow G_2$. First of all, the system parameters $(g \in G_1, Z = e(g, g) \in G_2)$ need to be randomly generated and distributed among the users and the owners. Then, each client needs to select a random

Table 1. The notation used in explanation of the proposed method

Symbol	Description
g	A generator for G_1
Z	$e(g,g)$
x_d	Secret Key of the data owner
x_u	Secret Key of the user
g^{x_d}	Data owner public key
g^{x_u}	User public key
r	Random Number
q	Large prime Number
DO	Data Owner
TPA	Trusted Third Party
CSP	Cloud Service Provider

number as a secret key and generate her public key based on this random number, for example, $(pk_u = g^{x_u}, sk_u = x_u)$ for each user and $(pk_d = g^{x_d}, sk_d = x_d)$ for each data owner.

(2) *Data encryption and key delegation.* Assume that the DO wants to share a file $F \in G_2$ among users. The Do generates a random number r and a unique large prime number q for each file. Then, the DO encrypts F by: $C = (Z^{rq}.F, g^{r.\frac{x_d}{q}})$. Then, the owner outsources the encrypted file (C) as well as a list of the authorized users to the vehicular cloud and delegates the management of the file to the CSP. Finally, the DO makes the TPA responsible for adding a new user for this file by sending a blind version of her secret key $(\frac{q^2}{x_d})$ to the TPA.

(3) *Data re-encryption.* If a new user requests the CSP to access the encrypted file (C), firstly, the CSP has to check whether the new user has eligibility to access data. After confirming that, the CSP asks the TPA to generate the re-encryption key based on the user's public key by:

$$rekey = pk_u^{(\frac{q^2}{x_d})} = g^{\frac{x_u \cdot q^2}{x_d}}$$

Up on receiving the re-encryption key, the CSP re-encrypts the outsourced file (C) by using the following equation:

$$C' = (Z^{rq}.F, re - encrypt(g^{r.\frac{x_d}{q}}, g^{\frac{x_u \cdot q^2}{x_d}}))$$

$$re - encrypt(g^{r.\frac{x_d}{q}}, g^{\frac{x_u \cdot q^2}{x_d}}) = e(g^{r.\frac{x_d}{q}}, g^{\frac{x_u \cdot q^2}{x_d}}) = Z^{rqx_u}$$

(4) *Data Decryption.* After obtaining the re-encrypted file $C' = (Z^{rq}.F, Z^{rqx_u})$, the user is able to decrypt the file by:

$$F = \frac{Z^{rq}.F}{(Z^{rqx_u})^{\frac{1}{x_u}}}$$

Remark 1. It is important to mention that although the transferred parameters between DO and TPA is blinded, we can generate a session key to encrypt and decrypt the data by using the public key of the DO (pk_d) and the public key of the TPA (pk_T).

Remark 2. All of the communications between DO and TPA, DO and CSP, and User and CSP are performed by using the existing RSU and OBU.

5   Related Work

Most of the existing methods for secure data sharing have been proposed for cloud and mobile cloud computing. In this section, we make an overview on some of the proposed method based on proxy re-encryption and focus on their advantages and disadvantages.

Proxy re-encryption (PRE) is a cryptosystem, which can be used to turn a ciphertext encrypted under one key into an encryption of the same plaintext under another key by using a proxy. Blaze et al. [21] was the first to propose a PRE scheme without having to learn the plaintext and secret key based on the ElGamal cryptosystem [24]. Although this scheme is semantically secure under the Decision Diffie-Hellman assumption in G, it suffers from several issues, such as bidirectionality, collusion, and re-encryption key generation process.

Ivan and Dodis [22] presented a unidirectional PRE approach on the basis of standard public key cryptosystems in which Alices secret key is divided in two parts $sk_a = sk_1 + sk_2$ and distributed between Proxy and Bob. Although this method addressed the bidirectional problem of the first PRE scheme, it needs a pre secret-sharing, which enforces Bob to store the additional secret key.

Ateniese *et al.* [23] solved the aforementioned problems and designed a unidirectional proxy re-encryption method by using the bilinear maps. To prevent the collision attack, the authors considered a master key security without requiring the pre-sharing of secret keys between parties.

Tysowski *et al.* [25] extended the Ateniese method [23] and presented a manager-based re-encryption scheme for mobile cloud computing based on the bilinear maps. However, this method has several drawbacks, such as: considering a manager as a trusted entity to generate the public key and secret key of all other parties, and requiring the re-encryption task by changing the group membership.

We propose a first proxy re-encryption method for sharing data securely in vehicular-based cloud computing. In this method, all parties are able to generates their public and private keys. One of the main contributions of this method is that the new user can access the outsourced data even if the data owner is unavailable. This is because the data owner delegated the management of data access control to the TPA by using a blinded key results in preserving the privacy of data.

6   Conclusion and Future Work

Secure data sharing is one of the important issues in vehicular ad hoc networks. Although the vehicles are able to directly share the data using V2V communication in vehicular networks, this technique is inefficient. Recently, researchers have introduced the vehicular cloud computing, which can provide several benefits for users, such as data sharing. In this paper, we presented a secure data sharing method for vehicular-based cloud computing using a proxy re-encryption technique. When the DO encrypts the file and outsources it in the vehicular cloud, the data access management is delegated to the TPA by a blind version of her key. This method also enables a new user to request the CSP for accessing the encrypted data even if the data owner is unavailable. Future work is in progress to consider trust management in the proposed framework.

References

1. Hartenstein, H., Laberteaux, K.P.: A tutorial survey on vehicular ad hoc networks. IEEE Commun. Mag. **46**(6), 164–171 (2008)
2. Al-Sultan, S., Al-Doori, M.M., Al-Bayatti, A.H., Zedan, H.: A comprehensive survey on vehicular ad hoc network. J. Netw. Comput. App. **37**, 380–392 (2014)
3. Toor, Y., Muhlethaler, P., Laouiti, A., Fortelle, A.D.L.: Vehicle ad hoc networks: applications and related technical issues. IEEE Commun. Surv. Tut. **10**, 74–88 (2008)
4. Dimitrakopoulos, G., Demestichas, P.: Intelligent transportation systems. IEEE Veh. Tech. Mag. **5**, 77–84 (2010)
5. Olariu, S., Hristov, T., Yan, G.: The Next Paradigm Shift: From Vehicular Networks to Vehicular Clouds Mobile Ad Hoc Networking. Wiley, Hoboken (2013)
6. Whaiduzzaman, M., Sookhak, M., Gani, A., Buyya, R.: A survey on vehicular cloud computing. J. Netw. Comput. App. **40**, 325–344 (2014)
7. Abuelela, M., Olariu, S.: Taking VANET to the clouds. In: Proceedings of the 8th International ACM Conference on Advances in Mobile Computing and Multimedia, pp. 6–13, New York (2010)
8. Yan, G., Wen, D., Olariu, S., Weigle, M.C.: Security challenges in vehicular cloud computing. IEEE Trans. Intel. Transp. Syst. **14**, 284–294 (2013)
9. Faouzi, N.-E.E., Leung, H., Kurian, A.: Data fusion in intelligent transportation systems: progress and challenges a survey. Inf. Fusion. **12**, 4–10 (2011)
10. Zeadally, S., Hunt, R., Chen, Y.-S., Irwin, A., Hassan, A.: Vehicular ad hoc networks (VANETS): status, results, andchallenges. Tele Commun. Syst. **50**, 217–241 (2012)
11. Harri, J., Filali, F., Bonnet, C.: Mobility models for vehicular ad hoc networks: a survey and taxonomy. IEEE Commun. Surv. Tut. **11**, 19–41 (2009)
12. Bitam, S., Mellouk, A., Zeadally, S.: Bio-inspired routing algorithms survey for vehicular ad hoc networks. IEEE Commun. Surv. Tut. **17**, 843–867 (2015)
13. Sookhak, M., Talebian, H., Ahmed, E., Gani, A., Khan, M.K.: A review on remote data auditing in single cloud server: taxonomy and open issues. J. Netw. Comput. App. **43**, 121–141 (2014)
14. Sookhak, M., Gani, A., Talebian, H., Akhunzada, A., Khan, S.U., Buyya, R., Zomaya, A.Y.: Remote data auditing in cloud computing environments: a survey, taxonomy, and open issues. ACM Comput. Surv. **47**, 65:1–65:34 (2015)

15. Sookhak, M., Akhunzada, A., Gani, A., Khan, M.K., Anuar, N.B.: Towards dynamic remote data auditing in computational clouds. Sci. World J., 1–12 (2014)
16. Sookhak, M., Gani, A., Khan, M.K., Buyya, R.: Dynamic remote data auditing for securing big data storage in cloud computing. Inf. Sci. (2015, in Press)
17. Yousafzai, A., Gani, A., Noor, R.M., Sookhak, M., Talebian, H., Shiraz, M., Khan, M.K.: Cloud resource allocation schemes: review, taxonomy, and opportunities. Knowl. Inf. Syst, 1–35 (2016)
18. Shiraz, M., Sookhak, M., Gani, A., Shah, S.A.A.: A study on the critical analysis of computational offloading frameworks for mobile cloud computing. J. Netw. Comput. App. **47**, 47–60 (2015)
19. Hussain, R., Abbas, F., Son, J., Oh, H.: TIaaS: secure cloud-assisted traffic information dissemination in vehicular ad hoc networks cluster. In: 13th IEEE/ACM International Symposium on Cloud and Grid Computing (CCGrid), pp. 178–179, Delft (2013)
20. Arif, S., Olariu, S., Wang, J., Yan, G., Yang, W., Khalil, I.: Datacenter at the airport: reasoning about time-dependent parking lot occupancy. IEEE Trans. Parallel Distrib. Syst. **23**, 2067–2080 (2012)
21. Blaze, M., Bleumer, G., Strauss, M.: Divertible protocols and atomic proxy cryptography. In: International Conference on the Theory and Application of Cryptographic Techniques, pp. 127–144 (1998)
22. Ivan, A., Dodis, Y.: Proxy cryptography revisited. In: Proceedings of 10th Annual the Network and Distributed System Security Symposium (NDSS), pp. 1–20, California (2003)
23. Ateniese, G., Fu, K., Green, M., Hohenberger, S.: Improved proxy re-encryption schemes with applications to secure distributed storage. ACM Trans. Inf. Syst. Secur. **9**, 1–30 (2006)
24. ElGamal, T.: A public key cryptosystem and a signature scheme based on discrete logarithms. In: Blakley, G.R., Chaum, D. (eds.) CRYPTO 1984. LNCS, vol. 196, pp. 10–18. Springer, Heidelberg (1985). doi:10.1007/3-540-39568-7_2
25. Tysowski, P.K., Hasan, M.A.: Re-encryption-based key management towards secure and scalable mobile applications in clouds. In: IACR Cryptology ePrint Archive, pp. 1–10 (2011)

Distributed Collaborative Beamforming for Real-World WSN Applications

Slim Zaidi$^{(\boxtimes)}$, Bouthaina Hmidet, and Sofiène Affes

INRS-EMT, Université du Québec, Montreal, QC H5A 1K6, Canada
{zaidi,hmidet,affes}@emt.inrs.ca

Abstract. In this paper, we consider a collaborative beamformer (CB) design that achieves a dual-hop communication from a source to a receiver in highly-scattered environments, through a wireless sensor network (WSN) comprised of K independent and autonomous sensor nodes. The weights of the considered CB design at these nodes, derived to maximize the received signal-to-noise ratio (SNR) subject to constraint over the nodes total transmit power, have expressions that inevitably depend on some form of the channel state information (CSI). Only those requiring the local CSI (LCSI) available at their respective nodes lend themselves to a truly distributed implementation. The latter has the colossal advantage of significantly minimizing the huge overhead resulting otherwise from non-local CSI (NLCSI) exchange required between nodes, which becomes prohibitive for large K and/or high Doppler. We derive the closed-form expression of the SNR-optimal CB (OCB) and verify that it is a NLCSI-based design. Exploiting, however, the polychromatic (i.e., multi-ray) structure of scattered channels as a superposition of L impinging rays or chromatics, we propose a novel LCSI-based distributed CB (DCB) design that requires a minimum overhead cost and, further, performs nearly as well as its NLCSI-based OCB counterpart. Furthermore, we prove that the proposed LCSI-based DCB outperforms two other DCB benchmarks: the monochromatic (i.e., single-ray) DCB and the bichromatic (i.e., two-ray) DCB (B-DCB).

Keywords: Distributed collaborative beamforming (CB, DCB) · Relaying · MIMO · Scattering · Device/machine-2-device/machine (D2D/M2M) communications · Wireless sensor networks (WSN)

1 Introduction

Due to its strong potential in establishing a reliable and energy-efficient communication in wireless sensor networks (WSN) applications, collaborative beamforming (CB) has garnered the attention of the research community [1–12]. The so far proposed CB designs could be broadly categorized as either the local

Work supported by the CRD, DG, and CREATE PERSWADE <www.create-perswade.ca> Programs of NSERC and a Discovery Accelerator Supplement Award from NSERC.

© ICST Institute for Computer Sciences, Social Informatics and Telecommunications Engineering 2017
Y. Zhou and T. Kunz (Eds.): ADHOCNETS 2016, LNICST 184, pp. 316–329, 2017.
DOI: 10.1007/978-3-319-51204-4_26

CSI (LCSI)-based (i.e., distributed) CB namely the monochromatic DCB (M-DCB) and the bichromatic DCB (B-DCB), or the non-local CSI (NLCSI)-based (i.e., non-distributed) CB namely the optimal CB. When designing M-DCB, authors in [1–12] ignored scattering present in almost all real-world scenarios but very few ones with both practical and investigation values in which they have consequently assumed a simple monochromatic (i.e., single-ray) channel. In scattered channels, however, said to be polychromatic (i.e., multi-ray) and characterized by the angular spread (AS) [13–17], due to channel mismatch, the performance of M-DCB slightly deteriorates in areas where the AS is small and becomes unsatisfactory when it grows large [18–23]. In contrast, B-DCB in [22,23] which accounts for scattering by an efficient two-ray approximation of the polychromatic channel at relatively low AS not only outperforms M-DCB, but also achieves the optimal performance at small to moderate AS values in lightly-to moderately-scattered environments. Nevertheless, its performance substantially deteriorates at large in highly-scattered environments [22,23]. OCB which is able to achieve optimal performance even in highly-scattered environments is NLCSI-based and cannot be implemented in WSNs due to its distributed nature [24]. Indeed, the often independent and autonomous sensors must estimate and broadcast their own channels at the expense of an overhead that becomes prohibitive for a large number of nodes and/or high Doppler [24,25]. The aim of this work is then to design a novel DCB technique implementation that requires a minimum overhead cost and, further, is able to achieve optimal performance for any AS values, thereby pushing farther the frontier of the DCB's real-world applicability range to include highly-scattered environments.

In this paper, we consider an OCB design whose weights are derived to maximize the received SNR subject to constraint over the nodes' total transmit power, to achieve a dual-hop communication from a source to a receiver in highly-scattered environments, through a WSN comprised of K independent and autonomous sensor nodes. We verify that the direct implementation of the so-obtained OCB is NLCSI-based. Exploiting, the polychromatic structure of scattered channels, we propose a novel DCB LCSI-based implementation that requires a minimum overhead cost and, further, performs nearly as well as its NLCSI-based OCB counterpart. Furthermore, we prove that the proposed LCSI-based DCB always outperforms M-DCB which is designed without accounting for scattering and that it is more robust against scattering than B-DCB whose performance substantially deteriorates in highly-scattered environments.

The rest of this paper is organized as follows. The system model is described in Sect. 2. Section 3 derives the power-constrained SNR-optimal CB design in closed-form and verifies that its direct implementation is NLCSI-based. Our novel DCB implementation is proposed in Sect. 4. Section 5 analyzes its performance while Sect. 6 verifies by computer simulations the theoretical results. Concluding remarks are given in Sect. 7.

Notation: Uppercase and lowercase bold letters denote matrices and column vectors, respectively. $[\cdot]_{il}$ and $[\cdot]_i$ are the (i,l)-th entry of a matrix and i-th entry of a vector, respectively. $\mathbf{I}_N$ is the N-by-N identity matrix. $(\cdot)^T$ and $(\cdot)^H$ denote the transpose and the Hermitian transpose, respectively. $\| \cdot \|$ is the 2-norm of a vector and $| \cdot |$ is the absolute value. $E\{\cdot\}$ stands for the statistical expectation and $(\xrightarrow{ep1}) \xrightarrow{p1}$ denotes (element-wise) convergence with probability one. $J_1(\cdot)$ is the first-order Bessel function of the first kind, $Ei(\cdot)$ is the exponential integral function, and $\odot$ is the element-wise vector product.

2   System Model

Consider a WSN comprised of K single-antenna sensor nodes uniformly and independently distributed on the disc $D(O, R)$. A source S and a receiver Rx are located in the same plane containing $D(O, R)$, as illustrated in Fig. 1. Due to high pathloss attenuation, we assume that there is no direct link from S to Rx. Let (r_k, ψ_k) denote the polar coordinates of the k-th node and (A_s, ϕ_s) denote those of the source. The latter is assumed, without loss of generality, to be at $\phi_s = 0$ and to be located relatively far from the nodes, i.e., $A_s \gg R$.

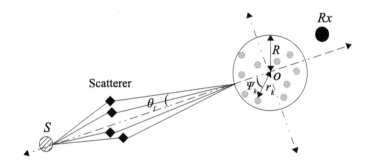

Fig. 1. System model.

Furthermore, the following assumptions are considered throughout the paper:

(A1) The source is scattered by a given number of scatterers located in the same plane containing $D(O, R)$. The latters generate from the transmit signal L rays or "spatial chromatics" (with reference to their angular distribution) that form a polychromatic propagation channel [15–18]. The l-th ray or chromatic is characterized by its angle deviation θ_l from the source direction ϕ_s and its complex amplitude $\alpha_l = \rho_l e^{j\varphi_l}$ where the amplitudes ρ_l, $l = 1, \ldots, L$ and the phases φ_l, $l = 1, \ldots, L$ are independent and identically distributed (i.i.d.) random variables, and each phase is uniformly distributed over $[-\pi, \pi]$. The θ_l, $l = 1, \ldots, L$ are i.i.d. zero-mean random variables with a symmetric probability density function (pdf) $p(\theta)$ and variance σ_θ^2 [13,15,16]. All θ_ls, φ_ls, and ρ_ls are mutually independent. All rays have equal power $1/L$ (i.e., $E\{|\alpha_l|^2\} = 1/L$). Note that the standard deviation σ_θ is commonly known as the angular spread (AS) while $p(\theta)$ is called the scattering or angular distribution. We are particularly interested here in addressing highly-scattered environments.

(A2) The nodes' forward channels to the receiver $[\mathbf{f}]_k$, $k = 1, \ldots, K$ are zero-mean unit-variance circular Gaussian random variables [9,11].

(A3) The source signal s is narrow-band with unit power while noises at the nodes and the receiver are zero-mean Gaussian random variables with variances $\sigma_v{}^2$ and $\sigma_n{}^2$, respectively. The source signal, noises, and the nodes' forward channels are mutually independent [9,11,12,26].

(A4) The k-th node is aware of its own coordinates (r_k, ψ_k), its forward channel $[\mathbf{f}]_k$, its backward channel $[\mathbf{g}]_k$, and the wavelength λ while being oblivious to the locations and the forward channels of *all* other nodes in the network [1–5,11,12].

Due to A1 and the fact that $A_s \gg R$, it can be shown that the backward channel gain from the source to the k-th node can be represented as

$$[\mathbf{g}]_k = \sum_{l=1}^{L} \alpha_l e^{-j\frac{2\pi}{\lambda} r_k \cos(\theta_l - \psi_k)}. \tag{1}$$

Obviously, in the conventional scenario where the scattering effect is neglected (i.e., $\sigma_\theta \longrightarrow 0$) to assume a monochromatic plane-wave propagation channel, we have $\theta_l = 0$ and, hence, $[\mathbf{g}]_k$ can be reduced to $[\mathbf{g}_1]_k = e^{-j(2\pi/\lambda)r_k \cos(\psi_k)}$, the well-known steering vector in the array-processing literature [1–12,19–25].

As can be observed from (1), the summation of L chromatics causes a variation, with a particular channel realization, of the received power at the k-th node. The channel is then said to experience a form of fading. When L is large, according to the Central Limit Theorem, the distribution of the channel gain $[\mathbf{g}]_k$ approaches a Gaussian. Since, according to A1, $\mathrm{E}\{\alpha_l\} = 0$ for $l = 1, \ldots, L$, then $[\mathbf{g}]_k$ is a zero-mean Gaussian random variable and, hence, its magnitude is Rayleigh distributed. Therefore, when L is large enough (practically in the range of 10), the channel from the source to the k-th node is nothing but a Rayleigh channel. It can also be observed from (1) that we did not take into account any line-of-sight (LOS) component in our channel model. If this were the case, $[\mathbf{g}]_k$'s distribution would approach a non-zero mean Gaussian distribution and the channel would become Rician.

A dual-hop communication is established from the source S to the receiver Rx. In the first time slot, S sends its signal s to the nodes while, in the second time slot, each node multiplies its received signal by a properly selected beamforming weight and forwards the resulting signal Rx. The received signal at the latter is given by

$$r = s\mathbf{w}^H \mathbf{h} + \mathbf{w}^H (\mathbf{f} \odot \mathbf{v}) + n, \tag{2}$$

where $\mathbf{w} \triangleq [w_1 \ldots w_K]$ is the beamforming vector with w_k being the k-th node's beamforming weight, $\mathbf{h} \triangleq \mathbf{f} \odot \mathbf{g}$ with $\mathbf{f} \triangleq [[\mathbf{f}]_1 \ldots [\mathbf{f}]_K]^T$ and $\mathbf{g} \triangleq [[\mathbf{g}]_1 \ldots [\mathbf{g}]_K]^T$, and $\mathbf{v}$ and n are the nodes' noise vector and the receiver noise, respectively. Several CB designs exist in the literature but we are only concerned herein by the one which maximizes the received signal-to-noise ratio (SNR) subject to constraint over the nodes' total transmit power [26].

3   Power-Constrained SNR-Optimal CB

Let $\mathbf{w}_O$ denote the power-constrained SNR-optimal CB, or simply OCB, which satisfies the following optimization problem:

$$\mathbf{w}_O = \arg\max \xi_{\mathbf{w}} \quad \text{s.t.} \quad P_T \le P_{\max}, \tag{3}$$

where $\xi_{\mathbf{w}}$ is the achieved SNR using $\mathbf{w}$ and $P_T = (1 + \sigma_v^2) \|\mathbf{w}\|^2$ is the nodes' total transmit power. From (2), $\xi_{\mathbf{w}}$ is given by

$$\xi_{\mathbf{w}} = \frac{P_{\mathbf{w},s}}{P_{\mathbf{w},n}}, \tag{4}$$

where $P_{\mathbf{w},s} = |\mathbf{w}^H \mathbf{h}|^2$ and $/P_{\mathbf{w},n} = \sigma_v^2 \mathbf{w}^H \mathbf{\Lambda} \mathbf{w} + \sigma_n^2$ are, respectively, the desired and noise powers with $\mathbf{\Lambda} \triangleq \operatorname{diag}\{|[\mathbf{f}]_1|^2 \dots |[\mathbf{f}]_K|^2\}$. Note that $\mathbf{w}_O$ should satisfy the constraint in (3) with equality. Otherwise, one could find $\epsilon > 1$ such that $\mathbf{w}_\epsilon = \epsilon \mathbf{w}_O$ verifies $(1 + \sigma_v^2) \|\mathbf{w}_\epsilon\|^2 = P_{\max}$. In such a case, since $d\xi_{\mathbf{w}_\epsilon}/d\epsilon > 0$ for any $\epsilon > 0$, the SNR achieved by $\mathbf{w}_\epsilon$ would be higher than that achieved by $\mathbf{w}_O$ contradicting thereby the optimality of the latter. As such, (3) could be rewritten as

$$\mathbf{w}_O = \arg\max \frac{\mathbf{w}^H \mathbf{h}\mathbf{h}^H \mathbf{w}}{\sigma_v^2 \mathbf{w}^H \tilde{\mathbf{\Lambda}} \mathbf{w}} \quad \text{s.t.} \quad (1 + \sigma_v^2) \|\mathbf{w}\|^2 = P_{\max}, \tag{5}$$

where $\tilde{\mathbf{\Lambda}} = \mathbf{\Lambda} + \beta \mathbf{I}$ and $\beta = \sigma_n^2 (1 + \sigma_v^2) / (\sigma_v^2 P_{\max})$. It is straightforward to show that the OCB solution of (5) is

$$\mathbf{w}_O = \left(\frac{P_{\max}}{K (1 + \sigma_v^2)\,\eta}\right)^{\frac{1}{2}} \tilde{\mathbf{\Lambda}}^{-1} \mathbf{h}, \tag{6}$$

where $\eta = \left(\mathbf{h}^H \tilde{\mathbf{\Lambda}}^{-2} \mathbf{h}\right)/K$. Nevertheless, according to (6), OCB is a NLCSI-based design since the computation of its beamforming weight $[\mathbf{w}_O]_k$ at the k-th node depends on information unavailable locally, namely $[\mathbf{g}]_k$, $k = 1, \dots, K$ and $[\mathbf{f}]_k$, $k = 1, \dots, K$ as well as $P_{\max}/K$ and $\sigma_n^2/P_{\max}$. In order to implement $\mathbf{w}_O$ in the considered WSN, each node should then estimate its backward channel and broadcast it over the network along with its forward channel. This process results in an undesired overhead which becomes prohibitive especially for large K and/or high backward channel's Doppler, resulting thereby in substantial throughput losses [24]. Therefore, OCB is unsuitable for implementation in WSNs, unless relatively exhaustive overhead exchange over the air were acceptable or if $\mathbf{w}_O$ were to be implemented in conventional beamforming, i.e., over a unique physical terminal that connects to a K-dimensional distributed antenna system (DAS).

4   Proposed DCB Implementation

In order to reduce the excessively large implementation overhead incurred by the NLCSI-based OCB, we resort to substitute η with a quantity that could be locally computed by all nodes at a negligible overhead cost. This quantity must also well-approximate η to preserve the optimality of the solution in (6). In this paper, we propose to use $\eta_D = \lim_{K \to \infty} \eta$ in lieu of η. First, we show that

$$\eta = \frac{1}{K} \sum_{k=1}^{K} \frac{|[\mathbf{f}]_k|^2}{\left(|[\mathbf{f}]_k|^2 + \beta\right)^2} \sum_{l=1}^{L} \sum_{m=1}^{L} \alpha_l \alpha_m^* e^{j4\pi \sin\left(\frac{\theta_l - \theta_m}{2}\right) z_k},$$

(7)

where $z_k = (r_k/\lambda) \sin\left((\theta_l + \theta_m)/2 - \psi_k\right)$. Using the strong law of large numbers and the fact that r_k, ψ_k and $[\mathbf{f}]_k$ are all mutually statistically independent, we have

$$\eta_D = \lim_{K \to \infty} \eta \xrightarrow{\text{p1}} \rho_1 \sum_{l=1}^{L} \sum_{m=1}^{L} \alpha_l \alpha_m^* \Delta \left(\theta_l - \theta_m\right),$$

(8)

where $\rho_1 = \mathrm{E}\left\{|[\mathbf{f}]_k|^2 / \left(|[\mathbf{f}]_k|^2 + \beta\right)^2\right\} = -(1 + \beta)e^\beta \mathrm{Ei}(-\beta) - 1$ and $\Delta(\phi) = \mathrm{E}\left\{e^{j4\pi \sin(\phi/2) z_k}\right\}$. Note that to derive the closed-form expression of $\Delta(\phi)$, the z_k's pdf $f_{z_k}(z)$ which is closely related to the nodes' spatial distribution is required. In this paper, we are only concerned by the main distributions frequently used in the context of collaborative beamforming that are: Uniform distribution and Gaussian distribution. It can be shown that [1,2]

$$f_{z_k} = \begin{cases} \frac{2\lambda}{R\pi}\sqrt{1 - \left(\frac{\lambda}{R} z\right)^2}, & -\frac{R}{\lambda} \le z \le \frac{R}{\lambda} \quad \text{Uniform} \\ \frac{\lambda}{\sqrt{2\pi}\sigma} e^{-\frac{(\lambda z)^2}{2\sigma^2}}, & -\infty \le z \le \infty \quad \text{Gaussian} \end{cases},$$

(9)

where σ^2 random variables corresponding to the nodes' cartesian coordinates. Using (9) we obtain

$$\Delta(\phi) = \begin{cases} 2\frac{J_1\left(4\pi \frac{R}{\lambda} \sin(\phi/2)\right)}{4\pi \frac{R}{\lambda} \sin(\phi/2)}, & \phi \ne 0 \quad \text{Uniform} \\ 1, & \phi = 0 \\ e^{-8\left(\pi \frac{\sigma}{\lambda} \sin(\phi/2)\right)^2}, & \text{Gaussian} \end{cases}.$$

(10)

Substituting η with η_D in (6), we introduce

$$\mathbf{w}_P = \left(\frac{P_{\max}}{K(1 + \sigma_v^2)\eta_D}\right)^{\frac{1}{2}} \tilde{\mathbf{\Lambda}}^{-1}\mathbf{h},$$

(11)

the beamforming vector of our proposed DCB. From (11), in contrast with $[\mathbf{w}_O]_k$, the k-th node's beamforming weight $[\mathbf{w}_P]_k$ solely depends on the forward and backward channels $[\mathbf{f}]_k$ and $[\mathbf{g}]_k$, respectively, which can be locally estimated. Therefore, according to (11), the proposed beamformer is a LCSI-based design that requires only a negligible overhead that does not grow neither with K nor with the Doppler, namely $P_{\max}/K$, $\sigma_n^2/P_{\max}$, and R or σ depending on the nodes' spatial distribution. Consequently, the proposed LCSI-based DCB is much more suitable for a distributed implementation over WSN than its NLCSI-based OCB counterpart. Furthermore, we will prove in the sequel that it performs nearly as well as the latter even for a relatively small number of nodes. We will also compare it with two other LCSI-based DCB benchmarks, namely M-DCB and the recently developed B-DCB. The former's design ignores scattering and assumes a monochromatic channel and, hence, its CB solution reduces from (11) to

$$\mathbf{w}_{\mathrm{M}} = \left(\frac{P_{\max}}{K\left(1 + \sigma_v^2\right)\rho_1}\right)^{\frac{1}{2}} \tilde{\mathbf{\Lambda}}^{-1}\mathbf{a}(0), \tag{12}$$

where $\mathbf{a}(\phi) \triangleq [[\mathbf{a}(\theta)]_1 \dots [\mathbf{a}(\theta)]_K]^T$ with $[\mathbf{a}(\theta)]_k = [\mathbf{f}]_k e^{-j(2\pi/\lambda)r_k \cos(\theta - \psi_k)}$. In turn, the B-DCB design whose CB solution is given by

$$\mathbf{w}_{\mathrm{BD}} = \left(\frac{P_{\max}}{K\left(1 + \sigma_v^2\right)\rho_1}\right)^{\frac{1}{2}} \frac{\tilde{\mathbf{\Lambda}}^{-1}\left(\mathbf{a}(\sigma_\theta) + \mathbf{a}(-\sigma_\theta)\right)}{\left(1 + \Delta\left(2\sigma_\theta\right)\right)}, \tag{13}$$

relies on a polychromatic channel's approximation by two chromatics at $\pm\sigma_\theta$ when the latter is relatively small. Note that from (12) and (13) both $\mathbf{w}_{\mathrm{M}}$ and $\mathbf{w}_{\mathrm{BD}}$ depends on the information commonly available at each node and, hence, are also suitable for a distributed implementation in WSNs.

5   Performance Analysis of the Proposed DCB

Let $\bar{\xi}_{\mathbf{w}} = \mathrm{E}\{P_{\mathbf{w},s}/P_{\mathbf{w},n}\}$ be the achieved average SNR (ASNR) using the CB vector $\mathbf{w}$. Note that the expectation is taken with respect to r_k, ψ_k and $[\mathbf{f}]_k$ for $k = 1, \dots, K$ and α_l and θ_l for $l = 1, \dots, L$. Since to the best of our knowledge, $\bar{\xi}_{\mathbf{w}}$ for $\mathbf{w} \in \{\mathbf{w}_{\mathrm{P}}, \mathbf{w}_{\mathrm{O}}, \mathbf{w}_{\mathrm{M}}\}$ is untractable in closed-form thereby hampering a its study rigorously, we propose to adopt the average-signal-to-average-noise ratio (ASANR) $\tilde{\xi}_{\mathbf{w}} = \mathrm{E}\{P_{\mathbf{w},s}\}/\mathrm{E}\{P_{\mathbf{w},n}\}$ as a performance measure instead to gauge the proposed DCB against its benchmarks.

5.1   Proposed DCB vs M-DCB

Following derivation steps similar to those in [22, Appendix A] and exploiting the fact that, according to A1, we have

$$\mathrm{E}\{\alpha_l^*\alpha_m\} = \begin{cases} 0 & l \neq m \\ \frac{1}{L} & l = m \end{cases}, \tag{14}$$

we obtain $\mathrm{E}\{P_{\mathbf{w}_{\mathrm{P}},s}\} = \frac{P_{\max}}{(1+\sigma_v^2)\rho_1}\left(\rho_2 + (K-1)\rho_3^2\right)$ where $\rho_2 = \mathrm{E}\{|[\mathbf{f}]_k|^4/(|[\mathbf{f}]_k|^2 + \beta)^2\} = 1 + \beta + \beta(2 + \beta)e^\beta\mathrm{Ei}(-\beta)$ and $\rho_3 = \mathrm{E}\{|[\mathbf{f}]_k|^2/\left(|[\mathbf{f}]_k|^2 + \beta\right)\} = 1 + \beta e^\beta\mathrm{Ei}(-\beta)$. Furthermore, to derive $\mathrm{E}\{P_{\mathbf{w}_{\mathrm{P}},n}\}$, one must first take the expectation only over the r_ks, ψ_ks and $[\mathbf{f}]_k$s yielding to

$$E\left\{P_{\mathbf{w}_P,n}\right\} = E_{\alpha_l,\theta_l}\left\{\frac{\sigma_v^2 P_{\max}\rho_2 \sum_{l,m=1}^{L} \alpha_l \alpha_m^* \Delta\left(\theta_l - \theta_m\right)}{\left(1+\sigma_v^2\right)\eta_D}\right\} + \sigma_n^2$$

$$= \sigma_v^2 \frac{P_{\max}\rho_2}{\left(1+\sigma_v^2\right)\rho_1} + \sigma_n^2. \tag{15}$$

It directly follows from the latter results that the achieved ASANR using the proposed DCB is

$$\tilde{\xi}_{\mathbf{w}_P} = \frac{\rho_2 + (K-1)\rho_3^2}{\sigma_v^2\left(\rho_2 + \beta\rho_1\right)}. \tag{16}$$

As can be observed from (16), $\tilde{\xi}_{\mathbf{w}_P}$ linearly increases with the number of nodes K. More importantly, from the latter result, $\tilde{\xi}_{\mathbf{w}_P}$ does not depend on the AS σ_θ meaning that the proposed DCB's performance is not affected by the scattering phenomenon even in highly-scattered environments where σ_θ is large.

Now, let us focus on the achieved ASANR $\tilde{\xi}_{\mathbf{w}_M}$ using M-DCB. Following the same approach above, one can prove that

$$\tilde{\xi}_{\mathbf{w}_M} = \frac{\rho_2 + (K-1)\rho_3^2 \int_\Theta p(\theta)\Delta^2\left(\theta\right)d\theta}{\sigma_v^2\left(\rho_2 + \beta\rho_1\right)}, \tag{17}$$

where Θ is the span of the pdf $p(\theta)$ over which the integral is calculated[1]. Since $\Delta\left(0\right) = 1$ regardless of the nodes spatial distribution, it follows from (16) and (17) that when there is no scattering (i.e., $\sigma_\theta = 0$), $\tilde{\xi}_{\mathbf{w}_M} = \tilde{\xi}_{\mathbf{w}_P}$. In such a case, indeed, $\mathbf{w}_P = \mathbf{w}_M \sum_{l=1} \alpha_l / \sqrt{\sum_{l=1} \alpha_l \sum_{m=1} \alpha_m^*}$ and, hence, $P_{\mathbf{w}_P,s} = P_{\mathbf{w}_M,s} \sum_{l=1} \alpha_l \sum_{m=1} \alpha_m^*$. Since according to (14) $E\{\sum_{l=1} \alpha_l \sum_{m=1} \alpha_m^*\} = 1$, we have $E\{P_{\mathbf{w}_P,s}\} = E\{P_{\mathbf{w}_M,s}\}$. Furthermore, it is straightforward to show that $P_{\mathbf{w}_P,n} = P_{\mathbf{w}_M,n}$ when $\sigma_\theta = 0$ and, therefore, M-DCB achieves the same ASANR as the proposed DCB when there is no scattering. This is in fact expected since the assumption of monochromatic channel made when designing the monochromatic solution is valid in such a case. Nevertheless, assuming that the nodes's spatial distribution and the scattering distribution $p(\theta)$ are both Uniform, it can be shown for relatively small AS that [27]

$$\tilde{\xi}_{\mathbf{w}_M} \simeq \frac{\rho_2 + (K-1)\rho_3^2 {}_3F_4\left(\frac{1}{2},2,\frac{3}{2};\frac{3}{2},2,2,3,-12\pi^2\left(\frac{R}{\lambda}\right)^2\sigma_\theta^2\right)}{\sigma_v^2\left(\rho_2 + \beta\rho_1\right)}, \tag{18}$$

where ${}_3F_4\left(\frac{1}{2},2,\frac{3}{2};\frac{3}{2},2,2,3,-12\pi^2(R/\lambda)^2 x^2\right)$ is a decreasing function of x whose peak is reached at 0 known as hypergeometric function. It can be inferred from (18), that the ASANR achieved by the M-DCB decreases when the AS σ_θ and/or R/λ increases. This is in contrast with the proposed DCB whose ASANR remains constant for any σ_θ and R/λ. Therefore, the proposed DCB is more robust against scattering than M-DCB whose design ignores the presence of scattering thereby resulting in a channel mismatch that causes severe ASANR deterioration.

---

[1] In the Gaussian and Uniform distribution cases, $\Theta = [-\inf,+\inf]$ and $\Theta = [-\sqrt{3}\sigma_\theta,+\sqrt{3}\sigma_\theta]$, respectively.

5.2   Proposed DCB vs B-DCB

It can be shown that the achieved ASANR using B-DCB is given by [22]

$$\tilde{\xi}_{\mathbf{w}_{BD}} = \frac{2\rho_2 + \frac{(K-1)\rho_3^2}{(1+\Delta(2\sigma_\theta))} \int_\Theta p(\theta)\left(\Delta(\theta+\sigma_\theta) + \Delta(\theta-\sigma_\theta)\right)^2 d\theta}{\sigma_v^2\left(\rho_2+\beta\rho_1\right)\left(1+\Delta(2\sigma_\theta)\right)}. \tag{19}$$

It follows from (19) that when there is no scattering (i.e., $\sigma_\theta = 0$), $\tilde{\xi}_{\mathbf{w}_{BD}}$ boils down as expected to its maximum level $\tilde{\xi}_{\mathbf{w}_P}$, regardless the nodes spatial distribution. Since, as has been shown in [22,23], B-DCB is able to achieve its maximum ASANR level for small to moderate AS values such as in lightly- to moderately-scattered environments, it turns out that the proposed DCB and its B-DCB counterpart achieve the same ASANR in such environments. However, when σ_θ is large such as in highly-scattered environments, using the fact that $\Delta(2\sigma_\theta) \simeq 0$ for large σ_θ, one can easily show that $\lim_{K\to\infty} \tilde{\xi}_{\mathbf{w}_{BD}}/\tilde{\xi}_{\mathbf{w}_P} = \int_\Theta p(\theta)\left(\Delta(\theta+\sigma_\theta) + \Delta(\theta-\sigma_\theta)\right)^2 d\theta$. Since the right-hand side (RHS) of the latter equality is a decreasing function of σ_θ, the ASANR gain achieved by the proposed DCB against B-DCB increases with the latter. Consequently, in highly-scattered environments where the AS is large, the proposed DCB outperforms B-DCB whose performance deteriorates due to the channel mismatch.

5.3   Proposed DCB vs OCB

As $P_{\mathbf{w}_O,s}$ and $P_{\mathbf{w}_O,n}$ are a very complicated functions of several random valuables, it turns out that it is impossible to derive the ASANR $\tilde{\xi}_{\mathbf{w}_O}$ in closed-form. However, a very interesting result could be obtained for large K. Indeed, one can show that

$$\lim_{K\to\infty}\frac{\tilde{\xi}_{\mathbf{w}_O}}{\tilde{\xi}_{\mathbf{w}_P}} = \frac{(\rho_2+\beta\rho_1)\,\mathrm{E}\left\{\frac{1}{\eta_D}\left(\lim_{K\to\infty}\frac{\mathbf{h}^H\tilde{\mathbf{\Lambda}}^{-1}\mathbf{h}}{K}\right)^2\right\}}{\rho_3^2\left(\mathrm{E}\left\{\frac{1}{\eta_D}\lim_{K\to\infty}\frac{\mathbf{h}^H\tilde{\mathbf{\Lambda}}^{-1}\mathbf{\Lambda}\tilde{\mathbf{\Lambda}}^{-1}\mathbf{h}}{K}\right\}+\beta\right)}$$
$$\xrightarrow{p1}\frac{\frac{(\rho_2+\beta\rho_1)}{\rho_1}\mathrm{E}\left\{\left(\sum_{l,m=1}^L \alpha_l\alpha_m^*\Delta(\theta_l-\theta_m)\right)\right\}}{\frac{\rho_2}{\rho_1}+\beta} = 1, \tag{20}$$

where the third line exploits (14) while the second exploits the law of large numbers by which we can prove that $\lim_{K\to\infty}\mathbf{h}^H\tilde{\mathbf{\Lambda}}^{-1}\mathbf{h}/K = \rho_3\sum_{l,m=1}^L \alpha_l\alpha_m^*\Delta(\theta_l-\theta_m)$ and $\lim_{K\to\infty}\mathbf{h}^H\tilde{\mathbf{\Lambda}}^{-1}\mathbf{\Lambda}\tilde{\mathbf{\Lambda}}^{-1}\mathbf{h}/K = \rho_2\sum_{l,m=1}^L \alpha_l\alpha_m^*\Delta(\theta_l-\theta_m)$. For large K, the latter result proves that the proposed LCSI-based DCB is able to achieve the same ASANR as the NLCSI-based OCB and, therefore, is able to reach optimality for any AS value. This further proves the efficiency of the proposed DCB.

Using the same method as in (20), one can easily show that $\lim_{K\to\infty}\tilde{\xi}_{\mathbf{w}}/\tilde{\xi}_{\mathbf{w}} \xrightarrow{p1} 1$ for $\mathbf{w}\in\{\mathbf{w}_P,\mathbf{w}_O,\mathbf{w}_M\}$. Therefore, all the above results hold also for the ASNR as K grows large.

6   Simulation Results

All the empirical average quantities, in this section, are obtained by averaging over 10^6 random realizations of all random variables. In all simulations, the number of rays or chromatics is $L = 10$ and the noises' powers σ_n^2 and σ_v^2 are 10 dB below the source transmit power $p_s = 1$ power unit on a relative scale. We also assume that the scattering distribution is uniform (i.e., $p(\theta) = 1/(2\sqrt{3}\sigma_\theta)$) and that α_ls are circular Gaussian random variables. For fair comparisons between the Uniform and Gaussian spatial distributions, we choose $\sigma = R/3$ to guarantee in the Gaussian distribution case that more than 99% of nodes are located in $D(O, R)$.

Figure 2 plots the empirical ASNRs and ASANRs achieved by $\mathbf{w} \in \{\mathbf{w}_O, \mathbf{w}_P, \mathbf{w}_M\}$ as well as the analytical ASANRs achieved by $\mathbf{w}_P$ and $\mathbf{w}_M$ versus K for $\sigma_\theta = 20$ (deg) and $R/\lambda = 1, 4$. The nodes' spatial distribution is assumed to be Uniform in Fig. 2(a) and Gaussian in Fig. 2(b). From these figures, we confirm that the analytical $\tilde{\xi}_{\mathbf{w}_P}$ and $\tilde{\xi}_{\mathbf{w}_M}$ match perfectly their empirical counterparts. As can be observed from these figures, the proposed DCB outperforms M-DCB in terms of achieved ASANR. Furthermore, the ASANR gain achieved using the proposed DCB instead of the latter substantially increases when R/λ grows large. Moreover, from Figs. 2(a) and (b), the achieved ASANR using the proposed LCSI-based DCB fits perfectly with that achieved using NLCSI-based OCB, which is unsuitable for a distributed implementation in WSNs, when K is in the range of 20 while it looses only a fraction of a dB when K is in the range of 5. This proves that the proposed DCB is able to reach optimality when K is large enough. It can be also verified from these figures that $\tilde{\xi}_{\mathbf{w}_P}$ and $\tilde{\xi}_{\mathbf{w}_B}$ perfectly match $\bar{\xi}_{\mathbf{w}_P}$ and $\bar{\xi}_{\mathbf{w}_M}$, respectively, for $K = 20$. All these observations corroborate the theoretical results obtained in Sect. 5.

Figure 3 displays the empirical ASNRs and ASANRs achieved by $\mathbf{w} \in \{\mathbf{w}_O, \mathbf{w}_{BD}, \mathbf{w}_P, \mathbf{w}_M\}$ as well as the analytical ASANRs achieved by $\mathbf{w}_P$ and $\mathbf{w}_M$ versus the AS for $K = 20$ and $R/\lambda = 1$. It can be observed from this figure that the ASANR achieved by M-DCB decreases with the AS while that achieved by the proposed beamformer remains constant. This corroborates again the theoretical results obtained in Sect. 5. Furthermore, we observe from Fig. 3 that B-DCB achieves the same ASNR as the proposed DCB when the AS is relatively small such as in lightly- to moderately-scattered environments. Nevertheless, in highly-scattered environments where the AS is large (i.e., $\sigma_\theta \geq 20$ deg), the proposed DCB outperforms B-DCB whose performance further deteriorates as σ_θ grows large. This is expected since the two-ray channel approximation made when designing B-DCB is only valid for small σ_θ. Moreover, it can be noticed from Figs. 3(a) and (b), that the ASNR gain achieved using the proposed DCB instead of M-DCB and B-DCB can reach until about 6.5 (dB) and 4 (dB), respectively. From these figures, we also observe that the curves of $\bar{\xi}_{\mathbf{w}_P}$ and $\bar{\xi}_{\mathbf{w}_O}$ are indistinguishable. As pointed out above, this is due to the fact that both OCB and the proposed DCB constantly reach optimality.

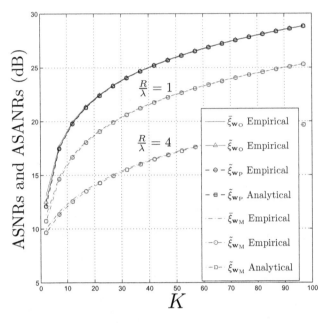

(a) Uniform distribution

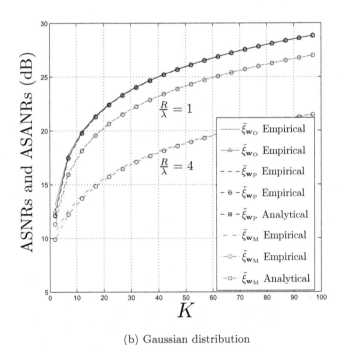

(b) Gaussian distribution

Fig. 2. The empirical ASNRs and ASANRs achieved by $\mathbf{w} \in \{\mathbf{w}_O, \mathbf{w}_P, \mathbf{w}_M\}$ as well as the analytical ASANRs achieved by $\mathbf{w}_P$ and $\mathbf{w}_M$ versus K for $\sigma_\theta = 20$ (deg) and $R/\lambda = 1, 4$ when the nodes' spatial distribution is (a): Uniform and (b): Gaussian.

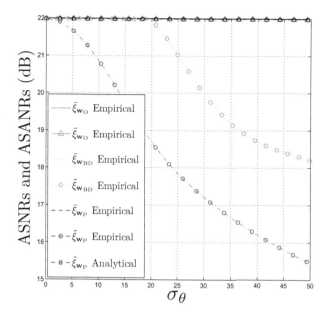

(a) Uniform distribution

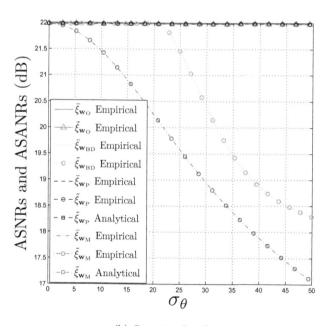

(b) Gaussian distribution

Fig. 3. The empirical ASNRs and ASANRs achieved by $\mathbf{w} \in \{\mathbf{w}_O, \mathbf{w}_{BD}, \mathbf{w}_P, \mathbf{w}_M\}$ as well as the analytical ASANRs achieved by $\mathbf{w}_P$ and $\mathbf{w}_M$ versus σ_θ for $K = 20$ and $R/\lambda = 1$ when the nodes' spatial distribution is (a): Uniform and (b): Gaussian.

7 Conclusion

In this paper, we considered a power-constrained SNR-optimal CB design that achieves a dual-hop communication from a source to a receiver, through a WSN comprised of K independent and autonomous sensor nodes. We verified that the direct implementation of this CB design is NLCSI-based. Exploiting, the polychromatic structure of scattered channels, we proposed a novel LCSI-based DCB implementation that requires a minimum overhead cost and, further, performs nearly as well as its NLCSI-based OCB counterpart. Furthermore, we proved that the proposed DCB implementation always outperforms M-DCB which is designed without accounting for scattering and that it is more robust to scattering than B-DCB whose performance substantially deteriorates in highly-scattered environments.

References

1. Ochiai, H., Mitran, P., Poor, H.V., Tarokh, V.: Collaborative beamforming for distributed wireless ad hoc sensor networks. IEEE Trans. Sign. Process. **53**, 4110–4124 (2005)
2. Ahmed, M.F.A., Vorobyov, S.A.: Collaborative beamforming for wireless sensor networks with Gaussian distributed sensor nodes. IEEE Trans. Wirel. Commun. **8**, 638–643 (2009)
3. Huang, J., Wang, P., Wan, Q.: Collaborative beamforming for wireless sensor networks with arbitrary distributed sensors. IEEE Commun. Lett. **16**, 1118–1120 (2012)
4. Zarifi, K., Ghrayeb, A., Affes, S.: Distributed beamforming for wireless sensor networks with improved graph connectivity and energy efficiency. IEEE Trans. Sign. Process. **58**, 1904–1921 (2010)
5. Ahmed, M.F.A., Vorobyov, S.A.: Sidelobe control in collaborative beamforming via node selection. IEEE Trans. Sign. Process. **58**, 6168–6180 (2010)
6. Mudumbai, R., Barriac, G., Madhow, U.: On the feasibility of distributed beamforming in wireless networks. IEEE Trans. Wirel. Commun. **6**, 1754–1763 (2007)
7. Mudumbai, R., Brown, D.R., Madhow, U., Poor, H.V.: Distributed transmit beamforming: challenges and recent progress. IEEE Commun. Mag. **47**, 102–110 (2009)
8. Han, Z., Poor, H.V.: Lifetime improvement in wireless sensor networks via collaborative beamforming and cooperative transmission. IET Microw. Antennas Propag. **1**, 1103–1110 (2007)
9. Dong, L., Petropulu, A.P., Poor, H.V.: A cross-layer approach to collaborative beamforming for wireless ad hoc networks. IEEE Trans. Sign. Process. **56**, 2981–2993 (2008)
10. Godara, L.C.: Application of antenna arrays to mobile communications, part II: Beam-forming and direction-of-arrival considerations. Proc. IEEE **85**, 1195–1245 (1997)
11. Zarifi, K., Zaidi, S., Affes, S., Ghrayeb, A.: A distributed amplify-and-forward beamforming technique in wireless sensor networks. IEEE Trans. Sign. Process. **59**, 3657–3674 (2011)
12. Zarifi, K., Affes, S., Ghrayeb, A.: Collaborative null-steering beamforming for uniformly distrubuted wireless sensor networks. IEEE Trans. Sign. Process. **58**, 1889–1903 (2010)

13. Astly, D., Ottersten, B.: The effects of local scattering on direction of arrival estimation with MUSIC. IEEE Trans. Sign. Process. **47**, 3220–3234 (1999)
14. Shahbazpanahi, S., Valaee, S., Gershman, A.B.: A covariance fitting approach to parametric localization of multiple incoherently distributed sources. IEEE Trans. Sign. Process. **52**, 592–600 (2004)
15. Souden, M., Affes, S., Benesty, J.: A two-stage approach to estimate the angles of arrival and the angular spreads of locally scattered sources. IEEE Trans. Sign. Process. **56**, 1968–1983 (2008)
16. Bengtsson, M., Ottersten, B.: Low-complexity estimators for distributed sources. IEEE Trans. Sign. Process. **48**, 2185–2194 (2000)
17. Besson, O., Stoica, P., Gershman, A.B.: Simple and accurate direction of arrival estimator in the case of imperfect spatial coherence. IEEE Trans. Sign. Process. **49**, 730–737 (2001)
18. Amar, A.: The effect of local scattering on the gain and beamwidth of a collaborative beampattern for wireless sensor networks. IEEE Trans. Wirel. Commun. **9**, 2730–2736 (2010)
19. Zaidi, S., Affes, S.: Distributed beamforming for wireless sensor networks in local scattering environments. In: Proceedings of IEEE VTC 2012-Fall, Québec City, Canada, 3–6 September (2012)
20. Zaidi, S., Affes, S.: Distributed collaborative beamforming with minimum overhead for local scattering environments. In: Proceedings of IEEE IWCMC 2012, Cyprus, 27–31 August (2012, Invited Paper)
21. Zaidi, S., Affes, S.: Spectrum-efficient distributed collaborative beamforming in the presence of local scattering and interference. In: Proceedings of IEEE GLOBECOM 2012, Anaheim, CA, USA, 3–7 December (2012)
22. Zaidi, S., Affes, S.: Distributed collaborative beamforming in the presence of angular scattering. IEEE Trans. Commun. **62**, 1668–1680 (2014)
23. Zaidi, S., Affes, S.: Distributed collaborative beamforming design for maximized throughput in interfered and scattered environments. IEEE Trans. Commun. **63**, 4905–4919 (2015)
24. Zaidi, S., Affes, S.: SNR and throughput analysis of distributed collaborative beamforming in locally-scattered environments. Wireless Commun. Mobile Comput. **12**, 1620–1633 (2012, Invited Paper). Wiley Journal
25. Zaidi, S., Affes, S.: Analysis of collaborative beamforming designs in real-world environments. In: Proceedings of IEEE WCNC 2013, Shanghai, China, 7–10 April (2013)
26. Havary-Nassab, V., Shahbazpanahi, S., Grami, A., Luo, Z.-Q.: Distributed beamforming for relay networks based on second-order statistics of the channel state information. IEEE Trans. Signal Process. **56**, 4306–4316 (2008)
27. Zaidi, S., Hmidet, B., Affes, S.: Power-constrained distributed implementation of SNR-optimal collaborative beamforming in highly-scattered environments. IEEE Wirel. Commun. Lett. **4**, 457–460 (2015)

A Gateway Prototype for Coalition Tactical MANETs

Mazda Salmanian[1(✉)], William Pase[2], J. David Brown[1],
and Chris McKenzie[3]

[1] Defence Research and Development Canada, Ottawa, Canada
{mazda.salmanian,david.brown}@drdc-rddc.gc.ca
[2] Armacode Inc., Ottawa, Canada
bill@armacode.com
[3] MIC, Ottawa, Canada
chris@mckenzieic.com

Abstract. Mobile ad hoc networks (MANETs) are well suited for tactical groups whose operations require that the network adapt to dynamic topology changes without the aid of centralized infrastructures. As more coalition forces deploy in tactical operations, their networks require inter-connectivity for sharing broadcast, multicast and unicast traffic from coalition applications such as situational awareness and sensor data. Inter-MANET connectivity, however, should not be at the cost of compromising a national MANET's sovereignty in terms of radio devices, subnet address space, and communication and routing protocols. In this paper, we describe our implementation of a gateway application that enables coalition tactical MANETs to inter-connect based on role names instead of IP addresses while keeping their national radios and networking sovereign and while protecting their private network addresses. We exhibit results and learned lessons from our laboratory experiments where we tested our gateway application in several MANET formations to ensure the functionality of relay connectivity, domain name service and network address translation features. The gateway application has potential to serve as a prototype for the future development of secure interoperability policies, service level agreements and standards at the tactical edge, e.g., for future NATO standardization agreements (STANAGs).

Keywords: MANET · Gateway · Interoperability · Domain name service · Network address translation · Wireless · Mobile ad hoc network Coalition networking

1 Introduction

As more national tactical forces are deployed in international coalition operations, they are expected to be interoperable to share basic application traffic such as situational awareness and sensor data. Future national forces are expected to take advantage of mobile ad hoc networking (MANET) technology, which adapts network topologies to tactical operations and enables dispersed nodes to pervasively inter-connect. The self-configuration benefits of MANET technology come from routing protocols that are

© ICST Institute for Computer Sciences, Social Informatics and Telecommunications Engineering 2017
Y. Zhou and T. Kunz (Eds.): ADHOCNETS 2016, LNICST 184, pp. 330–341, 2017.
DOI: 10.1007/978-3-319-51204-4_27

used in conjunction with wireless radios and networking protocols. In a coalition deployment, national forces deploy with their own tactical radios and networking protocols and require gateways to share common application data with other nations who may not be equipped with the same radios and protocols. Currently, NATO's (North Atlantic Treaty Organization) protected core networking (PCN) concept relies on gateways in strategic enterprise networks [1] to facilitate quality of service (QoS) and policy negotiations; however, its approach is not yet relevant at the tactical edge due to the dynamic inter-networking challenges of MANETs. The authors in [2] present an architecture for secure interoperability between coalition tactical MANETs that promotes keeping national tactical radios and networking protocols sovereign. The architecture includes a gateway node designed for national MANETs that inter-connects with peer coalition gateways for secure information exchange.

In this paper, we describe such a prototype gateway application constructed on an Android device using two IEEE 802.11 (WiFi) radios, one internal and one external to the Android device. The gateway application is designed to relay the traffic between the two radios, connecting intra-MANET traffic flows on a national radio to inter-MANET traffic shared between the gateways on the coalition radio, e.g., NATO narrowband waveform (NBWF) [3–5]. In other words, each nation's MANET connects to another nation's MANET by its own gateway and without changing its service set identification (SSID) or subnet address. The implementation also includes domain name service (DNS) functions that enable role-based connectivity between national MANET nodes, simplifying the prerequisite requirement of possessing the internet protocol (IP) addresses of a destination node before information sharing. In addition, the gateway application protects the private network addresses of a national MANET by mapping them to designated public network addresses via network address translation (NAT).

We present results from our laboratory experiments where we tested our gateway application in several MANET formations with WiFi radios. We demonstrate that our prototype application inter-connects autonomous MANETs using sovereign radios and networking protocols, including frequency assignments, routing protocols and network addresses.

The rest of the paper is organized as follows. We present the basic inter-connectivity and networking of MANETs under several scenarios in Sect. 2 where we recommend a frequency assignment scheme for gateway radios given our measured metrics. In Sect. 3, we present results and learned lessons from our DNS testing trials. We present similar results in Sect. 4 for testing the NAT function. We provide a summary in Sect. 5.

2   Basic Connectivity

In this section, we describe experiments we conducted to examine the inter-connectivity and networking of MANETs across gateways with three scenarios. In the first scenario, a MANET from nation A (MANET-A) and a MANET from nation B (MANET-B) inter-connect using gateways. In the second scenario, MANET-B connects MANET-A and MANET-C (from nation C), acting as a relay network between nations A and C. In the third scenario, the MANETs of the three nations inter-connect using their own local gateways while the gateways form a MANET of their own.

As mentioned earlier, the gateway application on a node routes the traffic between the two radios on the gateway: one used for inter-MANET connections and one for intra-MANET traffic. We use the internal WiFi radio of the Android device (with the gateway application) to connect to the local intra-MANET side of the gateway. We enable the gateway device to also drive an external radio tuned to a shared gateway channel that connects to the inter-MANET side of the gateway. The operation of the second radio is made possible by a customized Android operating system, Cyanogenmod 12.1 with a modified Kali Linux Nethunter kernel made for Nexus 5 (phone) and Nexus 7 (tablet) devices [6]. The external radio used throughout our experiments is the TP-Link^{TM} [7].

To simplify the experiments, we assign different radio frequency (RF) channels to represent different nations; the national subnet address and the MANET's SSID are other distinguishing features. We use the non-overlapping North American channel selection i.e., channels 1, 6 and 11 in accordance with [8].

Our experiments were performed in a laboratory under controlled static conditions. Mobility trials are planned for future work. The MANETs use the optimized link state routing protocol (OLSR) [9] for multi-hop connectivity. To enforce multi-hop connections and to avoid cross-talk between co-channel non-adjacent nodes, we used a separate customized application that generates 'iptables' commands and creates firewall rules. These rules permit testing of topologies that would otherwise require physically separating devices to restrict connectivity.

We used an iPerf [10] application to generate TCP and UDP (transmission control and user datagram protocols) traffic for the experiments and measured throughput to evaluate the effects of network changes in the scenarios.

Given an SSID, a common WiFi channel (via TP-Link) and a shared subnet address, the gateways discover one another as nodes of a MANET using the OLSR signaling. While gateway assignment is pre-determined in our experiments, future MANETs could use algorithms such as [11] to dynamically assign the gateway role to a node.

A gateway's primary task is to relay traffic from one of its radios to another, serving two SSIDs and subnets. The relay function in our gateway application is implemented by configuration settings of 'iptables' commands and OLSR routing table instructions. The gateway nodes are equipped with two OLSR instances, one for each RF interface; there is no cross-talk between the two OLSR instances.

Scenario 1. In this first scenario, depicted in Fig. 1, commander of nation A (CMDR-A) connects to commander of nation B (CMDR-B) under several cases, detailed in Table 1. In this stage of the experiment (basic connectivity) we assume that the commanders know one another's IP addresses. As shown in the figure, MANET-A is on subnet 192.168.11.xx and MANET-B is on subnet 192.168.12.xx.

The two phones in Fig. 2, which represent nodes 1 and 6 of Fig. 1, are CMDR-A and CMDR-B, respectively with addresses 192.168.11.10 and 192.168.12.11. The two tablets in Fig. 2 represent nodes 3 and 4[1] of Fig. 1 as GW-A (gateway of MANET-A) and GW-B (gateway of MANET-B), respectively, each with two radios, one internal

---

[1] Notations for nodes 2 and 5 are reserved for multi-hop configurations of this scenario.

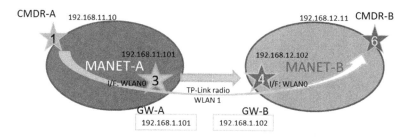

Fig. 1. The configuration of two MANETs in Scenario 1. Nodes 3 and 4 are gateways with two radios tuned according to use cases in Table 1. The internal radio interface (I/F) is denoted WLAN0, while the external radio interface is denoted WLAN1. MANETs are partitioned with three parameters: The MANET's SSID, the operating WiFi channel and the subnet address. The subnet address of the gateway MANET is 192.169.1.xx.

Table 1. Channel assignments of the radios in Scenario 1.

Use case	MANET-A	GW-A	GW-B	MANET-B	Description
0	11	11/11	11/11	11	One flat MANET, base measure
1	11	11/11	11/11	11	Two MANETs, two GWs, all @ Ch.11
2	11	11/11	11/06	06	Two MANETs, one GW @ Ch.11/06
3	01	01/11	11/06	06	Two MANETs & GWs, Ch. 01/11/06

and one TP-Link external to the tablet. The external radios for inter-gateway communication share subnet 192.168.1.x. Traffic packets are generated by the iPerf application at CMDR-A node destined for CMDR-B's IP address. The packets are sent to GW-A by default due to their unknown destination subnet address in MANET-A; every MANET node is assigned a default gateway. The packets are then forwarded to GW-B where their subnet addresses are recognized and where they are forwarded to CMDR-B.

The use cases tabulated in Table 1 differ in the RF channels assigned to the MANET radios and to the gateways. For example in use case 3, GW-A is assigned channel 01 (intra-MANET radio) and channel 11 (inter-MANET radio). In addition:

- Case 0 also differs from the other use cases in its network address assignments. In this use case we make a baseline measure of throughput from a flat (same subnet) multi-hop network. Here, we expect co-channel interference to contribute to collisions and loss of packets.
- Case 1 examines the potential overhead of the gateway application and its introduction to two MANETs separated by subnet addresses but not by radio frequency. All nodes, including the gateways' internal and external radios, are tuned to channel

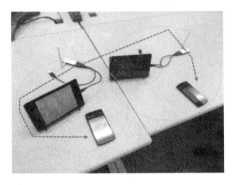

Fig. 2. The laboratory configuration of two MANETs in Scenario 1. The two MANETs are depicted in yellow ellipses. The two tablets (representing nodes 3 and 4 of Fig. 2) are gateways with two radios, one internal and one external to the tablet. The external radio is the TP-Link. (Color figure online)

11, as they were in case 0. The subnet addresses of the two MANETs are 192.168.11. xx and 192.168.12.xx, and that of the gateway MANET is 192.168.1.xx.

- Case 2 is designed to measure the effect of reduced co-channel interference. One MANET is assigned to operate on channel 6; its gateway's internal radio is also tuned to channel 6 while the rest of the nodes operate on channel 11.
- Case 3 offers the environment in which we measure potential improvements in throughput by dedicating channel 11 to inter-gateway communication. Both gateways' external radios are tuned to channel 11 while the MANETs are tuned to channels 1 and 6, respectively.

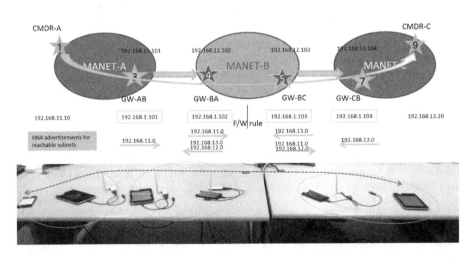

Fig. 3. The configuration of three MANETs in Scenario 2. Nodes 4 and 5 are gateways of MANET-B depicted in our laboratory experiment with a Nexus 7 tablet and a Nexus 5 phone, respectively. Their external radios are the TP-Link.

Scenario 2. In Scenario 2 (single use case), MANET-B provides connectivity between MANETs A and C so that CMDR-A establishes a multi-hop connection to CMDR-C. This scenario is depicted in Fig. 3.

Scenario 2 continues the theme of minimizing co-channel interference; MANETs A, B and C are assigned to operate on channels 1, 6 and 1 respectively, while the gateways operate on one common channel 11. This test ensures that an iPerf-generated packet destined from CMDR-A to CMDR-C is relayed correctly through the gateways in MANET-B, leaving subnet 192.168.11.x through 192.168.12.x for 192.168.13.x. MANET-B has two gateways: a firewall rule forces them to establish their peer-to-peer connection via their internal radios under subnet 192.168.12.x, leaving their TP-Links on channel 11 under subnet 192.168.1.x to serve connections to each of their neighbouring MANETs. In this scenario, the OLSR on the gateway nodes is configured (via our gateway application) to advertise host network association (HNA) messages so that subnets can be found for route discovery. The HNA messages advertise network routes (subnet IDs) in the same way topology control (TC) messages advertise host routes. We avoid (and recommend against) using multiple interface declaration (MID) messages that advertise all the internal routes to other subnets; this ensures that internal routes are kept private on a multi-national scenario.

Scenario 3. In Scenario 3, the traffic flow is from CMDR-A to CMDR-C. This scenario configuration is depicted in Fig. 4.

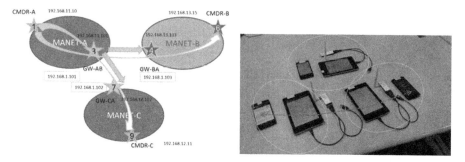

Fig. 4. The configuration of three MANETs in Scenario 3. Nodes 3, 7 and 4 are gateways that form a MANET of their own; these gateways are depicted in our laboratory experiment with Nexus 7 tablets. Their external radios are the TP-Link.

Scenario 3 is similar to Scenario 2 in that MANETs A, B and C are assigned to operate on channels 1, 6 and 1 respectively, while the gateways operate on one common channel 11. In Scenario 3, however, each MANET is assigned one gateway whereas in Scenario 2, MANET-B had two gateways.

Laboratory Results. We present our experimental results and summarize learned lessons of our three scenarios. To put the throughput results in perspective, we first present the commercial specifications of our link connections. The specifications are:

- TP-Link to the tablet/phone connection is USB 2.0 (universal serial bus) with maximum data rate up to 480 Mbps.

- Internal radios of the tablets are (802.11b/g/n) rated up to 150 Mbps.
- Internal radios of the phones are (802.11a/b/g/n/ac) rated up to 300 Mbps.
- TP-Link to TP-Link connection (802.11n) is rated up to 150 Mbps; this is the main inter-gateway connection common for interoperability in our scenarios and the relative bottleneck among the links.

The TCP and UDP throughput results are presented in Figs. 5 and 6, respectively. Each scenario (and use case) is run ten times; each time (i.e., every point on the graph) is an averaged value over a 30-second iPerf transmission. The TCP packets are transmitted at the highest possible rate for assured reception; the iPerf UDP packets are transmitted at 50 Mbps load.

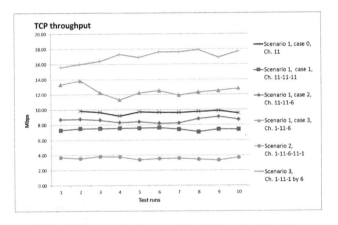

Fig. 5. TCP throughput results of Scenarios 1 to 3.

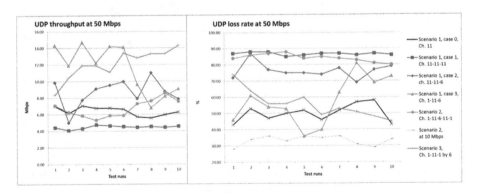

Fig. 6. UDP throughput and loss rate results of Scenarios 1 to 3.

The TCP throughput increases of Scenario 1 cases 1 to 3 are well noted and result from the reduction of co-channel interference. The throughput of Scenario 1 case 0, however, is higher than those of cases 1 and 2. The MANET in Scenario 1 case 0 is a flat merged network using channel 11; with no gateway, it presumably has the most co-channel interference among the scenarios. This throughput observation indicates that there is overhead associated with the gateway application managing a connection in a two-MANET formation partitioned by subnet addresses (Scenario 1 cases 1 and 2). We also presume that the carrier sense multiple access (CSMA) of the 802.11 radios had the opportunity to optimize access time to channel 11.

The channel transition of the iPerf traffic for Scenario 1 case 3 is 1-11-6 whereas that of Scenario 3 is 1-11-1 next to a MANET operating on channel 6. Our results show that Scenario 3 produced higher average throughput than Scenario 1 case 3. It is possible that co-channel sensing of CSMA influences transmission times such that optimum throughput performance may be achieved, e.g., by performing bulk acknowledgements.

In the UDP throughput results shown in Fig. 6, we note that the data points are not as stable as those of TCP in Fig. 5 because UDP is a connectionless protocol and does not re-transmit lost or erroneous packets. Here, the throughput of Scenario 1 case 1 is lower than Scenario 2; both have similar loss rates. This observation indicates that co-channel interference in Scenario 1 case 1 reduces the throughput of UDP packets such that it is lower than hopping over two additional gateways in Scenario 2 where there is no co-channel interference.

The HNA feature of OLSR is implemented per (radio) interface in order to enforce control over the direction of advertised subnets and to avoid circular routes. This allows us to dedicate a gateway to a neighbouring MANET – the external radio of the gateway that connects to the neighbour MANET advertises a specific subnet.

A MANET providing a connection service between two MANETs (e.g., Scenario 2) must advertise subnet addresses inside its network as well as outside. Every MANET node is assigned a default gateway; these internal HNA advertisements help the default gateway of a node route the message to the subnet of the destination address. These internal HNA messages are not needed in a MANET that has only one gateway serving multiple neighbours, such as the formation in Scenario 3.

3   Domain Name Service

Domain name service (DNS) is an Internet standard protocol made of a collection of request for comments (RFC) documents [12] published by the Internet Engineering Task Force. The messaging exchange in the protocol allows a node to address its destination by a name, instead of by the numerical representation of the IP destination address. In this work, we use the role name of a MANET node such as CMDR. Earlier, in the basic connectivity section, we had assumed that the two commanders in Fig. 1 were in possession of one another's IP addresses. Now, the IP address of the destination does not have to be known a priori at the source. With the DNS feature, CMDR-A can address CMDR-B by its role name and its domain name, e.g., cmdr.b representing the commander's role name in MANET-B domain.

The DNS protocol message flow and its trigger by CMDR-A are explained below with reference to Fig. 1. A traffic message is formed with cmdr.b as the destination address. The message is queued until DNS software resolves the 'RoleName.domain' and maps it to an IP address. A DNS signaling message is sent to GW-A, the default gateway of the source node, CMDR-A. The DNS signaling message is forwarded to

GW-B because domain '.b' is identified with GW-B in GW-A. At GW-B, the DNS software resolves the 'RoleName.domain' to the IP address mapping of CMDR-B. GW-B responds to GW-A with the IP address for CMDR-B. Then, GW-A forwards the signaling message to CMDR-A's queue where the original traffic message is stored. Subsequently, the traffic message is sent from 192.168.11.10 to 192.168.12.11.

In Fig. 7, we present a screen shot of our DNS experiment where cmdr.ca generates a 'ping' command to cmdr.uk whose address is successfully resolved to 192.168.12.11. In the experiment, the '.ca' and '.uk' represent Canada and United Kingdom domains, respectively replacing CMDR-A and CMDR-B. The DNS query (signaling message) is triggered per 'traffic flow', i.e., as needed. The result of a query is cached for a configurable time-to-live (TTL). In our Android implementation, we use a Linux DNS server (DNSmasq) and replace the Android default DNS, i.e., Google, because it requires connection to the Internet[2]. The DNS server is currently implemented on the gateway nodes; however, it can be implemented on any node in the MANET, so long as a traffic source can route DNS queries to the DNS server node.

Fig. 7. A 'ping' command from cmdr.ca (CMDR-A in Fig. 1) is translated to cmdr.uk's IP address (CMDR-B) via the DNS server implemented on GW-A and GW-B. Screen shots of our customized application on Nexus 5 phones representing CMDR-A (left) and CMDR-B (right) show the CMDRs' default gateway and DNS server addresses as 192.168.11.101 and 192.168.12.102, respectively.

Figure 7 also shows screen shots of our customized application on Nexus 5 phones representing CMDR-A (left) and CMDR-B (right) of Fig. 1. As mentioned earlier, in our customized application, every node is assigned a default gateway, and now, in this stage, a default DNS server. It is this customized application that enables a node to control its radio interfaces, act as or assign a gateway or a DNS server. The screen shots of CMDR nodes show the CMDRs' default gateway and DNS server addresses as 192.168.11.101 and 192.168.12.102, respectively. In our experiments, we successfully

---

[2] DNSmasq allows us local control to add and to resolve domain names.

performed end-to-end 'ping' and 'traceroute' commands on Scenarios 1, 2, and 3, testing the DNS function in all aforementioned MANET configurations, but for brevity we do not show their screen captures here.

4 Network Address Translation

We extend the features of our custom application to include network address translation (NAT) such that the DNS query response returns a configurable public IP address of a MANET node's RoleName (in the query) not its actual private IP address. The private IP address of the destination node is not only kept sovereign within its MANET domain, it is further protected by the NAT feature such that it is not shared outside of the MANET domain.

We have chosen to implement this feature on the gateway node which is at the edge of the MANET as the final trusted node that releases traffic packets to external domains. An outgoing traffic packet from a source node is sent with a private source IP address and a public destination IP address, acquired by the DNS protocol. When this packet arrives at the default gateway (of the source node), its private source IP address is mapped to a public source IP address. This process is referred to as a source NAT.

Analogously, an incoming traffic packet that arrives at the gateway carries public source and destination IP addresses. At the gateway, by a process referred to as a destination NAT, the packet's public destination IP address is mapped to a private destination IP address.

We present Fig. 8 as evidence of our NAT implementation. The Wireshark^{TM} flow capture of Fig. 1 configuration is shown where cmdr.ca (CMDR-A) generates a 'ping' command to cmdr.uk (CMDR-B) whose address is successfully resolved to 10.168.12.11. This address is the public address of cmdr.uk as it is in 10.x.x.x network address space, not the 192.x.x.x network address space where the private address resides. The flow capture shows the implementation of both source and destination NATs as the private addresses are protected and hidden.

Source and destination NAT techniques are employed and executed in the Linux kernel. We use the Netfilter feature of the Linux kernel to implement NAT in our application. The application generates source and destination NAT rules, S-NAT and D-NAT respectively, such that the IP Table gets reconfigured with 'iptables' commands.

It is important to note that the gateway node itself, as a member of the MANET, must subject its internal (radio) traffic to the NAT rules. It is also important to note that with our scheme, the subnet address space of the public domain becomes limited as the number of nodes increases and their addresses are mapped to a public address. In a coalition operation, the public subnet addresses must be managed and even pre-assigned to the participating nations.

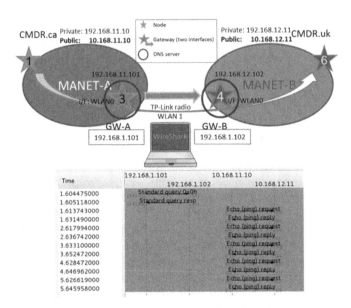

Fig. 8. A 'ping' command from cmdr.ca is translated to cmdr.uk's public IP address via the DNS server and NAT implemented on GW-A and GW-B. Screen shot of Wireshark captured messages shows the gateway addresses as 192.168.1.101 and 192.168.1.102, and the public addresses of the two commanders as 10.168.11.10 and 10.168.12.11, respectively in subnet 10.x.x.x. not 192.x.x.x.

5 Summary

Gateways provide traffic relay service between two autonomous MANETs. In this work we presented experimental results performed on three MANET scenarios. We have shown that even though a gateway is an extra relay in the MANET, there is a noticeable improvement in throughput when the frequency assignments are managed in a coalition deployment such that co-channel interference is reduced. Frequency management, however, should be weighed against the inevitable exposure to jamming as more frequency allocations (different channels) mean more targets.

We implemented the relay functionality along with DNS and NAT in a customized Android application. We implemented the gateway discovery function by taking advantage of the OLSR protocol and its HNA messages. The gateway application manages configuration settings such that MANETs can be partitioned by RF radio channels, subnet addresses and SSIDs.

From a security perspective, we showed that MANETs can operate using sovereign private subnet addresses and only share their assigned public subnet addresses. The public address space assignments in a coalition deployment, however, require careful planning to avoid public-address collisions between network gateways.

In this work, we have realized the architecture that was detailed in [2]. For brevity, we have omitted the prototyping results of the encryption strategy as they are out of scope of this work.

In a MANET, there is no "physical" way to ensure that a high-assurance mobile exchange point is connected to the perimeter of the network. Appropriate architectures and algorithms may employ multiple exchange points or allow for a single mobile exchange point to serve as the "logical" gateway between two networks. Along that theme, we are considering the following areas in our future research: policy enforcement at the tactical edge, distributed DNS and load balancing between multiple gateways. These new architectures will ensure that network connectivity between two partners will be more robust (i.e., survivable) despite mobility and changes to topology.

Acknowledgment. The authors wish to thank Ms. Susan Watson for her valuable time and suggestions for this work.

References

1. Lies, M., Dahlberg, D., Steinmetz, P., Hallingstad, G., Calvez, P.: The Protected Core Networking (PCN) Interoperability Specification (ISPEC), NATO Communications and Information Agency. Technical report 2013/SPW008905/13 (2013)
2. Salmanian, M., Brown, J.D., Watson, S., Song, R., Tang, H., Simmelink, D.: An architecture for secure interoperability between coalition tactical MANETs. In: 2015 IEEE Military Communications Conference, MILCOM 2015 (2015)
3. North Atlantic Treaty Organization (NATO), Standardization Aggrement (STANAG) 5631/AComP-5631, Narrowband Waveform Physical Layer, 1 edn. Ratification Draft (2015)
4. North Atlantic Treaty Organization (NATO), Standardization Aggrement (STANAG) 5632/AComP-5632, Narrowband Waveform Link Layer, 1 edn. Ratification Draft (2015)
5. North Atlantic Treaty Organization (NATO), Standardization Aggrement (STANAG) 5633/AComP-5633, Narrowband Waveform Network Layer, 1 edn. Ratification Draft (2015)
6. Kali NetHunter Documentation (2016). https://github.com/offensive-security/kali-nethunter/wiki
7. TP-Link. http://www.tp-link.com/en/products/details/TL-WN722N.html
8. IEEE Computer Society, IEEE Standard for information technology - telecommunications and information exchange between systems local and metropolitan area networks - specific requirements: Part 11: Wireless LAN Medium Access Control (MAC) and Physical Layer (PHY) Specifications, IEEE Std 802.11™ (2012)
9. Clausen, T., Jacquet, P.: Optimized Link State Routing Protocol (OLSR), RFC 3626, IETF (2003)
10. iPerf - The network bandwidth measurement tool (2016). https://iperf.fr/
11. Wong, S.H.Y., Chau, C.K., Lee, K.W.: Managing interoperation in multi-organization MANETs by dynamic gateway assignment. In: 12th IFIP/IEEE International Symposium on Integrated Network Management (IM 2011) and Workshops (2011)
12. Domain Name System (2016). https://en.wikipedia.org/wiki/Domain_Name_System

A Prototype Implementation of Continuous Authentication for Tactical Applications

J. David Brown[1(✉)], William Pase[2], Chris McKenzie[3],
Mazda Salmanian[1], and Helen Tang[1]

[1] Defence Research and Development Canada, Ottawa, Canada
{david.brown, mazda.salmanian,
helen.tang}@drdc-rddc.gc.ca
[2] Armacode Inc., Ottawa, Canada
bill@armacode.com
[3] MIC, Ottawa, Canada
chris@mckenzieic.com

Abstract. Recent advances in wireless and computing technology have led to accelerated efforts to equip soldiers at the tactical level with sophisticated handheld communications devices to share situational awareness data. An important consideration is how to secure these devices, and how to ensure that the users of the devices have not been compromised. This paper presents the details of prototyping activity we conducted in which two commercial biometric devices were integrated with a handheld communication device to perform continuous user authentication. We discuss the design of the prototype, its performance, and lessons learned that apply to future efforts at implementing continuous authentication in a military-focused setting.

Keywords: Continuous authentication · Mobile ad hoc networks · Biometric authentication

1 Introduction

The concept of a networked soldier has been a part of military doctrine for years [1–4]. With recent advances in commercial and military technology, in the near future dismounted soldiers will be equipped with handheld networked computing devices to provide geographic situational awareness and to provide communications and information sharing capabilities during tactical operations. To ensure the integrity and confidentiality of the data on the communications devices, some form of user authentication is required such that a user can authenticate to his or her device. Additionally, it is desired that beyond a one-time user-to-device authentication, the trust relationship between user and device can be maintained through some form of continuous user authentication. The need for this continuous authentication is especially relevant in tactical operations, where users may be operating in contested or dangerous environments and are faced with a significant risk of device loss or capture by an adversary.

© ICST Institute for Computer Sciences, Social Informatics and Telecommunications Engineering 2017
Y. Zhou and T. Kunz (Eds.): ADHOCNETS 2016, LNICST 184, pp. 342–353, 2017.
DOI: 10.1007/978-3-319-51204-4_28

In this paper, we discuss the design, implementation, and evaluation of a prototype in which we integrated two commercial biometric devices with a smartphone operating in a mobile ad hoc network (MANET). The first biometric device measured the user's electrocardiogram (ECG) reading and the second measured the user's pulse. While many of the more common authentication techniques currently being explored in the literature may have significant value for commercial applications, this paper discusses how those techniques are less applicable in a military setting. Specifically, the contribution of this paper is twofold:

(1) the paper details the military-specific constraints that place serious real-world limitations on the types of continuous authentication techniques that would be useful and feasible in a tactical edge scenario; and

(2) the paper demonstrates—with prototyping, integration, and experimentation—the feasibility of a practical method for continuous user authentication that could operate under (a subset of) the constraints imposed by a military tactical networking use case.

A variety of techniques for continuous user authentication have been proposed in the academic literature, where these are often based on monitoring one or more of a user's biometric features. Initial variants of this work considered the case of a user sitting at a desktop computer and focused on monitoring data such as keystroke timing—see, for instance [5] for an early discussion of this style of continuous authentication, or [6, 7] for more recent iterations on this theme. Exploring user mouse dynamics was seen as a promising technique for the desktop user as well, as reported in [8]. For mobile devices such as smartphones, however, keyboard and mouse techniques are not applicable, and recent studies have focused on using elements such as touch-screen interactions or leveraging the outputs from on-board gyroscopes or cameras to develop a reliable biometric. Promising results have been obtained using touch-screen interaction for continuous authentication, looking at how a user swipes, "pinches to zoom", or performs other touch gestures; these gestures are compared to a stored template as in [9] or to data learned dynamically during the session as in [10]. The possibility of using a smartphone's on-board motion sensors to identify users and/or perform continuous authentication is explored in [11], but it is clear that there is still more work to be done to make such a system robust. In [12], the authors combine motion sensing with images observed by the on-board camera in an attempt to increase the reliability of a motion-based system.

Despite the promise of the continuous authentication techniques proposed in [5–12], their applicability to a tactical military environment is limited. Keyboard- or typing-based techniques are not useful since dismounted tactical users are unlikely to have keyboards. Touch screen-based techniques have limited value since users are not expected to be interacting frequently enough with the device for this technique to ensure "continuity" (i.e., there will be large gaps in time where the user will not touch a screen); additionally, users will likely be wearing gloves, which will reduce the sensitivity of any such algorithms. Examining output from the onboard gyroscopes still requires more robustness; furthermore, in a military setting, users can be expected to move in unpredictable ways, which could complicate generating a "template" of user behaviour. The high tempo and unpredictable nature of tactical operations necessitates a non-intrusive method of performing continuous authentication.

In this work, we consider two commercial sensors worn on the user's wrist; one measures a user's ECG reading as a means of providing strong initial authentication, and the other continuously measures the user's pulse as a means of providing "continuity". The possibility of using ECG data as a means for user authentication has been a topic of study for some time. In 2001, [13] showed that ECG data collected from a 12-lead ECG on individuals at rest was sufficient to perform user identification, correctly identifying better than 45 individuals out of a group of 50. The effect of anxiety and stress on the accuracy of ECG identification was explored in [14], which showed that high-resolution ECG data provided a reliable means of user identification for individuals at rest and for those performing in high anxiety situations, with a better than 90% correct identification rate. Work by a number of researchers, including [15, 16], has focused on methods to improve the accuracy of performing user identification through ECG and has achieved success rates above 95%. Taken together, ECG and pulse measurements are a potentially attractive biometric combination since they also provide a proof-of-life indicator, which is vital in contested military environments.

The remainder of this paper is organized as follows. In Sect. 2, we discuss the implementation details of our prototype, including the devices we used, how they were integrated with a custom smartphone application (app), and how they—in combination with the app—performed continuous authentication. Section 3 details the testing and evaluation of the prototype; we show the results of experiments run to measure the average time to detect a "lost device" and the average time to detect a "compromised" user. In Sect. 4 we provide discussion and lessons learned from this exercise, including notes on how such technology could be modified in order to be better suited for a tactical operations use case. We sum up with a brief conclusion in Sect. 5.

2 Prototype Implementation

We developed a prototype implementation that provides continuous authentication functionality for the particular use case of a dismounted soldier operating in a tactical environment. Our basic assumptions were that the user was equipped with a portable communications device consisting of a small graphical user interface display and a radio for short-range communication. In this use case, it is desirable (and practical) for the user to authenticate to the device only once—upon power up at the beginning of the mission. Thereafter, the device should remain active and should not "lock" even if the user has not interacted with the device for several minutes; it is impractical for a user in a tactical setting to re-authenticate once a mission has begun. In the absence of locking or re-authentication, a continuous authentication system should detect if the user has lost the device (i.e., detect a dropped/stolen device) or if the user has been compromised (i.e., detect loss of life).

For our prototype, we used a Nexus 5 smartphone with an external 802.11 transceiver to serve as our tactical communications device. The smartphone communicated using ad hoc 802.11 with other similarly-configured devices as part of a mobile ad hoc network. We integrated two external devices with the smartphone: (1) the "Nymi band", a commercial wristband produced by Nymi Inc. (see [17]) that measures a wearer's ECG reading; and (2) the "MIO LINK", a commercial wristband produced by

MIO Global (see [18]) that measures a wearer's heart rate. The outputs of the Nymi band and the MIO LINK were fused by a custom Android app we created to detect a lost device or a compromised user.

Out of the box, the Nymi band operates as follows. A user completes an enrollment process (only required on first use) in which the user wears the wristband and uses a company-provided "companion app" to assign a password and provide training data of the user's ECG reading (performed by the Nymi wristband). Whenever the user subsequently puts on the Nymi band, the user must complete an "activation" step, which consists of the Nymi band running an ECG measurement and validating (through the "companion app") that the correct user is wearing the band. Once the band is activated the user no longer needs to re-activate the band unless the band is removed. Note that this is an important security feature provided by the Nymi—the fact that wristband removal is detected ensures that (in most use cases) the user who authenticated is still the user wearing the wristband. At this point, the activated band can unlock appropriately provisioned devices. Nymi also provides an API for the Nymi band to communicate with custom applications; at the time this work was completed, we used the Nymi software development kit (SDK) version 2.0 and our smartphone communicated with the wristband over Bluetooth.

The MIO LINK device is not tied to a particular user and thus does not require enrollment. The device simply measures a user's heartrate when worn on the user's wrist. The heart rate signal is broadcast using Bluetooth. We paired our smartphone with the MIO LINK and polled the heart rate signal over the Bluetooth channel. Note that if multiple users were operating in close proximity, the individual MIO LINK devices would be each be paired to a unique smartphone, thus avoiding any inadvertent crosstalk.

We created a custom smartphone app that took the inputs from both the Nymi band and MIO LINK. The process flow of the continuous authentication app is shown in Fig. 1. Initially the app ensures that a live user exists and is within range of the smartphone by checking the pulse rate from the associated MIO LINK. If the pulse rate is greater than zero, then the app will query the Nymi band. If the user had previously activated the Nymi band and has not yet removed the wrist strap, then the Nymi band will authenticate the user to the device. Note that this authentication is essentially confirming that an authenticated user put on the band at some earlier time, ran an ECG authentication (to "activate" the band) and has not yet removed it; this authentication step is not re-running the ECG, it is merely performing an integrity check that the band has not been removed.

Following the Nymi band integrity check, initial authentication is complete and the app enters a "continuity" checking phase. In this phase, the app continuously polls the MIO LINK (once per second) and confirms the presence of the MIO LINK signal. If the signal is absent for more than 3 s, the app concludes that the smartphone has been physically separated from the MIO LINK; the inference is that the smartphone has been lost and the app de-authenticates the user. If the signal is present but the measured pulse rate is zero, the inference is that the user has been compromised (potentially deceased) and the app de-authenticates the user. If the signal is present and the pulse is non-zero the user remains authenticated. Note that the need for the second device in addition to the Nymi band—the MIO LINK in this case—is to check for continuity of the user's proximity and proof-of-life. The Nymi band can confirm that an authenticated user has

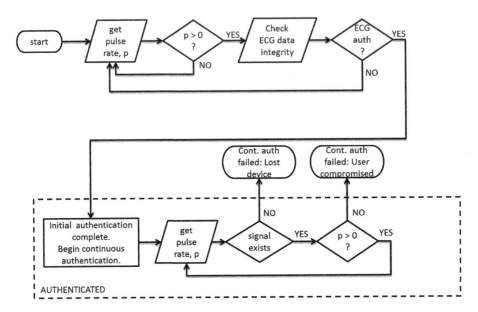

Fig. 1. Process flow of continuous authentication app. The app obtains pulse rate data from a commercial heart rate monitor and obtains ECG data integrity from a commercial ECG reader.

not removed the band and thus can "unlock" a device, but it does not continually update the unlocked device with a proof-of-life signal.

When a de-authentication event occurs, the app logs the event and immediately sends a message to the commander using the communication channels available in the MANET (where it is assumed that the IP address of a commander or administrator node is provisioned ahead of time). The message to the commander indicates the ID of the smartphone that was de-authenticated along with a "reason code" for the de-authentication—either that the phone was lost or that the user was compromised.

Figure 2 provides screenshots of the continuous authentication app. In Fig. 2(a), the app shows that the user has successfully authenticated with the Nymi band, as shown by the two green "OK" circles on the left[1]. The user's pulse signal is present (Status is "OK") and the heart rate is non-zero (it is 74 beats per minute here). Thus, this user has successfully completed initial authentication and continuity has also been maintained, leading to an overall "Authentication Status" of "OK". In Fig. 2(b), while the user has completed initial authentication, at some point the signal from the heart rate monitor was lost for more than 3 s, leading to an invalid status (shown as "—"). This results in an "Authentication Status" of "NO" and generates a message of "Phone lost". This message is timestamped, logged by the app, and sent to the user with the IP specified on the display (in this case the commander with IP 192.168.10.17).

---

[1] Note that the two green circles in the app are labeled "provision" and "validation". These are terms and concepts used in the Nymi SDK 2.0 package. We used the "provision" and "validation" buttons to trigger the Nymi band to communicate with our app and confirm initial authentication.

(a)   (b)

Fig. 2. Screenshot of continuous authentication app. The app detects the initial authentication from the ECG and the user continuity from the heart rate. In (a), initial authentication is successful and the heart rate signal is present and is non-zero. In (b), initial authentication is successful but the heart rate signal has been lost so the "Authentication Status" is set to "NO" and a message is automatically forwarded to the commander. (Color figure online)

3 Testing and Evaluation Results

We performed two simple experiments to evaluate the performance of our continuous authentication prototype. In the first experiment we measured the time required for the system to detect a lost device (i.e., absence of MIO LINK signal) and in the second experiment we measured the time to detect a compromised user (i.e., pulse rate of zero).

For the first experiment we conducted a series of 100 trials[2]. The procedure followed for each trial is described below, where the participants involved are a user (who wears the equipment) and an experimenter (who records the data):

(1) User dons the Nymi band and MIO LINK;
(2) User completes the "activation" step for the Nymi band by measuring his/her ECG reading; experimenter ensures the activation is successful (i.e., Nymi band has recognized user);
(3) Experimenter ensures the MIO LINK is functioning properly (i.e., blue light on wristband is flashing approximately once per second);

---

[2] Note that since the trials involved testing on human subjects—including measuring and recording pulse rates and ECG readings—all testing was conducted with the approval of the DRDC Human Research Ethics Committee.

(4) User opens the custom continuous authentication app on the smartphone and authenticates to the smartphone (this involves pressing two buttons on the app to tell the app to begin accepting outputs from the Nymi and from the MIO LINK);

   NOTE: At this point the user is authenticated to the smartphone and the app will detect a dropped phone or a compromised user.

(5) User holds the smartphone, while the experimenter monitors the app for at least one minute and ensures that Authentication Status remains listed as "OK";

(6) Experimenter logs a "begin test" event on the app; then the user immediately places the phone on the ground and walks away at a comfortable walking pace;

(7) User continues walking, while the experimenter monitors the app until the Authentication Status changes to "NO" (at which point user can return);

(8) Experimenter notes the distance travelled by the user at the instant the app changes the Authentication Status to "NO"; experimenter ensures the reason-code logged by the app for loss of authentication is due to "Phone lost".

In Fig. 3 we plotted the cumulative distribution function (cdf) of the distance travelled by the user before the lost device was detected. We note that a lost device is detected after the user has travelled a median of 13.2 mand 95% of all lost devices are detected within 16.4 m. The detection sensitivity depends upon the range of the BLE (Bluetooth Low Energy) signal connecting the smartphone to the MIO LINK; presumably a tighter detection range could be obtained using a lower power signal.

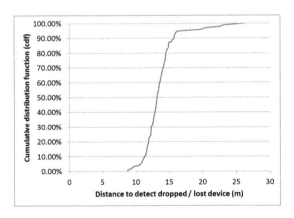

Fig. 3. Performance of continuous authentication prototype to detect a lost device showing the cdf of the distance a user has travelled before detection.

For the second experiment we conducted a series of 100 trials, where once again the participants involved were a user and an experimenter. The procedure for each trial was as follows:

NOTE: Steps 1 to 5 are identical to those followed in the first experiment.

(6) Experimenter logs a "begin test" event on the app; then the user immediately removes the MIO LINK from his/her wrist;

(7) The experimenter monitors the app until the Authentication Status changes to "NO";

(8) Experimenter notes the time between the "begin test" log timestamp and the log timestamp at which the app changes the Authentication Status to "NO"; experimenter ensures the reason-code logged by the app for loss of authentication is due to "user health compromised".

The cumulative distribution function (cdf) for the time to detect a user health compromise (i.e., a zero pulse rate from the MIO LINK as opposed to a loss of signal) as measured over 100 trials is shown in Fig. 4. We note that a heart rate of zero beats per minute was detected after a median time of 6.5 s, and 95% of all cases were detected within 9.4 s.

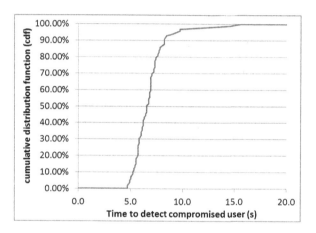

Fig. 4. Performance of continuous authentication prototype to detect user health compromise (i.e., a heart rate of zero beats per minute).

4   Discussion

In implementing the prototype app using the two commercial wristbands, it became clear that the devices were being used for two distinctly separate functions. The first, authentication, dealt with the user's identity and was being performed by the Nymi band; the second, continuity, dealt with the ongoing validity of the user's identity and was performed by the MIO LINK. Together these two sources of data provided an implementation of continuous authentication applicable to a military tactical network use case.

A corollary to this realization is that either function could be performed by other devices (or indeed by a single device), and that continuous authentication can be achieved by separating the "user authentication" function from the "user continuity" function. In fact, the continuity function does not need to continually validate a user's identity—it must simply validate continuity of the presence (and life) of the user who performed the initial authentication. This is an important distinction from typical studies of continuous authentication, which generally focus on biometrics that can identify a user in an ongoing manner based on current activity that matches a stored user profile or set of features (see, for example, [9–12]). In our use case, there is no expectation that a user should ever become separated from his/her device, and there is no expectation that a device should ever "lock" once a user has authenticated. As long as the device can trace the continuity of the initial authentication, this continuity does not require an additional authentication component.

In our prototype, initial authentication is currently performed by the Nymi band, but it could be replaced by another authentication device including a fingerprint scanner, a secure token, or even a simple password. We note that the Nymi band supports password authentication as an alternative to the ECG reading, and we have found that the password authentication was generally simpler to perform. The primary advantage offered by the Nymi device is the presence of an "integrity" measure, whereby the Nymi band detects if an authenticated user has removed the band. For most commercial applications this is adequate to ensure a "continuous authentication" and to unlock a device. Unfortunately the Nymi band (in its current form) does not offer a continued proof-of-life detection such as a heart rate monitor, nor does it continue to beacon out its signal once initial authentication has taken place and the target device is unlocked—thus our use of the MIO LINK to provide "continuity". Continuity and proof-of-life could, of course, be performed by other means such as a different heart rate monitor or an alternate monitor such as body temperature, body sounds, body movement, etc.

In Fig. 5, we propose a process flow for continuous authentication using a single device (instead of the two devices in our prototype) that meets the requirements of our tactical military use case. A single device is mechanically linked to a user (e.g., using a wristband) and the user performs an initial authentication with the device (e.g., using a biometric, token, or password). Once initial authentication is complete, the device will maintain its "integrity" so long as the mechanical linkage is intact—that is, if the mechanical linkage is broken, the device will detect the mechanical break (as the Nymi does) and will indicate this in its communications with the smartphone. The device measures the user's pulse (or other proof-of-life, though pulse is simple and reliable) and sends the pulse signal along with an "integrity" signal at frequent periodic intervals. Three conditions would result in a de-authentication event: (1) the integrity signal is invalid, (2) the entire signal is absent, or (3) the pulse is zero. Each of these conditions would result in a different reason code for de-authentication to be presented to a commander.

The process flow in Fig. 5 is similar to the flow from Fig. 1, but includes the important addition of an integrity check with every transmission of the pulse signal. Our prototype included the integrity check only at initial authentication and then relied on the pulse signal alone to provide continuity. While the omission of the integrity check in our prototype may appear reasonable since a pulse is already a "continuous" signal, it leaves open the vulnerability of the system to an adversary that removes the heart rate monitor from our user and replaces it on his/her own wrist in the short window of time before the loss of pulse is detected (e.g., in less than the median detection time shown in Fig. 4). If an integrity check is included with each instance of

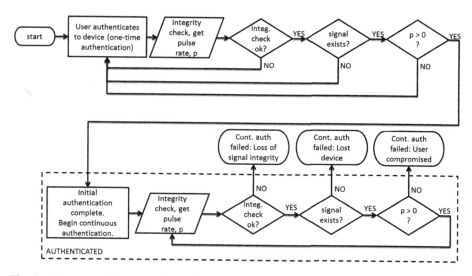

Fig. 5. Recommended process flow for continuous authentication for tactical operations. Once initial (strong) authentication is complete, the primary goal of the continuous authentication system is to ensure proof-of-life and continuity of user, which can be achieved using a heart rate monitor (for proof-of-life) and a mechanical linkage (e.g., a wristband) that detects removal to ensure the integrity of the initial authentication.

the heart rate signal, the integrity check will fail in the case described above due to the loss of mechanical linkage, thus this vulnerability is averted[3].

In developing and using our prototype, we arrived at a number of other conclusions that are of interest to a military application of continuous authentication. These are summarized below:

- In a contested tactical network environment, we envision users operating as part of a MANET to communicate with the other members of their unit (e.g., Platoon, Section, Squad, etc.), where each user's node is equipped with a form of continuous authentication. Should any user's device experience a continuous authentication failure, the commander of the unit would be notified. The course of action the commander should follow at this point is not obvious. From a technical standpoint it is not difficult to lock out or disable the device. However, from an operational standpoint, this may not be the most desirable immediate course of action—perhaps the user is still there but has experienced equipment difficulties; perhaps an adversary has control of the device but it is preferable to observe what the adversary does with the device before it is disabled, etc. Ultimately, we maintain that a failure of continuous authentication that alerts the commander will empower the commander to take action; the commander could use another communication channel to determine the user's status, could request a re-authentication, could revoke the

---

[3] Note that in our prototype we have made the tacit assumption that the BLE connection between the wristbands and the smartphone is secure. To ensure the security of this connection is beyond the scope of this paper.

device's encryption key, could remotely zeroize the node, or could defer the decision until more information is available, depending upon mission constraints.

- The inclusion of a "reason code" for a failure of continuous authentication is vital in order to assist a commander in determining a sensible course of action in response. For instance, if the reason for failure is deemed a "lost device", it might be reasonable to remotely lock the device, but at the same time allow the device to continue broadcasting situational awareness messages in order to more easily locate and recover the device. If the reason for failure is a "user compromise", the commander may wish to direct immediate effort (including medical expertise) to the last known location of the user.

- For nodes that are part of a MANET, it is possible that certain nodes will be disconnected from the commander at any given time. Should a node fail continuous authentication while disconnected from the commander, the commander may miss the notification of the failure. We suggest that it may be desirable for nodes to periodically transmit continuous authentication status information to the commander as part of their standard situational awareness updates. In this fashion, the commander will be notified of the failure as soon as the node returns in range and the commander receives a situational awareness update.

- In our prototype, the two wristbands communicated with the smartphone using Bluetooth Low Energy. Relying on a wireless connection (commercial or otherwise) for continuous user authentication is problematic as it may be vulnerable to jamming and other simple denial of service techniques. In addition to simple jamming, BLE (or other standard protocols) may be vulnerable to spoofing at the protocol level. A wired connection to other body sensors is more robust, however this presents problems as well since additional wiring to mechanical linkages and body sensors may be cumbersome. The choice ultimately depends upon envisioned use cases and adversarial capabilities.

5   Conclusion

In this paper, we presented a prototype implementation of a continuous authentication system for a military tactical network use case in which two commercial biometric wristbands were tethered to a smartphone in a MANET; the prototype system detects if a user loses the smartphone or becomes compromised. For our use case, we observed that continuous authentication can be achieved by a device that monitors the continued presence of the user following an initial strong authentication. It is not necessary for the continued presence to validate the identity of the user, but merely to confirm the continuity of the user from initial login. We recommend the use of an authentication device that contains a mechanical linkage (e.g., a wristband) that can perform initial authentication and also detect the continued presence of a user (e.g., through a heart rate monitor). With such a device, any loss of continuity or integrity—either through a loss of heart rate signal or a loss of mechanical linkage—would result in a de-authentication event. Developing or implementing such a device is feasible today using existing commercial products and could increase the usability and security of mobile devices for tactical applications.

References

1. Directorate of Land Concepts and Design, Department of National Defence, Government of Canada: Land Operations 2021: Adaptive dispersed operations, the force employment concept for Canada's army of tomorrow (2007)
2. Canadian Army Land Warfare Centre, Department of National Defence, Government of Canada: No man's land: tech considerations for Canada's future army (2012)
3. Alberts, D.S., Garstka, J.J., Stein, F.P.: Network Centric Warfare. CCRP Publication Series, August 1999. http://dodccrp.org/files/Alberts_NCW.pdf
4. Alberts, D.S., Hayes, R.E.: Power to the Edge. CCRP Publication Series, June 2003. http://www.dodccrp.org/files/Alberts_Power.pdf
5. Monrose, F., Rubin, A.D.: Keystroke dynamics as a biometric for authentication. Future Gener. Comput. Syst. **16**, 351–359 (2000)
6. Hossain, M.S., Balagani, K.S., Phoha, V.V.: New impostor score based rejection methods for continuous keystroke verification with weak templates. In: Proceedings of 2012 IEEE Fifth International Conference on Biometrics: Theory, Applications and Systems, September 2012
7. Roth, J., Liu, X., Metaxas, D.: On continuous user authentication via typing behavior. IEEE Trans. Image Process. **23**(10), 4611–4624 (2014)
8. Ahmed, A.A.E., Traore, I.: A new biometric technology based on mouse dynamics. IEEE Trans. Dependable Secure Comput. **4**(3), 165–179 (2007)
9. Feng, T., Liu, Z., Kwon, K.-A., Shi, W., Carbunar, B., Jiang, Y., Nguyen, N.: Continuous mobile authentication using touchscreen gestures. In: Proceedings of 2012 IEEE Conference on Technologies for Homeland Security, November 2012
10. Buduru, A.B., Yau, S.S.: An effective approach to continuous user authentication for touch screen smart devices. In: Proceedings of IEEE International Conference on Software Quality, Reliability and Security (QRS), August 2015
11. Zhao, X., Feng, T., Xu, L., Shi, W.: Mobile user identity sensing using the motion sensor. In: Proceedings of SPIE DSS, Baltimore, MD, May 2014
12. Crouse, D., Han, H., Chandra, D., Barbello, B., Jain, A.K.: Continuous authentication of mobile user: fusion of face image and inertial measurement unit data. In: Proceedings of 2015 IEEE International Conference on Biometrics (ICB), May 2015
13. Biel, L., Pettersson, O., Philipson, L., Wide, P.: ECG analysis: a new approach in human identification. IEEE Trans. Instrum. Meas. **50**(3), 808–812 (2001)
14. Israel, S.A., Irvine, J.M., Cheng, A., Wiederhold, M.D., Wiederhold, B.K.: ECG to identify individuals. Pattern Recogn. **38**(1), 133–142 (2005)
15. Wang, Y., Agrafioti, F., Hatzinakos, D., Plataniotis, K.N.: Analysis of human electrocardiogram for biometric recognition. EURASIP J. Adv. Signal Process. **2008**, 1–11 (2008)
16. Singh, Y.N., Gupta, P.: Biometric method for human identification using electrocardiogram. In: Tistarelli, N., Nixon, M.S. (eds.) ICB 2009. LNCS, vol. 5558, pp. 1270–1279. Springer, Heidelberg (2009). doi:10.1007/978-3-642-01793-3
17. Nymi corporate website: www.nymi.com. Accessed 16 May 2016
18. MIO LINK brochure website: http://www.mioglobal.com/Mio-Link-heart-rate-wristband/Product.aspx. Accessed 16 May 2016

Workshop on Convergence of Airborne Networking, Wireless Directional Communication Systems, and Software Defined Networking

On Delay Tolerant Airborne Network Design

Shahrzad Shirazipourazad, Arun Das$^{(\boxtimes)}$, and Arunabha Sen

Computer Science and Engineering Program, School of Computing, Informatics and
Decision System Engineering, Arizona State University, Tempe, USA
{sshiraz1,arun.das,asen}@asu.edu

Abstract. Mobility pattern of nodes in a mobile network has significant impact on the connectivity properties of the network. Due to its
importance in civil and military environments, and due to the several
complex issues present in this domain, one such mobile network that has
drawn attention of researchers in the past few years is Airborne Networks
(AN). Since the nodes in an airborne network (AN) are heterogeneous
and mobile, the design of a reliable and robust AN is highly complex and
challenging. This paper considers a persistent backbone based architecture for an AN where a set of Airborne Networking Platforms (ANPs)
such as aircrafts, UAVs and satellites, form the backbone of the AN. As
ANPs may be unable to have end-to-end paths at all times due to the
limited transmission ranges of the ANPs, the AN should be delay tolerant and be able to transmit data among ANPs within a bounded time.
In this paper we propose techniques to compute the minimum transmission range required by the ANPs in such delay tolerant airborne networks.

Keywords: Airborne Networks · Airborne Networking Platform · Delay
tolerant networks

1 Introduction

An Airborne Network (AN) is a mobile ad-hoc network that utilizes a heterogeneous set of physical links (RF, Optical/Laser and SATCOM) to interconnect a
set of terrestrial, space and highly mobile Airborne Networking Platforms (ANPs)
such as satellites, aircrafts and Unmanned Aerial Vehicles (UAVs). Airborne networks can benefit many civilian applications such as air-traffic control, border
patrol, and search and rescue missions. The design, development, deployment and
management of a network with mobile nodes is considerably more complex and
challenging than a network of static nodes. This is evident by the elusive promise of
the Mobile Ad-Hoc Network (MANET) technology where despite intense research
activity over the past years, mature solutions are yet to emerge [1,2]. One major
challenge in the MANET environment is the unpredictable movement pattern of
the mobile nodes and its impact on the network structure. In case of an AN, there
exists considerable control over the movement pattern of the mobile platforms.
For instance, to realize the functional goals of an AN, Air Force personnel can

© ICST Institute for Computer Sciences, Social Informatics and Telecommunications Engineering 2017
Y. Zhou and T. Kunz (Eds.): ADHOCNETS 2016, LNICST 184, pp. 357–368, 2017.
DOI: 10.1007/978-3-319-51204-4_29

specify the controlling parameters of the network, such as the *location, flight path* and *speed* of the ANPs that form the backbone of the AN. Such control provides designers an opportunity to develop a topologically stable network even when the network nodes are highly mobile.

It is increasingly being recognized in the networking research community that the level of *reliability* needed for continuous operation of an AN may be difficult to achieve through *a completely mobile, infrastructure-less network* [3]. In order to enhance *reliability* and *scalability* of an AN, Milner *et al.* in [3] suggested the formation of a *backbone network* with ANPs. In order to deal with the reliability and scalability issues of an AN, we consider an architecture for an AN where a set of ANPs form the *backbone* of the AN. This set of ANPs may be viewed as *mobile base stations* with *predictable and well-structured flight paths* and the combat aircrafts on a mission as *mobile clients*.

It is desirable that such a backbone network remain connected at all times even though the topology of the network may change with the movement of the ANPs. Such continuous network connectivity can be achieved if the transmission range of the ANPs is sufficiently large. However, a large transmission range also implies high energy usage. Accordingly, one would like to know the smallest transmission range for the ANPs which ensures connectivity at all times. In [4], the authors precisely address this problem and propose techniques to find the smallest transmission range to ensure that the backbone network remains connected at all times. The authors define the *critical transmission range* (CTR) as the minimum transmission range of the ANPs to ensure that the dynamic network formed by the movement of the ANPs remains connected at all times, and present algorithms to compute the CTR when the flight paths are known.

Due to the critical nature of ANs, the ANPs may be subject to adversarial attacks such as Electromagnetic Pulse attacks or network jamming. Such attacks can impact specific geographic regions at specific times and if an ANP is within the fault region during the time of attack, it will be rendered inoperable. In [5], the authors consider the scenario where some of the AN nodes fail due to a region fault, i.e., the failed nodes are confined to a *geographic region*. The authors define a *critical transmission range in faulty scenario (CTR_f)* as the smallest transmission range necessary to ensure that the surviving nodes of the AN remain connected irrespective of the location of the fault and the time of the fault. The authors study this problem in [5] and propose techniques to compute the CTR_f.

It may be noted that in previous problems studied in [4,5] the backbone network is required to be connected at all times. Accordingly, techniques were proposed to compute CTR and CTR_f in [4,5]. However, it may not be possible to equip the transmitters of the ANPs with transmission ranges at least as large as the CTR or CTR_f. In such a scenario, the backbone network may be forced to operate in a disconnected mode for some amount of time. It is also conceivable that the data to be transmitted through the ANPs may be tolerant to some amount of delay. Hence, ANPs may not need to have end-to-end paths at all times but should be able to transmit data to each other within a bounded time.

These requirements lead us to study the problem of computation of critical transmission range in delay tolerant airborne networks. More specifically, the critical transmission range in delay tolerant network (CTR_D) is defined as the minimum transmission range necessary to ensure that every pair of nodes in the backbone network can transmit at least one bit of data with each other within a bounded time. In this paper we formulate the problem for computing CTR_D and propose techniques to compute CTR_D. To the best of our knowledge this problem has not been studied before.

The rest of the paper is organized as follows: In Sect. 2 we present the related works, in Sect. 3 the AN architecture considered in this study is detailed, in Sect. 4 the connectivity problem in delay tolerant ANs is formulated and solution techniques proposed, finally in Sect. 5 we present our experiments.

2 Related Works

Due to the Joint Aerial Layer Networking (JALN) activities of the U.S. Air Force, design of a robust and resilient ANs has received considerable attention in the networking research community in recent years. It has been noted that purely mobile ad-hoc networks (i.e., networks without infrastructure) have limitations with respect to reliability, data transmission, communication distance and scalability [3,6]. Accordingly, the authors of [3,6] have suggested the introduction of a mobile wireless backbone network where the nodes serve as mobile base stations (analogous to cellular telephony or the Internet backbone), in which topology of the dynamic backbone network can be managed through the control of movement patterns and transmission ranges of the backbone nodes.

Although there have been several studies on various aspects of mobile ad-hoc networks, most of these studies consider infrastructure-less networks, whereas the focus of this study is on ANs with a backbone infrastructure. Noted among the studies on infrastructure-less mobile ad-hoc networks is topology control in MANETs [6–9]. The goal of these studies is to assign power values to the nodes to keep the network connected while reducing energy usage. The authors of [7,8] have proposed distributed heuristics for power minimization in mobile ad-hoc networks, but have offered no guarantees on their worst case performance. Santi in [9] studied the minimum transmission range required to ensure network connectivity in mobile ad-hoc networks. He proved that the critical transmission range for connectivity (CTR) is $c\sqrt{\frac{\ln n}{\pi n}}$ for some constant c where the mobility model is obstacle free and nodes are allowed to move only within a bounded area. In these studies the mobility patterns are not known unlike the problem studied in this paper where it is assumed that the flight paths of the ANPs are known. Also, this paper studies the computation of the minimum transmission range in a delay tolerant setting that has not been studied in previous studies.

As there may be times that the networks may have to operate in a disconnected mode, the last few years have seen considerable interest in the networking research community in delay tolerant network (DTN) design [10]. The authors of [11] survey challenges in enhancing the survivability of mobile wireless networks.

It is mentioned in [11] that one of the aspects that can significantly enhance network survivability is the design of end-to-end communication in environments where the path from source to destination is not wholly available at any given instant of time. In this design, adjusting the transmit power of the nodes plays an important role. Existing DTN research mainly focuses on routing problem in DTNs [12,13]. The paper [14] provides a survey on routing algorithms for DTN. For such algorithms to be effective, every pair of nodes should be able to communicate with each other within a bounded period of time. Papers such as [15,16] have studied the problem of topology control in DTNs. In these papers, the time evolving network is modeled by a space-time graph and it is assumed that the this graph is initially connected and the problem is to find the minimum cost connected subgraph of the original graph. To the best of our knowledge, no studies exist that study the computation of the minimum transmission range of nodes in DTNs such that the time evolving network is connected over time.

3   System Model and Architecture

As mentioned previously, the level of *reliability* needed for continuous operation of an AN may be difficult to achieve through *a completely mobile, infrastructure-less network*, and if possible, a *backbone network* with ANPs should be formed to enhance reliability. It may be noted that in [5] the authors present the system model and architecture of the AN considered in this paper. We summarize that description in this section to preserve the completeness of our presentation.

In order to achieve the goal of reliability, an architecture of an AN is proposed where a set of ANPs form a backbone network and provide reliable communication services to combat aircraft on a mission. In this architecture, the nodes of the backbone networks (ANPs) may be viewed as *mobile base stations* with *predictable and well-structured flight paths* and the combat aircrafts on a mission as *mobile clients*. A schematic diagram of this architecture is shown in Fig. 1. In the diagram, the black aircrafts are the ANPs forming the infrastructure of the AN (although in Fig. 1, only aircrafts are shown as ANPs, UAVs/satellites can also be considered as ANPs). It is assumed that the ANPs follow a circular flight path. The circular flight paths of the ANPs and their coverage area (shaded spheres with ANPs at the center) are also shown in Fig. 1. Thick dashed lines indicate the communication links between the ANPs. The figure also shows three fighter aircrafts on a mission passing through a space known as an *air corridor*, where network coverage is provided by ANPs 1 through 5. As the fighter aircrafts move along their trajectories, they pass through the coverage area of multiple ANPs and there is a smooth hand-off from one ANP to another when the fighter aircrafts move from the coverage area of one ANP to that of another. At points P1, P2, P3, P4, P5 and P6 of Fig. 1, the fighter aircrafts are connected to the ANPs (4), (2, 4), (2, 3, 4), (3), (1, 3) and (1), respectively.

In this paper, it is assumed that two ANPs can communicate with each other whenever the distance between them does not exceed the specified threshold (transmission range of the on board transmitter). We are aware that in an

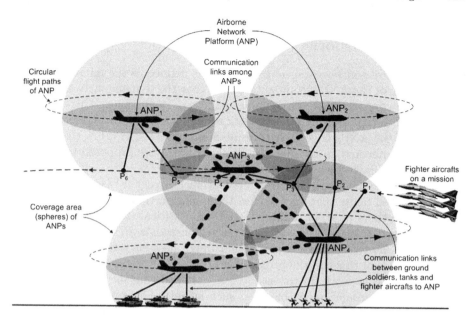

Fig. 1. A schematic view of an Airborne Network

actual airborne network deployment, successful communication between two airborne platforms does not depend only on the distance, but also on various other factors such as (i) the line of sight between the platforms [17], (ii) changes in the atmospheric channel conditions due to turbulence, clouds and scattering, (iii) the banking angle, the wing obstruction and the dead zone produced by the wake vortex of the aircraft [18] and (iv) Doppler effect. Moreover, the transmission range of a link is not a constant and is impacted by various factors, such as transmission power, receiver sensitivity, scattering loss over altitude and range, path loss over propagation range, loss due to turbulence and the transmission aperture size [18]. However, the distance between the ANPs remains an important parameter in determining whether communication between the ANPs can take place, and as the goal of this study is to understand the basic and fundamental issues of designing an AN with twin invariant properties of coverage and connectivity, such simplifying assumptions are necessary and justified. Once the fundamental issues of the problem are well understood, factors (i)–(iv) can be incorporated into the model to obtain more accurate solutions.

For simplicity of the analysis, two more assumptions are made. We assume that (i) all ANPs are flying at the same altitude thus restricting the problem to a two-dimensional plane, and (ii) they follow a circular flight path. Although these assumptions are made for this study to simplify our analysis, our techniques remain valid even when the ANP altitudes are different and the flight paths are irregular. The techniques proposed in this paper are equally applicable to any scenario as long as the flight paths are periodic and are a priori known.

4   Computation of Critical Transmission Range in Delay Tolerant Airborne Networks CTR_D

In [4] the *critical transmission range* (CTR) was defined as the minimum transmission range of the ANPs required to ensure that the dynamic network formed by the movement of the ANPs remains connected at all times. This definition of CTR implies that although the network topology is dynamic and may change with time, if the ANPs have their transmission ranges set to CTR, the network will remain connected at all times. In [4] algorithms were proposed to compute the CTR when flight paths of the ANPs are known.

As ANPs may be susceptible to faults, specifically region based or geographically correlated faults that may render ANPs in a particular geographic region to fail, the authors of [5] introduce the notion of critical transmission range in faulty scenario (CTR_f). For a given fault region radius R, the authors of [5] define CTR_f as the smallest transmission range necessary to ensure network connectivity at all times in the presence of at most one region fault of radius R anywhere in the network. In this study it is assumed that all nodes within the fault radius R become inoperable after such a fault. The authors of [5], given region fault radius R, propose techniques to compute the CTR_f that allows the network to remain connected at all times after a region failure of radius R, irrespective of the location of the fault and the time of failure.

As noted above, the CTR and CTR_f computations of [4,5] respectively ensure that the network always remains connected in a non faulty, and faulty scenario. However, due to resource constraints of the AN, it may not be possible to equip the ANPs with radios that have coverage of radius CTR or CTR_f. Therefore, in this situation the backbone network cannot remain connected at all times. On the other hand, based on the type of data that should be transmitted between ANPs, data transmissions may be tolerant to some amount of delay. Hence, ANPs may not require to have end-to-end paths between all ANPs at all times, but instead they should be able to transmit data to each other within some limited time through intermediate nodes across different network topologies over time. In this section we investigate the problem of computation of minimum transmission range in such delay tolerant airborne networks.

We consider that the trajectories and the distance function $s_{ij}(t)$ of the network nodes are periodic over time. As a consequence, the network topologies are periodically repeated. However, periodicity is not an underlying assumption and our results can be utilized in non-periodic scenario as well as long as the node trajectories for the whole operational duration of a network are given. In [5] the authors present how to compute the *link lifetime timeline* and accordingly the network topologies caused by ANPs mobility in a time period when all inputs are given. We represent the set of topologies in a periodic cycle starting from time t_0 (current time) by the set $\mathcal{G} = \{G_1, G_2, \ldots, G_l\}$. Each network topology G_i exists for a time duration of T_i.

In Fig. 2, an example of a dynamic graph with two topologies G_1 and G_2, one periodic cycle is shown. G_1 and G_2 last for T_1 and T_2 time units respectively. It can be observed that there is no end-to-end path from A to C in either G_1

or G_2. However, A can transmit data to B in G_1, and B can forward it to C in G_2. In this situation we say that A can reach C through a temporal path with delay equal to the lifetime of G_1, i.e. T_1; and that the temporal path is completed in G_2. We define a *temporal path* from node s to d to be a set of tuples $\{(t_1, (v_1, v_2)), (t_2, (v_2, v_3)), \ldots, (t_k, (v_{k-1}, v_k))\}$ such that $v_1 = s, v_k = d$, $v_i \in V$, and for every tuple $(t_i, (v_i, v_{i+1}))$, edge (v_i, v_{i+1}) is active at time t_i, and $t_i \geq t_{i-1}$ for all $1 \leq i \leq k$. Moreover, without loss of generality, we assume that t_i corresponds to the starting time of a topology in $\mathcal{G}$. Then, the path delay is defined to be $t_k - t_0$ where t_0 is the starting time of G_1 in the first periodic cycle. We note that all path delays are computed with respect to starting point t_0 but we later show that we can modify the starting point to any time.

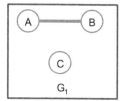 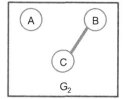

Fig. 2. A dynamic graph with two topologies G_1 and G_2

We note that the existence of a path from node i to j with some delay does not guarantee the existence of a path from j to i with the same delay. For example, in Fig. 2 the path from C to A has a delay of $T_1 + T_2$ while the path delay from A to C is equal to T_1. We say that a dynamic graph $G(t)$ is *connected with delay D* if there exists a temporal path from every node $i \in V$ to every node $j \in V \setminus \{i\}$ with delay smaller than D. It may be noted that, if the transmission range Tr is too small, ANPs may not be able to communicate with each other at all, i.e. there may be no temporal path of finite delay between the ANPs. We define *critical transmission range in delay tolerant network* (CTR$_D$) to be the minimum transmission range necessary to ensure that the dynamic graph is *connected with delay D*. We define *the connectivity problem in delay tolerant networks* as the problem of computation of CTR$_D$ given the delay threshold D, and the following input parameters:

1. a set of points $\{c_1, c_2, \ldots, c_n\}$ on a two dimensional plane (representing the centers of circular flight paths),
2. a set of radii $\{r_1, r_2, \ldots, r_n\}$ representing the radii of circular flight paths,
3. a set of points $\{p_1, p_2, \ldots, p_n\}$ representing the initial locations of the platforms, and
4. a set of velocities $\{v_1, v_2, \ldots, v_n\}$ representing the speeds of the platforms

In order to find the value of CTR$_D$, first we explain a technique to check whether a transmission range Tr is adequate for having a connected dynamic network with delay D. First, we determine the set of events $(L(tr))$ when the state of a link changes from active to inactive (and vice-versa), when the transmission range is set to Tr. The technique to build the set $L(tr)$ was proposed in [5] and we restate it here for the sake of completeness. From the specified input parameters (1) through (4) we first determine the lifetime (active/inactive intervals) of every link between every pair of nodes i and j by comparing $s_{ij}(t)$ with Tr and finding the time points that the state of a link changes. Let $L(Tr) = \{e_1, e_2, \ldots, e_l\}$ denote the set of events, or e_i's, when the state of a link changes when transmission range is Tr; let $L(tr)$ be sorted in increasing order of the time of the events. Hence, between two consecutive events e_i and e_{i+1} that occur at times t_i and t_{i+1} the set of active links is unchanged. Algorithm 1 summarizes this technique of computing $L(Tr)$.

Algorithm 1. Link Lifetime Computation

Input: (i) a set of points $\{c_1, c_2, \ldots, c_n\}$ representing the centers of circular flight paths, (ii) a set of radii $\{r_1, r_2, \ldots, r_n\}$ representing the radii of circular flight paths, (iii) a set of points $\{p_1, p_2, \ldots, p_n\}$ representing the initial locations of the platforms, (iv) a set of velocities $\{v_1, v_2, \ldots, v_n\}$ representing the speeds of the platforms.
Output: $L(Tr)$: an ordered set of events that the state of a link changes from active to inactive or inactive to active.

1: $L(Tr) \leftarrow \emptyset$
2: **for all** pairs i, j **do**
3:      Compute l to be the set of time points t such that $s_{ij}(t) = Tr$ (using equation (5) of [5]) over a period of time, to find the instances of times t where the state of the link (i, j) changes. If $s_{ij}(t) = Tr$ and is $s_{ij}(t)$ increasing at t, it implies that the link dies at t, and if $s_{ij}(t)$ decreasing at t, it implies that the link becomes active at t.
4:      **for all** $l_k \in l$ **do**
5:          Find the position of l_k in $L(Tr)$ using binary search and Add the event into $L(Tr)$. ($L(Tr)$ is sorted in increasing order)
6:      **end for**
7: **end for**

It has been shown in [5] that the time complexity of Algorithm 1 is $O(n^4)$, and $|L(Tr)| = O(n^2)$. Before describing the rest of the technique to check whether a transmission range Tr is adequate for having a connected dynamic network with delay D, we propose the following observation:

Observation 1. *For a given transmission range Tr, there is a temporal path from every node u to every node v with finite delay iff the superimposed graph $G_c = \{V, \bigcup_{i=1}^{l} E_i\}$, where E_i is the set of edges in G_i, is connected.*

Although a transmission range Tr may be enough to result in a connected superimposed graph G_c, it may not be sufficient for the existence of a temporal path between every pair of nodes with delay smaller than a threshold D even if D is as large as $\sum_{i=1}^{l-1} T_i$. Figure 3 depicts an AN with three topologies in one period.

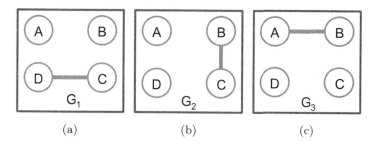

Fig. 3. A dynamic graph with three topologies G_1, G_2 and G_3

It can be observed that A cannot have a temporal path from A to D in the first period. Actually the fastest path includes edges (A, B) in G_3 in first period, (B, C) in G_2 in the second period and (C, D) in G_1 in the third period. Therefore, the path delay is $2(T_1 + T_2 + T_3)$. Generally, in the worst case in every period just a subpath (a set of consecutive edges) in one topology is used and therefore the maximum delay of a temporal path will be $D_{max} = (l-1)\sum_{i=1}^{l} T_i$. Hence, if $D \geq D_{max}$, examining the connectivity of G_c is enough to decide whether for a transmission range there exists a temporal path of delay smaller than D between every pair of nodes in the dynamic network.

We now present an algorithm that checks for a given value of transmission range Tr, whether a network is connected with delay D where $D < D_{max}$. Let $N(u)$ denote the set of nodes that are reachable from $u \in V$ with delay smaller than D. Initially $N(u) = \{u\}$. The algorithm starts by computing the connected components in every topology G_i. Let $C_i = \{C_{i,1}, C_{i,2}, \ldots C_{i,q_i}\}$ represent the set of connected components in G_i where $C_{i,j}$ is the set of nodes in the jth component of G_i, and $q_i = |C_i|$. Let g and h be the quotient and remainder of $\frac{D}{\sum_{i=1}^{l} T_i}$ respectively, and $t_0 + h$ is the time where the network topology is G_p for a p, $1 \leq p \leq l$. Therefore, the topologies in time duration t_0 to $t_0 + D$ includes G_1 to G_l for g number of cycles and G_1 to G_p in the last periodic cycle. Starting from the first topology G_1 in the first period, in each topology G_i, if a node v is in the same connected component with a node $w \in N(u)$, then v can be reachable from u through a temporal path which is completed in G_i; hence, $N(u)$ is updated to $N(u) \cup (\bigcup_{k:N(u)\cap C_{i,k}\neq\emptyset} C_{i,k})$. In this step the algorithm goes through all the topologies from t_0 to $t_0 + D$. In the end, if $N(u) = V$ for all $u \in V$, then the transmission range Tr is sufficient for having a connected network with delay D. In Algorithm 2 the steps of checking the connectivity of a dynamic graph with delay D is presented.

As shown in [5] the number of topologies, l in a single period of the dynamic topology is $O(n^2)$. Thus, the computation of the connected components of a graph $G_i = (V, E_i)$ using either breadth-first search or depth-first search requires a time complexity of $O(|V| + |E_i|) = O(n^2)$. Hence, Step 2–4 of Algorithm 2 takes $O(n^4)$. It may be noted that this algorithm is used for the case when $D < (l-1)\sum_{i=1}^{l} T_i$. Therefore, the number of periods $g < l - 1$, and $g = O(n^2)$. Computation of

Algorithm 2. Checking Connectivity of Airborne Network with delay D

Input: $\mathcal{G}(t) = \{G_1, G_2, \ldots, G_l\}$ and delay threshold D
Output: *true* if dynamic graph $\mathcal{G}(t)$ is connected with delay D; otherwise *false*.

1: Initialize $N(u) = \{u\}$ for every $u \in V$
2: **for all** topologies $G_i, 1 \leq i \leq l$
3:    Compute $C_i = \{C_{i,1}, C_{i,2}, \ldots C_{i,q_i}\}$, the set of connected components of G_i
4: **for all** periods 1 to g
5:    **for all** topologies $G_i, 1 \leq i \leq l$
6:       **for all** node $u \in V$
7:          $N(u) \leftarrow N(u) \cup (\bigcup_{k:N(u) \cap C_{i,k} \neq \emptyset} C_{i,k})$
8: **for all** topologies $G_i, 1 \leq i \leq p$ (the topologies in the last period)
9:    **for all** node $u \in V$
10:       $N(u) \leftarrow N(u) \cup (\bigcup_{k:N(u) \cap C_{i,k} \neq \emptyset} C_{i,k})$
11: **for all** node $u \in V$
12:    **if** $N(u) \neq V$, **return** *false*
13: **return** *true*

Step 5 is also $O(n^2)$ since $|N(u)|$ and the total size of all components in G_i is $O(n)$. Overall, it can be concluded that the time complexity of Algorithm 2 is $O(n^7)$.

Additionally, as all the delays are computed with respect to t_0, we can easily extend this technique to any t_i, by repeating Algorithm 2 for every $t_i, 1 \leq i \leq l$ where t_i is the starting time of topology G_i. Thus, this extension increases the complexity by a factor of $l = O(n^2)$.

Finally, similar to the computation of CTR and CTR$_f$ in [4,5], in order to compute CTR$_D$ a binary search can be carried out within the range $0 - Tr_{max}$ to determine the smallest transmission range that will ensure the AN is connected with delay D during the entire operational time. The binary search adds a factor of $\log Tr_{max}$ to the complexity of Algorithm 2 to compute CTR$_D$.

5   Simulations

The goal of our simulation is to compare critical transmission ranges in different scenarios of non faulty (CTR), faulty (CTR$_f$) and delay tolerant (CTR$_D$) to investigate the impact of various parameters, such as the number of ANPs, the region radius and delay, on the critical transmission range. In our simulation environment, the deployment area considered was a 1000×1000 square mile area. The centers of the orbits of the ANPs were chosen randomly in such a way that the orbits did not intersect with each other. In our simulation, we assumed that all the ANPs move at the same angular speed of $\omega = 20$ rad/h. Hence a period length is 0.1π h. One interesting point to note is that, in this environment where all the ANPs are moving at the same angular speed on circular paths, the value of CTR is independent of the speed of movement of the ANPs. This is true because changing the angular speed ω effects just the time at which the events take place, such as when a link becomes active or inactive. If we view the

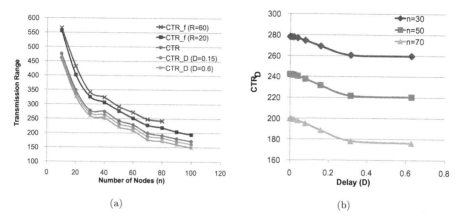

Fig. 4. (a) Transmission Range vs. Number of Nodes; (b) Transmission Range (CTR$_D$) vs. Delay

dynamic topology of the backbone network over one time period as a collection of topologies $\mathcal{G} = \{G_1, G_2, \ldots, G_l\}$, where G_i morphs into $G_{i+1}, 1 \leq i \leq l$ at some time, by increasing or decreasing the angular speed of all ANPs, we just make the transitions from G_i to G_{i+1} faster or slower, without changing the topology set $\mathcal{G}$. Similarly, the set of ANPs that fail due to failure of a region at a certain time remains unchanged.

In our first set of experiments we compute CTR, CTR$_f$ when $R = 20, 60$, and CTR$_D$ when $D = 0.5$ *period*, 2 *period* for different values of number of nodes, n. Figure 4(a) depicts the result of these experiments. In these experiments, for each value of n we conducted 30 experiments and the results were averaged over the 30 different random initial setups. We set *orbit radius* $= 10$. We observed that as expected, an increase in the number of nodes results in a decrease in CTR, CTR$_f$ and CTR$_D$. Moreover, CTR$_D \leq$ CTR $\leq$ CTR$_f$ for all instances. In all of the experiments, we computed CTR$_D$ with respect to all times (corresponding to beginning of a new topology) and not just t_0.

In the second set of experiments, we conducted experiments to investigate the impact of delay D on the value of CTR$_D$. Figure 4(b) depicts the results. We observe that when value of delay D is zero the value of CTR$_D$ is equal to CTR and by increasing delay, CTR$_D$ decreases and the interesting observation is that when delay becomes greater than 2 *period* the decrease in the value of CTR$_D$ is unnoticeable or even zero.

References

1. Burbank, J.L., Chimento, P.H., Haberman, B.K., Kasch, W.T.: Key challenges of military tactical networking and the elusive promise of MANET technology. IEEE Commun. Mag. **44**, 39–45 (2006)
2. Conti, M., Giardano, S.: Multihop ad-hoc networking: the reality. IEEE Commun. Mag. **45**, 88–95 (2007)

3. Milner, S.D., Thakkar, S., Chandrashekar, K., Chen, W.L.: Performance and scalability of mobile wireless base-station-oriented networks. ACM SIGMOBILE Mob. Comput. Commun. Rev. **7**(2), 69–79 (2003)

4. Shirazipourazad, S., Ghosh, P., Sen, A.: On connectivity of airborne networks with unpredictable flight path of aircrafts. In: Proceedings of the First ACM MobiHoc Workshop on Airborne Networks and Communications, pp. 1–6. ACM (2012)

5. Shirazipourazad, S., Ghosh, P., Sen, A.: On connectivity of airborne networks in presence of region-based faults. In: 2011-MILCOM 2011 Military Communications Conference, pp. 1997–2002. IEEE (2011)

6. Milner, S., Llorca, J., Davis, C.: Autonomous reconfiguration and control in directional mobile ad hoc networks. IEEE Circ. Syst. Mag. **9**(2), 10–26 (2009)

7. Ramanathan, R., Rosales-Hain, R.: Topology control of multihop wireless networks using transmit power adjustment. In: INFOCOM 2000 of Nineteenth Annual Joint Conference of the IEEE Computer and Communications Societies Proceedings, vol. 2, pp. 404–413. IEEE (2000)

8. Cabrera, J., Ramanathan, R., Gutierrez, C., Mehra, R.: Stable topology control for mobile ad-hoc networks. IEEE Commun. Lett. **11**(7), 574–576 (2007)

9. Santi, P.: The critical transmitting range for connectivity in mobile ad hoc networks. IEEE Trans. Mob. Comput. **4**, 310–317 (2005)

10. Fall, K.: A delay-tolerant network architecture for challenged internets. In: Proceedings of the 2003 Conference on Applications, Technologies, Architectures, and Protocols for Computer Communications, SIGCOMM 2003, pp. 27–34. ACM, New York (2003)

11. Sterbenz, J.P.G., Krishnan, R., Hain, R.R., Jackson, A.W., Levin, D., Ramanathan, R., Zao, J.: Survivable mobile wireless networks: issues, challenges, and research directions. In: Proceedings of the 1st ACM Workshop on Wireless Security, WiSE 2002, pp. 31–40. ACM, New York (2002)

12. Jain, S., Fall, K., Patra, R.: Routing in a delay tolerant network. In: Proceedings of the 2004 Conference on Applications, Technologies, Architectures, and Protocols for Computer Communications, SIGCOMM 2004, pp. 145–158. ACM, New York (2004)

13. Alonso, J., Fall, K.: A linear programming formulation of flows over time with piecewise constant capacity and transit times. Intel Research Technical report IRB-TR-03-007 (2003)

14. Cao, Y., Sun, Z.: Routing in delay/disruption tolerant networks: a taxonomy, survey and challenges. IEEE Commun. Surv. Tutorials **15**(2), 654–677 (2013)

15. Huang, M., Chen, S., Zhu, Y., Xu, B., Wang, Y.: Topology control for time-evolving and predictable delay-tolerant networks. In: 2011 IEEE 8th International Conference on Mobile Adhoc and Sensor Systems (MASS), pp. 82–91 (2011)

16. Huang, M., Chen, S., Zhu, Y., Wang, Y.: Cost-efficient topology design problem in time-evolving delay-tolerant networks. In: Global Telecommunications Conference (GLOBECOM 2010), pp. 1–5. IEEE (2010)

17. Tiwari, A., Ganguli, A., Sampath, A.: Towards a mission planning toolbox for airborne networks: optimizing ground coverage under connectivity constraints. In: IEEE Aerospace Conference, pp. 1–9, March 2008

18. Epstein, B., Mehta, V.: Free space optical communications routing performance in highly dynamic airspace environments. In: IEEE Aerospace Conference Proceedings (2004)

Enabling Dynamic Reconfigurability of SDRs Using SDN Principles

Prithviraj Shome[1], Jalil Modares[2], Nicholas Mastronarde[2(✉)],
and Alex Sprintson[1(✉)]

[1] Department of Electrical and Computer Engineering,
Texas A&M University, College Station, USA
{prithvirajhi,spalex}@tamu.edu
[2] Department of Electrical Engineering, University at Buffalo, Buffalo, USA
{jmod,nmastron}@buffalo.edu

Abstract. Dynamic reconfiguration and network programmability are active research areas. State of the art solutions use the Software Defined Networking (SDN) paradigm to provide basic data plane abstractions and programming interfaces for control and management of these abstractions; however, SDN technologies are currently limited to wired networks and do not provide the appropriate abstractions to support ever changing wireless protocols. On the other hand, the Software Defined Radio (SDR) paradigm enables complex signal processing functionality to be implemented efficiently in software, instead of on specialized hardware; however, SDR does not cater to the demand for adaptive radio network management with respect to changing channel conditions and policies. To overcome these limitations, we present CrossFlow, a principled approach for application development in SDR networks. CrossFlow defines fundamental radio port abstractions and an interface to manipulate them. It provides a flexible and modular cross-layer architecture using the principles of SDR and a mechanism for centralized control using the principles of SDN. Through the convergence of SDN and SDR, CrossFlow works towards providing a target independent framework for application development in wireless radio networks. We validate our design using proof-of-concept applications, namely, adaptive modulation, frequency hopping, and cognitive radio.

Keywords: Software-defined networking · Software-defined radios · Dynamic reconfigurability · OpenFlow · GNU radio

1 Introduction

With ever-changing wireless standards and protocols, there has been a conscious shift towards a programmatic approach for designing and implementing wireless radios. This has led to a tremendous interest in Software Defined Radios (SDR). SDR is a powerful concept in which filters, amplifiers, modulators and other complex signal processing blocks are realized in software, instead of on specialized hardware. As the task of signal processing is handed over to software, it

© ICST Institute for Computer Sciences, Social Informatics and Telecommunications Engineering 2017
Y. Zhou and T. Kunz (Eds.): ADHOCNETS 2016, LNICST 184, pp. 369–381, 2017.
DOI: 10.1007/978-3-319-51204-4_30

is possible to use general purpose hardware, connected to an RF front end, to create powerful and highly flexible radios.

While the SDR paradigm has revolutionized the design of wireless radios, it does not provide an efficient method to control a network of SDRs. Since SDRs can be reconfigured to provide a wide variety of radio functionalities, it would be highly desirable to have a consistent interface to expose the SDR's functional modules to the network application developer. As modules can be added, removed or changed any time, such an interface framework must be able to adapt to these changes, report events to the application, and allow control of various constituent modules while hiding their complexity from the application developer. This level of abstraction is necessary because, as the network grows and becomes more heterogeneous, it is impossible for the application developer to keep track of low-level details. Here, by the notion of heterogeneous networks, we take into consideration a network containing both wired and wireless devices. Hence, the architecture should enable network control, meet requirements of users, and abstract away the details of the implementation.

In order to provide abstractions taking into account the above considerations, we use the concept of Software Defined Networking (SDN). SDN defines abstractions that represent data plane components and the interface to control and manage these abstractions. These primitives (including asynchronous callback function event reporting) enable an application developer to obtain a logically centralized view of the network. The application developer can then dynamically adjust rules to reflect changing network conditions and requirements.

In this paper, we aim to integrate SDR and SDN to provide a principled approach for developing a consistent interface to manage underlying abstractions of SDRs. We build upon the abstract model presented in our previous paper [1], where we described a monolithic architecture for *wireless radio port* abstraction. In the current paper, we go beyond that and broaden the design space to provide a modular design, which is in line with the design principle of SDR. This also enables an integration of both wired and wireless networks which can be managed in a programmatic manner, thereby enabling development of key applications catering to a heterogeneous network. We call our platform CrossFlow. Some network applications that we envision can leverage CrossFlow include, but are not limited to, the following:

1. **Physical layer adaptation** including (i) *frequency hopping* to resist narrowband interference and prevent unauthorized interception; (ii) *transmission power control* to maintain a target link quality while reducing interference to other users and/or extending battery life; (iii) *adaptive modulation and coding* to trade-off throughput and communication reliability and adapt to channel conditions (e.g., pathloss and interference).
2. **Quality of service (QoS) provisioning** to provide QoS policies implemented through medium access control, throttling, admission control, scheduling, and error control techniques (e.g., ARQ and FEC).

3. **Adaptive routing** to allow a CrossFlow controller, with its global view of the network, to dynamically switch between different routing protocols depending on the network conditions and the application constraints.
4. **Self-healing network** to allow the CrossFlow controller to deploy fault management applications based upon self-healing mechanisms.
5. **Cross-layer control** to allow joint optimization of parameters, algorithms, and protocols at all layers of the protocol stack.

We use the generalized model of SDN introduced in [2] as a template for defining the abstractions and their features discussed above. We also build upon the concept of *wireless radio ports* as discussed in [3]. This abstraction is composed of a number of smaller abstractions, one for each processing block, so that fine-grained control of the processing capabilities of a radio device is provided to application developers without exposing its intricate details. This enables manipulation of critical physical, data link, and network layer properties through various well defined interfaces. Thus, using the architecture of CrossFlow, we can build applications across all layers of the network stack.

For validation purposes, we use the popular GNU Radio [4] framework, which provides a modular, open-source Digital Signal Processing (DSP) software environment for SDRs. GNU Radio modules are written in C++ and provide a mechanism to connect and manage data between them. A Python wrapper ties these blocks together to implement applications. We run GNU Radio on a host PC connected (via Ethernet) to a Universal Software Radio Peripheral (USRP) N210 device from Ettus Research, and we also run CPqd SoftSwitch software [5] as a separate module. CPqd SoftSwitch serves as a switch agent interacting between the SDN controller and GNU Radio modules. This is done through message extensions which we will discuss in subsequent sections. We also develop three proof-of-concept applications to validate our design principles: *frequency hopping*, *adaptive modulation*, and *cognitive radio*.

Our contributions can be summarized as follows.

1. We propose a framework that provides a uniform and consistent view of SDRs, so that a network of SDRs can be managed in an efficient manner.
2. We extend the SDN model with message extensions to provide support for wireless radio interfaces.
3. We provide sample applications using the framework for validation.

The rest of the paper is organized as follows. In Sect. 2, we review the related work done in this area. Section 3 describes the CrossFlow architecture with its SDN extensions. Section 4 describes a proof-of-concept implementation of three applications using our framework. Section 5 concludes the paper.

2   Background and Related Work

Software Defined Networking. Network reconfigurability is a major challenge in the networking industry. The explosion of mobile devices and cloud services has increased the need for on-demand installation of services and reconfiguration of flow rules according to changing traffic patterns. In addition, network elements like routers and switches have their own unique interfaces and as such management of network components is a source of concern for network application developers. As the network grows, this complexity increases exponentially and rolling out new services becomes a tedious and complicated process.

Software Defined Networking (SDN) is an architecture which tries to address these challenges by decoupling the control and forwarding functions. This enforces abstraction of underlying implementation and enables applications or network services to be developed using the abstractions. This simple and elegant design also provides applications a centralized view of the network. As a result, it has sparked tremendous research interest in providing a scalable, secure and programmatic approach towards the challenges discussed above. While SDN is a revolutionary approach, it is still mainly geared towards wired networks. Through our previous work, ÆtherFlow [3], we tried to provide a protocol independent approach for bringing wireless into the SDN model. In this paper, we go a step further and try to provide a mechanism for dynamic radio resource management for SDRs to obtain true network visibility in a heterogeneous network.

GNU Radio framework. GNU Radio [4] is a free and open-source framework that provides signal processing functionality to implement SDRs. The main constituents of the framework are basic blocks which perform distinct signal processing functions. GNU Radio enables the composition of these blocks to synthesize new radio functionality on general purpose hardware, but it is not suitable for developing applications to control a network of SDRs. This is because each block exposes its own set of interfaces which does not scale with increasing numbers of radios in the network. In this paper, we provide uniform interfaces to control and manage these processing block abstractions, so that an application developer does not need to handle every block's unique interface characteristics.

Aside from GNU Radio, the idea of providing a programmable wireless data plane has been implemented in [6,7]. These papers provide modular blocks and focus on real-time guarantees for processing signals. But, like GNU Radio, they do not provide any logical interface to control a network of such programmable devices. The paper [8] deals with centralized control of devices but it focuses mainly on LTE networks. Our paper is orthogonal to these works as we provide a mechanism for centralized control while making the exposed interfaces protocol independent. The combination of SDRs and SDN has recently been used in a variety of contexts [9–12]. In [11], SDR and SDN are used to create a testbed for LTE technologies while [9,10] focus on integration of SDR and SDN for 4G/5G technology. In [13], an SDR model for management of interference in dense heterogeneous networks is proposed while [12] developed a jamming architecture using SDN and SDR principles. These papers provide distinct solutions for various scenarios but do not provide a generic framework for handling various protocols in a principled manner.

The most closely related work to CrossFlow is the RcUBe framework [14], which provides structured abstractions for decision, control, data, and register

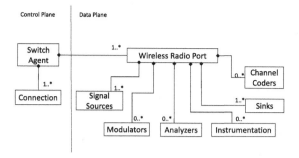

Fig. 1. Abstraction model of CrossFlow

planes of SDRs. A key difference between CrossFlow and RcUBe is that Cross-Flow allows SDRs to be managed by the same SDN controller as other network devices, thereby enabling unified control of a heterogeneous network.

3   CrossFlow Architecture and Design

In this section, we motivate and describe the architecture and design of Cross-Flow. In Sect. 3.1, we introduce the proposed data plane abstractions. Then, in Sect. 3.2, we describe how we extend the OpenFlow protocol to accommodate CrossFlow messages.

3.1   Data Plane Abstractions

We extend the data model proposed in [2] to create an abstraction model for the CrossFlow framework, which is displayed in Fig. 1. We build upon the *wireless radio port* concept proposed in [3] to create a new layer of abstractions. This layer of abstractions exhibits a composition or *has-a* relationship with the wireless radio port abstraction (i.e., the wireless radio port *has a* sink, modulator, or channel coder). This means that the blocks of this layer are the objects or members that comprise the wireless radio port. These blocks are derived from the most commonly used processing blocks in GNU Radio [4]. This abstract wireless radio port model serves the following design vision:

- It allows visibility into the signal processing blocks from an application point of view, without going into implementation details.
- It allows for the development of an event driven framework for radio operation.
- It could enable composition of blocks to implement new functionality. For future work, we envision that a network application could specify which blocks to connect for a specific wireless port instance, and the internal framework could handle the implementation.

In this paper, we focus on the first two bullets. Specifically, we develop an abstract interface to enable event-driven dynamic configuration of a fixed set of signal processing blocks at run-time. In order to change a signal processing block's parameters, the application needs to send a $<command, value>$ tuple in a message. For queries and receive event responses, it registers for events with each block and, when an event occurs, appropriate callbacks are invoked. One of the main requirements of the CrossFlow model is that each abstraction should implement four types of interfaces as proposed in both [2,3], namely, capabilities, configuration, statistics, and events. Thus far, CrossFlow provides the interfaces for a wireless radio port abstraction with two processing blocks: *Sink* and *Modulators*. However, we plan to extend it to include the other processing blocks shown in Fig. 1. The Sink abstraction allows the SDN controller to manage the signal sinks which can be a USRP device, file, or a socket, while the Modulators abstraction allows management of modulation schemes (e.g., BPSK, QPSK, and 8PSK).

The interfaces for *Sink* and *Modulators* are categorized as follows:

1. **Sink:**
   - **Capabilities**: The interface allows the SDN controller to query the capabilities of sinks such as: (i) Type of sink (USRP, socket, etc.); (ii) Channels supported; (iii) Center Frequency; and (iv) IP address.
   - **Configuration**: The interface allows the SDN controller to configure properties of signal sinks such as: (i) Gain; (ii) Frequency, and (iii) Sample rate.
   - **Statistics**: The interface allows the SDN controller to gather statistics for sinks such as the received signal strength indicator (RSSI).
   - **Events**: The interface allows the SDN controller to take decisions based upon events such as low or high RSSI.
2. **Modulators:**
   - **Capabilities**: The interface allows the SDN controller to query the properties of the modulator block such as: (i) Modulations supported; (ii) Current samples/symbol; and (iii) Use of a Gray code indicator.
   - **Configuration**: The interface allows the SDN controller to configure properties of the modulator block such as: (i) Choice of modulation scheme (e.g. BPSK, QPSK and 8PSK); (ii) Sample/symbol; and (ii) Use of a Gray code.
   - **Statistics**: The interface allows the SDN controller to gather statistics for the modulator block such as: (i) Signal to Noise Ratio (SNR) and (ii) Bit Error Rate (BER).
   - **Events**: The interface allows the SDN controller to take decisions based upon events in the modulator block such as: (i) Low or high SNR and (ii) Low or high BER.

3.2 Message Extensions

CrossFlow uses SDN design principles to control a network of configurable SDRs. As such, to enable control plane interactions between the SDN controller and the SDR, we had two options: either we could have implemented our own control protocol to enable their interactions or extend the existing OpenFlow [15] framework. This is because OpenFlow does not natively support wireless features. In order to enable a cleaner implementation, we decided to extend OpenFlow by using experimenter messages within the OpenFlow protocol, similar to ÆtherFlow. Experimenter messages are a part of the standard OpenFlow protocol which provides a mechanism for vendors to include propriety information within the protocol. This approach has two advantages. First, we do not need to implement a new protocol for control and data plane interactions. Second, since we are using experimenter messages to carry CrossFlow messages, the SDN controller does not need to perform special handling for these messages. This enables the controller to remain target independent and hence it can handle both wired and wireless devices. In the current version of CrossFlow, we define three messages:

- **Configuration message request:** Request for modification of parameters, such as gain, frequency, SNR threshold, and modulation scheme.
- **Statistics message request:** Request for statistics, such as SNR and BER.
- **Event message response:** Response for events, such as low SNR.

4   Proof-of-Concept Implementation

4.1   Illustrative CrossFlow Implementation

In this section, we describe our implementation of *adaptive modulation*, *frequency hopping* and *cognitive radio* applications using the CrossFlow framework. For illustration, we implement our model on a USRP N210 SDR from Ettus Research. We use the CPqD Softswitch [5] (`ofsoftswitch`) software as the switch agent in the SDN model. Its main functionality is to enable communication between GNU Radio and the python based Ryu SDN controller. As described in previous sections, the applications will send messages to the processing blocks, e.g., to configure them. The `ofsoftswitch` then forwards this request to a centralized `CrossFlow Hub` inside the GNU Radio domain.

There are four main components (blocks) in the illustrative CrossFlow module that we implement in GNU Radio, namely, the `CrossFlow Hub`, the Modulation Controller (Mod Controller for brevity), the SNR Monitor and the USRP Controller (see Fig. 2).

- The `CrossFlow Hub` is the interface between the Modulation (Mod for brevity) and USRP controllers in GNU Radio and the Ryu SDN controller. The `CrossFlow Hub` and the Ryu SDN controller communicate via packet data unit (PDU) socket. The `CrossFlow Hub` is responsible for receiving commands from `ofsoftswitch` (or any other compliant interface), interpreting the commands, and forwarding the commands to the appropriate controller block (i.e., the USRP or Mod controller in our implementation). It is also responsible for receiving information from different controller blocks and sending information to the Ryu SDN controller. The `CrossFlow Hub` has in/out ports to send commands and receive information to/from the GNU Radio controller blocks. It also has in/out PDU ports for interfacing with Socket PDU.

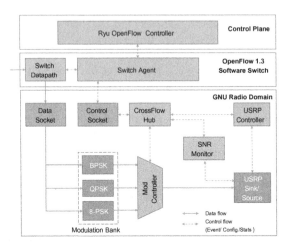

Fig. 2. Transmitter implementation diagram of CrossFlow with two processing blocks: Sink and Modulators

- The `Mod Controller` is responsible for receiving commands from the `CrossFlow Hub`, and selecting the appropriate modulation scheme from the modulation bank. For illustration, we include three modulation schemes (BPSK, QPSK, and 8PSK); however, thanks to the modular design, we can easily add more schemes. The Mod Controller can also feedback information to the Ryu SDN controller about the modulation scheme that is currently in use and the number of modulation schemes available in the modulation bank.
- The `SNR Monitor` is responsible for monitoring the SNR level and generating an event in case the SNR level falls below a certain threshold, which can be configured by the application. Currently the framework uses the existing SNR probe of GNU Radio, which supports four SNR estimators for M-PSK modulated signals. This monitoring block is also responsible for relaying the SNR statistics back to `CrossFlow Hub` in response to a SNR statistics query generated by the application.
- The `USRP Controller` is responsible for controlling different RF parameters of the USRP Transmitter/Receiver based on commands from the `CrossFlow Hub`. In our proof-of-concept implementation, we control the carrier frequency and the power of the signal. It can also feedback information to the `CrossFlow Hub` about the current RSSI, SNR, carrier frequency, power, etc.

Although our illustrative implementation only has three controllers (facilitating abstraction of the USRP Sink/RF implementation, SNR estimation, and the adaptive modulation implementation), additional controllers can be easily added to support new functionalities and abstractions.

Table 1. Variation of the SU's SNR and PER as a function of the PU's normalized transmission power. The SU transmits at a fixed rate of 1 Mbps.

PU's normalized TX power	SU's packet error rate	SU's signal-to-noise-ratio (dB)
0	0.15 %	5.8553
0.09	6.34 %	−0.2983
0.14	19.89 %	−0.9483

4.2 Example Applications

Frequency Hopping Application: Frequency hopping is a technique of transmitting radio signals by spreading the signal over a sequence of changing frequencies. It can be used against jamming and for protecting against unauthorized eavesdropping. In our implementation, the Ryu SDN controller simply issues a *GNU-CONFIG-FREQ* command with the desired frequency and pushes this configuration to the device. As shown in Fig. 2, the ofsoftswitch receives this command and forwards it to the GNU Radio domain. The centralized CrossFlow Hub inside the GNU Radio domain processes this request and issues appropriate commands to the USRP Controller, which ultimately signals the USRP block to tune to the requested frequency.

Adaptive Modulation Application: Adaptive Modulation is a technique where the modulation is changed according to the conditions of the channel. There are various estimators which are used for obtaining channel quality. These can be based on SNR, BER, or other environment specific parameters. Similar to the *frequency hopping* application, the Ryu SDN controller issues the *GNU-CONFIG-MOD* command with the appropriate modulation scheme (e.g., BPSK, QPSK, or 8PSK) and forwards the request to the device. The request ultimately reaches the Mod Controller, which is a multiplexer block that selects the requested modulation scheme as shown in Fig. 2.

Cognitive Radio Application: We build upon the frequency hopping application mentioned above to construct a cognitive radio application. Cognitive radio is a type of radio in which the device is aware of its environment and can dynamically change its operating parameters like transmission power, frequency, gain, etc., in response to the environmental conditions. In CrossFlow, we implement an application that can configure a radio device to switch channels based upon a low SNR event measured by the device. The experimental setup is shown in Fig. 3, where we have three USRP N210 devices that act as sender, receiver and noise source. In our setup, the sender and receiver are secondary users (SUs) and the noise source is the primary user (PU). We assume that the SUs are frequency agile and can operate at either 910 MHz or 915 MHz. All transmissions use a 2 MHz bandwidth. Meanwhile, we assume that the PU only operates at 910 MHz. The SUs are set 5 m apart.

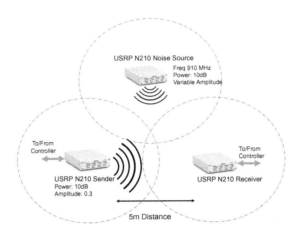

Fig. 3. Setup for cognitive radio application in CrossFlow

Using this setup, we conduct two tests: one to measure the effect of the PU's transmission power on the SU's packet error rate (PER) and SNR at the 910 MHz carrier frequency (measured over 1,000,000 packets transmitted by the SU at 1 Mbps while the PU operates at normalized transmission powers of 0.0, 0.09, 0.14, where 0.0 indicates that the PU is silent), and another to measure how quickly the cognitive radio can trigger and respond to a low SNR event (with normalized PU transmission powers 0.09, 0.17 and 0.22, and SU data rates of 256 Kbps, 512 Kbps and 1 Mbps).

Table 1 shows the PER and SNR values obtained in the first experiment where the PU and SU transmit on the 910 MHz carrier frequency at the same time. As expected, the SU's PER increases and SNR decreases as the PU's transmission power increases on the same channel.

In the second experiment, which demonstrates a simple cognitive radio application, we assume that the PU is initially silent and that the SU is initially transmitting in the spectrum "hole" at 910 MHz. Once the PU starts to transmit, the SU experiences a decrease in its SNR. When the SNR falls below a specified threshold (4 dB in our experiment), a low SNR event is triggered by the SU's SNR Monitor block and the event summary is sent to ofsoftswitch through the CrossFlow Hub. This request is then forwarded to the Ryu SDN controller using the event response message. The application, upon receiving this message, sends a GNU-CONFIG-FREQ command so that the SU changes its carrier frequency to 915 MHz to avoid the interference from the PU. The sequence of actions involved in changing the channel is similar to the sequence in the *frequency hopping* application described earlier.

In Fig. 4(a), we show the number of packets that are lost over the course of time required for the SNR to be sensed below the 4 dB threshold, for the receiver to generate the low SNR event, and for the Ryu SDN controller to respond by issuing the GNU-CONFIG-FREQ command, and finally for the transmitter to switch frequencies. We repeat this experiment three times for each combina-

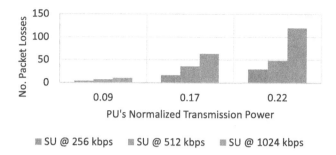

(a) Average number of packets lost while changing channels.

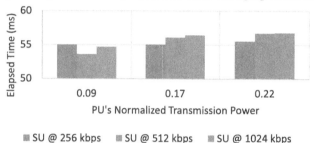

(b) Elapsed time to change channels.

Fig. 4. Packet losses incurred while the SU transitions to the unoccupied channel, and the associated transition delay, for di erent SU rates (kbps) and normalized PU transmission powers.

tion of SU data rate (256, 512, and 1024 kbps) and normalized PU transmission power (0.09, 0.17, and 0.22), and report the average of each group of three measurements in Fig. 4(a). In Fig. 4(b), we report the time that elapses over the aforementioned sequence of events. We note that this switching time is independent of the SU's data rate and the PU's transmission power.

5   Conclusion and Future Work

In this paper, we presented a framework for programming a network of software defined radios using software defined networking (SDN) principles. The framework we propose allows adaptive, flexible, and real-time (re)configuration of software defined radio interfaces from a network controller application. It streamlines the development of network applications by hiding the low level internal details of the signal processing pipeline. In order to validate our approach, we also provide three proof-of-concept applications: *frequency hopping*, *adaptive modulation* and *cognitive radio*. This shows that our design is viable and can be extended to introduce new capabilities.

One of the challenges that we need to consider is in-band control of the radio devices. Currently we implemented our design using an out-of-band wired control channel. The CrossFlow framework can easily be extended to enable in-band control of devices and it will be our next design goal. Another area of focus is the latency between controller and SDR framework. The issue can be mitigated by the introduction of distributed control module in SDR. The distributed control module will allow devices to take local decisions while the centralized controller is responsible for introducing policies and global management, thereby ensuring reduced latency.

Acknowledgments. This material is based upon work supported by the National Science Foundation under Grants No. 1422655, 1423322, by the AFOSR under contract No. FA9550-13-1-0008, and by the Air Force Research Laboratory under Grant No. FA8750-14-1-0073.

References

1. Shome, P., Yan, M., Najafabadi, J.M., Mastronarde, N., Sprintson, A.: CrossFlow: a cross-layer architecture for SDR using SDN principles. In: Proceedings of the IEEE Conference on Network Function Virtualization and Software Defined Networks (IEEE NFV-SDN), November 2015
2. Casey, C.J., Sutton, A., Sprintson, A.: TinyNBI: distilling an API from essential OpenFlow abstractions. In: Proceedings of the Third Workshop on Hot Topics in Software Defined Networking, ser. HotSDN 2014. New York, NY, USA, pp. 37–42. ACM (2014). http://doi.acm.org/10.1145/2620728.2620757
3. Yan, M., Casey, J., Shome, P., Sprintson, A., Sutton, A.: Aetherflow: principled wireless support in SDN. In: Proceedings of the ICNP Workshop on Control, Cooperation, and Applications in SDN protocols (CoolSDN 2015) (2015)
4. GNU Radio. http://gnuradio.org/redmine/projects/gnuradio/wiki
5. CPqD OpenFlow 1.3 Software Switch. http://cpqd.github.io/ofsoftswitch13/
6. Bansal, M., Schulman, A., Katti, S.: Atomix: a framework for deploying signal processing applications on wireless infrastructure. In: Proceedings of NSDI (2015)
7. Bansal, M., Mehlman, J., Katti, S., Levis, P.: Openradio: a programmable wireless dataplane. In: Proceedings of HotSDN (2012)
8. Gudipati, A., Perry, D., Li, L.E., Katti, S.: SoftRAN: software defined radio access network. In: Proceedings of the Second Workshop on Hot Topics in Software Defined Networks, ser. HotSDN 2013 (2013)
9. Cho, H.-H., Lai, C.-F., Shih, T., Chao, H.-C.: Integration of SDR and SDN for 5G. Access IEEE **2**, 1196–1204 (2014)
10. Sun, S., Kadoch, M., Gong, L., Rong, B.: Integrating network function virtualization with SDR and SDN for 4G/5G networks. Netw. IEEE **29**(3), 54–59 (2015)
11. Mancuso, V., Vitale, C., Gupta, R., Rathi, K., Morelli, A.: A prototyping methodology for SDN-controlled LTE using SDR (2014)
12. Corbett, C., Uher, J., Cook, J., Dalton, A.: Countering intelligent jamming with full protocol stack agility. Secur. Priv. IEEE **12**(2), 44–50 (2014)
13. Gupta, R., Bachmann, B., Kruppe, A., Ford, R., Rangan, S., Kundargi, N., Ekbal, A., Rathi, K., Asadi, A., Mancuso, V., et al.: LabVIEW based software-defined physical/MAC layer architecture for prototyping dense LTE Networks (2015)

14. Demirors, E., Sklivanitis, G., Melodia, T., Batalama, S.N.: Rcube: real-time recon-figurable radio framework with self-optimization capabilities. In: 12th Annual IEEE International Conference on Sensing, Communication, and Networking (SECON), pp. 28–36. IEEE (2015)
15. McKeown, N., Anderson, T., Balakrishnan, H., Parulkar, G., Peterson, L., Rexford, J., Shenker, S., Turner, J.: Openflow: Enabling Innovation in Campus Networks. ACM SIGCOMM Computer Communication Review **38**(2), 69–74 (2008)

Reliability, Throughput and Latency Analysis of an Aerial Network

Kamesh Namuduri[1(✉)] and Amjad Soomro[2]

[1] University of North Texas, Denton, USA
kamesh.namuduri@unt.edu
[2] Air Force Research Laboratory, Rome, NY, USA
amjad.soomro@af.us.mil

Abstract. An aerial network is a self-organized network formed by air-craft systems deployed for a mission-specific purpose. Aerial networks are characterized by dynamic topologies and frequent network connections and disconnections primarily due to the fast moving nature of the aircraft systems. This paper considers an abstract paradigm for an aerial network, modeling it as a dynamic graph, and analyzes its reliability, throughput and latency characteristics. It defines metrics to measure the performance of an aerial network in terms of reliability, throughput, and latency. The analysis is intended to provide insights into the performance characteristics of an aerial network as a function of topology and mobility. The insights derived from this analysis are expected to lead to strategies for improving the network performance in real-world applications. This analysis also provides a set of building blocks which would lead to a general framework for identifying reliable and critical paths in a network (Approved for public release: distribution unlimited.).

Keywords: Aerial network · Adhoc network · Reliability · Through-put · Latency

1   Introduction

An aerial network is formed by aircraft systems deployed in the air for a specific mission. The network may include manned and unmanned aircraft systems (UASs), ground vehicles, control stations and services, as illustrated in Fig. 1. In an aerial network, nodes may travel at high speeds, range extends to hundreds of miles, and network topology is constantly changing. The use of airborne networks has typically been restricted to military applications until now. In recent years, with the expanded roles of UASs and the desire to bring dynamic infrastructure to disaster areas, there has been increased interest in airborne network research. Airborne networks are envisioned to play an important role in Next Generation (NextGen) Air Transportation Systems.

© ICST Institute for Computer Sciences, Social Informatics and Telecommunications Engineering 2017
Y. Zhou and T. Kunz (Eds.): ADHOCNETS 2016, LNICST 184, pp. 382–389, 2017.
DOI: 10.1007/978-3-319-51204-4_31

1.1   Aerial versus Terrestrial Networks

Aerial networks can be viewed as a class of Mobile Ad-hoc Networks (or MANETs) consisting of nodes moving in and out of the network as they fly. Typically, MANETs are formed without any supporting infrastructure. Aerial networks differ from traditional MANETs and ground networks due to their extreme high speed, long distances, relative multi-path free operation, and dynamic rate adaptation. These differences solicit different operating characteristics and present challenging problems as well. In addition to operating conditions described above, NextGen air transportation systems envisioned by the U.S.'s Federal Aviation Administration (FAA) require enhanced safety and security. Although significant progress has been made in many dimensions of air transportation systems (e.g. airspace management, data link capability), there is a clear need for continued efforts to enhance the safety and security capabilities (e.g. sense and avoid, safe navigation, and secure communication) of NextGen air transportation systems. Aerial networking is an approach towards achieving these objectives.

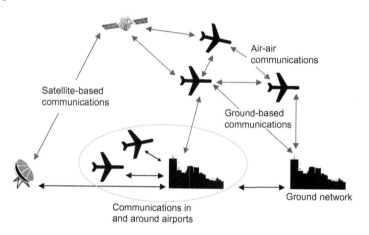

Fig. 1. An illustration of airborne network consisting of terrestrial, satellite, and RF links to connect to control stations on the ground and to other airplanes.

1.2   Aerial Network Protocols

Until recently, aerial networking is limited to military applications only. Within the civilian domain, aerial networking protocols are still emerging. It is envisioned that aerial networks will use Internet Protocol and require high-data-rate, low-latency, and self-configuration capabilities. The Air Force Research Laboratory (AFRL) has developed a proprietary, IP-based tactical targeting networking technology capable of communicating over 300 nautical miles for defense applications [1]. Aerial networks can also be designed to use a heterogeneous set of physical links (for example, radio-frequency and satellite communications) to interconnect terrestrial, space, and other airborne networks. Link 16, a TDMA-based secure, jam-resistant, high-speed digital data link that operates over-the-air in the L-band portion (969–1206 MHz) of the UHF spectrum, is another example. With Link 16, military aircraft as well as ships and ground control stations can exchange real-time information. Link 16 also supports the exchange of text messages and imagery data and provides two channels of digital voice (2.4 Kbit/s and/or 16 Kbit/s in any combination).

Successful deployment of aerial networks require comprehensive modeling and simulation of airborne networks. Modeling and simulation of airborne networks, in turn, require models of airborne vehicles, antenna propagation patterns, mobility models, terrain models, and weather patterns. Deployment of successful aerial networks also require the implementation of information assurance strategies and their integration with network management and planning tools.

1.3   Contributions

This paper considers an abstract paradigm for an aerial network, modeling it as a dynamic graph, and analyzes its reliability, throughput, and latency characteristics. It defines metrics to measure the performance of an aerial network. The analysis is intended to provide insights into the performance characteristics of an aerial network as a function of topology and mobility. The insights derived are expected to lead to strategies for improving the reliability, throughput and latency in real-world applications.

2   Aerial Network as a Dynamic Graph

This section outlines the mathematical preliminaries of random graphs that are necessary for the analysis presented here. An aerial network with its dynamic topology can best be represented as a random graph. We consider networks with two terminals: a source (s) node and a target (t) node and follow the notation used in [3]. Let $G=(V, E, P)$ represent a probabilistic graph with a set of nodes $v_i \in V$, a set of edges $e_{ij} \in E$, and a link failure probability matrix $p_{ij} \in P$. Let G_{st} (V, E_{st}, P_{st}) represent an overlay graph containing a path from s to t with its associated set of edges and probabilities (E_{st}, P_{st}). An overlay graph is created during the route discovery process (RDP); a process followed by a source node to find its target node. Although either nodes or links may fail in a network, the scope of this analysis is limited to networks with link failures only, i.e., nodes are assumed to be failure-free. An edge e_{ij} represents a link connecting two adjacent nodes v_i and v_j. A path between two nodes v_i and v_j that are not adjacent to each other in G is defined as a sequence of distinct links connecting the two nodes. Information flows from one node to another as long as there is a path connecting the two nodes. A $(s-t)$ cut divides the set of vertices V in the graph G_{st} (V, E_{st}, P_{st}) into two disjoint subsets S and T such that $s \in S$ and $t \in T$. $C_{st}(i)$ represents a cut-set indexed by i in the overlay graph connecting the two nodes s and t (Fig. 2).

Table 1. Terminology and notation used to represent a graph [3]

G(V,E,P)	A network with the set of nodes V, set of links E and link failure probability matrix P
s	A source node in the network G
t	A target node in the network G
S	The set of all source nodes
T	The set of all target nodes
NS	A set of network states
NS_i	Network state i
G_{st} (V, E_{st}, P_{st})	An overlay network that contains a path from s to t with its associated (V, E_{st}, P_{st})
n	Number of nodes, $|V|$
m	Number of edges, $|E_{st}|$
nc	Number of cut-sets in a graph
F_p	Probability that a network is disconnected
$s - t$	A cut in G_{st} (V, E_{st}, P_{st}) where $s \in S$ and $t \in T$
$C_{st}(i)$	i^{th} cut-set of G_{st} (V, E_{st}, P_{st})
c_{ij}	Capacity of the link e_{ij}
$c(S, T)$	Capacity of an $s - t$ cut in a static network
$c_p(S, T)$	Capacity of an $s - t$ cut in a probabilistic network
$R_{st}(G_{st})$	Reliability of route between s and t
R	Reliability of an entire network
$\Re_{st}$	Data flow between s and t
z_{st}	Probability that a data flow occurs between s and t
$l(i, j)$	Link latency between two nodes i and j
$L(s, t)$	Path latency between the a source node s and its target node t
F_p	Probability that the network gets disconnected

3 Network Reliability Analysis

This section presents the reliability analysis of an aerial network. It outlines the concept of reliability in the context of an aerial network and provides an approach to estimate the reliability of a path between a source node, s, and a terminal node, t [3].

An aerial network is characterized by fast moving aircraft systems. Thus, a link failure is primarily attributed to mobility of a node. A link failure may be temporary because an inactive link may become active again when the node comes back within the range of another node that is connected to the network. On the other hand, if a node fails, it will be removed from the aerial network. The topology of the network might change when a node is disconnected from one node and is connected back again possibly to a different node. Hence, it is reasonable to assign a probability of failure to every link in the network. A probabilistic graph is appropriate representation of an aerial network.

An overlay network is created while a node s is discovering a path to its destination t. Although the graphs, in general, may be directed, we consider undirected graphs for simplicity of analysis. The model can easily be extended to directed graphs as well. While the probability of failures may be different from one link to another, for simplification, it is assumed that the failure probabilities are same for all the links, i.e., $p_{ij} = p$ and they are independent of each other.

For illustration purpose, let us consider a benchmark graph from among those given in [2] shown in Fig. 2. It represents a typical overlay network created during a route discovery process.

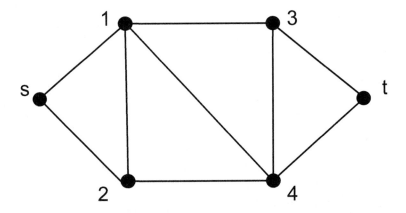

Fig. 2. An illustration of an overlay network [2]

Reliability is a performance measure for the overlay network created during RDP between two nodes. Network reliability can be computed as a function link failure probabilities and cut-sets in the corresponding graph. The problem of enumerating all cut-sets in a graph is an NP-hard problem, for which a solution is proposed in [2]. The graph shown in Fig. 2 has the following cut-sets:

$$C_{st}(1) = \{e_{s1}, e_{s2}\}$$
$$C_{st}(2) = \{e_{3t}, e_{4t}\}$$
$$C_{st}(3) = \{e_{s1}, e_{12}, e_{24}\}$$
$$C_{st}(4) = \{e_{13}, e_{14}, e_{24}\}$$
$$C_{st}(5) = \{e_{13}, e_{34}, e_{4t}\}$$
$$C_{st}(6) = \{e_{s2}, e_{12}, e_{13}, e_{14}\}$$
$$C_{st}(7) = \{e_{14}, e_{24}, e_{34}, e_{3t}\} \tag{1}$$

Assume that there are m physical links in a network between s and t, i.e., ($|E_{st}| = m$). Let p represent the probability of link failure. The failure probability of a network state NS with exactly i physical link failures, i.e., ($|NS| = i$) is $p^i(1 - p)^{m-i}$. Let N_i be the number of disconnected states NS with $|NS| = i$. Then, the probability that the network gets disconnected (F_p) is the sum of the probabilities over all disconnected states, i.e.,

$$F_p = \sum_{i=0}^{m} N_i p^i (1 - p)^{m-i} \tag{2}$$

Reliability of a two-terminal network is defined as the probability of having atleast one path between the two nodes [5]. When viewed as the complement of the network failure probability, it can be expressed as follows:

$$R_{st}(G_{st}) = 1 - \sum_{i=0}^{m} N_i p^i (1 - p)^{m-i}. \tag{3}$$

Network failure states (NS) can be completely characterized by cut-sets. With the use of cut-sets, reliability $R_{st}(G_{st})$ of the network $G_{st}(V, E_{st}, P_{st})$ [3] can be expressed in the following closed form:

$$R_{st}(G_{st}) = 1 - \sum_{i=nc}^{m} | C_{st}(i) | p^i (1 - p)^{m-i} \tag{4}$$

where $m = |E_{st}|$ is the cardinality of the edge set E_{st}, nc is the number of cut-sets, and $|C_{st}(i)|$ is the cardinality of cut-set with exactly i edges. The reliability of the entire overlay network can be defined as [3]

$$R = \frac{\sum_{s \in V} \sum_{t \in V, t \neq s} z_{st} R_{st}(G_{st})}{n(n - 1)} \tag{5}$$

where $n = |V|$, and z_{st} is the probability that a data flow occurs between the two nodes s and t.

4   Throughput Analysis

Throughput of a network can be estimated using cut-sets of a graph that represents the network. The concept of max-flow min-cut strategy to estimate the throughput of a network was first introduced in [4]. This section extends this concept to probabilistic networks.

Cut-set: An $s - t$ cut is a partition of V such that $s \in S, t \in T$, and S and T are disjoint subsets of V. The capacity of a $s - t$ cut is defined as follows:

$$c(S, T) = \sum_{(u,v) \in (S \times T), (i,j) \in E} c_{ij} d_{ij} \tag{6}$$

where c_{ij} is the capacity of the link e_{ij} and $d_{ij} = 1$ if $i \in S$ and $j \in T, 0$ otherwise. Minimum $s - t$ cut is obtained by minimizing $c(S, T)$.

Max-flow min-cut: The max-flow min-cut theorem suggests that the maximum amount of data passing from the source (s) to the target (t) in a network is equal to the amount of flow corresponding to the minimum $s - t$ cut [4].

Throughput for a probabilistic network needs to take the reliability of the links into account. In its simplistic form, throughput of an unreliable link can be obtained by multiplying the amount of flow on the link with its reliability. Thus, the capacity of an $s - t$ cut in a probabilistic network can be expressed as follows:

$$c_p(S, T) = \sum_{(u,v) \in (S \times T), (i,j) \in E} c_{ij}(1 - p_{ij})d_{ij} \tag{7}$$

where p_{ij} represents the failure probability of the link e_{ij}. The minimum $s - t$ cut for a probabilistic network will be different from a static network. Hence, the throughput of a probabilistic network will be different from that of a static network.

5   Latency Analysis

Link latency ($l_{i,j}$) is a parameter that characterizes an aerial communication link (i, j) between two nodes i and j. If a path consisting of n number of nodes exists between a source s node and its target (t) node, then, the path latency, $L(s, t)$, is the sum of the latencies corresponding to the sequence of links $\{(s, 1), (1, 2), \ldots, (i, i+1), \ldots, (n-1, t)\}$ that constitute the path (s–t).

$$L(s, t) = l_{s,1} + \sum_{i=1}^{n-2} l_{i,i+1} + l_{n-1,t} \tag{8}$$

Path latency can be viewed as the end-to-end delay between the source and target nodes assuming that there is no queuing delay. Latency is a deterministic parameter for a given path unlike throughput which is a function of the reliability of the communication links between the source and target nodes.

6   Discussion and Summary

Modeling an aerial network as a dynamic graph, this paper presents the basic concepts of reliability, throughput and latency as related to an aerial network with intermittent links. The objective of this analysis is to characterize an aerial network and to get insights into the performance of the network in terms of its capabilities and limitations. The analysis presented here is suitable for a scenario in which the topology of the network is known. It will be extended to a more general framework that includes strategies for identifying most reliable and most critical paths in the network and a study of topologies that maximize the network performance.

Acknowledgments. This material is based upon work supported by the Air Force Research Laboratory under the visiting faculty research program and the National Science Foundation under Grant No. 1622978.

References

1. TTNT.   https://www.rockwellcollins.com/Data/Products/Communications_and_Networks/Networks/Tactical_Targeting_Network_Technology.aspx.   Accessed   25 July 2016
2. Benaddy, M., Wakrim, M.: Cutset enumerating and network reliability computing by a new recursive algorithm and inclusion exclusion principle. Intern. J. Comput. Appl **45**, 22–25 (2012)
3. Caleffi, M., Ferraiuolo, G., Paura, L.: A reliability-based framework for multi-path routing analysis in mobile ad-hoc networks. Int. J. Commun. Netw. Distrib. Syst. **1**(4–6), 507–523 (2008)
4. Elias, P., Feinstein, A., Shannon, C.: A note on the maximum flow through a network. IRE Trans. Inf. Theor. **2**(4), 117–119 (1956)
5. Lee, K., Lee, H.-W., Modiano, E.: Reliability in layered networks with random link failures. IEEE/ACM Trans. Netw. (TON) **19**(6), 1835–1848 (2011)

Gender Assignment for Directional Full-Duplex FDD Nodes in a Multihop Wireless Network

Moein Parsinia[1], Qidi Peng[2], Sanjukta Bhowmick[3], John D. Matyjas[4], and Sunil Kumar[5(✉)]

[1] Computational Science Research Center, San Diego State University, San Diego, USA
mparsinia@mail.sdsu.edu
[2] Institute of Mathematical Sciences, Claremont Graduate University, Claremont, CA, USA
qidi.peng@cgu.edu
[3] Computer Science Department, University of Nebraska at Omaha, Omaha, USA
sbhowmick@unomaha.edu
[4] Air Force Research Laboratory, Rome, NY, USA
john.matyjas@us.af.mil
[5] Electrical and Computer Engineering Department, San Diego State University, San Diego, USA
skumar@mail.sdsu.edu

Abstract. The frequency-division duplex (FDD) nodes use two separate frequency bands (separated by a guard band) for transmission and reception, thus enabling the full-duplex (FD) communication. On the other hand, the use of directional FDD nodes in multihop wireless network offers the advantages of larger transmission range, better link reliability, and spatial reuse, resulting in a much higher throughput and superior interference mitigation. However, the multihop FDD communication partitions the nodes in two classes (or genders) wherein the nodes of the same class (or gender) in a neighborhood cannot communicate with each other. This can seriously impact the availability of neighboring nodes for communication, and lead to disconnected nodes (or regions) in the network. In this paper, an algorithm is presented to assign the appropriate genders to these nodes in a multi-hop network such that each node is able to communicate with its multiple 1-hop neighbors, located in different directions. Our simulation results demonstrate that approximately half of the neighbors of each node are of the opposite gender and they are distributed in different directions, thus enabling robust, multipath, and high throughput communication in the network.

Keywords: Frequency Division Duplex (FDD) · Full-duplex communication · Gender assignment · Frequency assignment · Directional antenna · Graph coloring · Multihop network

© ICST Institute for Computer Sciences, Social Informatics and Telecommunications Engineering 2017
Y. Zhou and T. Kunz (Eds.): ADHOCNETS 2016, LNICST 184, pp. 390–401, 2017.
DOI: 10.1007/978-3-319-51204-4_32

1   Introduction

Multihop wireless networks are commonly used for the range extension and throughput improvement in ad hoc and peer-to-peer networks, with applications in sensor networks and emergency/disaster communication. Recently, the multihop communication (in the form of relaying) is also being used in wide-area cellular systems [1–3]. The multihop transmission can use time division duplex (TDD) or frequency division duplex (FDD) modes, both having their pros and cons. FDD uses two distinct frequency channels for transmitting and receiving the packets at the same time, while TDD operates on only one frequency channel where the transmission and reception of the packet(s) is interleaved in time. Since FDD does not need a guard time interval to separate the transmission from the reception, it is more efficient in wide area scenarios with long transmission links which introduce non-negligible signal propagation delays [4]. The full-duplex and/or half-duplex FDD modes have been proposed for the mobile radio systems like 3GPP LTE [3], GSM/GPRS [5], UMTS [6], WiMAX [7]. Recently, the FDD has also been proposed for 5G networks [8].

In an FDD system, the available frequency band (F) is divided in two distinct bands (F1 and F2), with a guard band to separate them. The presence of guard band can allow a node to cancel the self-interference by sufficiently attenuating the received side-band power from its own transmission by using RF duplex filter. The node can therefore transmit the data on F1 and receive on F2 at the same time, leading to the full-duplex communication. In FDD, the F1 as well as F2 frequency bands can be grouped into blocks of contiguous channels. The available spectrum can thus be allocated to support the multiple users simultaneously on different bands. Since the transmissions on F1 and F2 bands in FDD systems are generally continuous, the receiver can feedback the most recent channel information to the transmitter. Thus, the delay is reduced for the channel information feedback, medium access control, and retransmission, which can enhance the system throughput [4].

Use of Directional Communication: The increasing use of high resolution multimedia applications by a rapidly increasing number of mobile users is making the wireless resources scarce. The proliferation of 5G devices would further aggravate this problem. As a result, the designers are considering the use of above 6 GHz frequency bands (e.g., 12–18 GHz, 28 GHz, 60 GHz and 72–73 GHz bands), as larger RF spectrum bandwidth is available at these frequencies. However, the attenuation experienced by the signals (i.e., the path losses caused by the atmospheric path loss) is much higher at higher frequencies. So higher frequency links require more power. Use of directional antennas can be very helpful in realizing high data rate links as they increase the transmission range, besides offering spatial reuse [9,10]. The directional antennas are also more resistant to the interference as any unwanted signal which is not in the direction of the antenna beam has little impact on the node. Moreover, the use of multiple beam directional antennas (MBDA) in a wireless node allows the simultaneous transmission or reception on different beams using the same channel. As a result,

Fig. 1. FDD communication: (a) A node can communicate only with another node of opposite gender; (b) Two nodes of the same gender cannot communicate with each other.

the system capacity can be enhanced considerably. Fortunately, the size of these antennas also becomes more manageable at higher frequency bands.

In this paper, we consider a multihop wireless network consisting of the FDD nodes equipped with MBDA, which are operating in full-duplex mode. However, this partitions the nodes available in the network in two classes (or *genders*) as described below. If node A uses F1 for signal transmission and F2 for reception (we denote it as the **male node**), it can establish a link to communicate with another node B only if the latter receives the signal at F1 and transmits at F2 (we denote it as the **female node**). Node A cannot communicate with another node of the same gender (i.e., male node) that also transmits at F1 and receives at F2 (see Fig. 1). This requires that the frequency bands F1 and F2 in FDD nodes be chosen carefully for the transmission and reception such that a node could effectively use its multiple beams to establish links with its multiple 1-hop neighbors, in order to simultaneously communicate with them. *We call this problem as the gender assignment of the node.*

When multiple nodes are available in a 1-hop neighborhood of a given node, the gender should be assigned such that this node can establish communication links with multiple neighboring nodes spread in different directions. This can help in establishing the routes which pass through the nodes in different parts of the network while avoiding the congested network nodes and regions. In fact, the gender should be assigned such that each node can establish links with approximately half of its 1-hop neighbors in different directions. This strategy allows almost all the nodes in the graph to have communication links with nearly half of their 1-hop neighbors.

2   Proposed Distributed Gender Assignment Algorithm

We consider solving the gender assignment problem as a simple undirected graph coloring problem. Recall that a simple undirected graph $G = (V, E)$ consists of a set of vertices V and a set of edges E. Each edge $e \in E$ connects two vertices $u, v \in V$. This edge is thus represented as $e = \{u, v\}$. For two vertices $u, v \in V$, u is called a neighbor of v if u and v are joined by an edge e (i.e., $\{u, v\} \in E$). Throughout this paper, we use graph to mean simple undirected graph for short.

We can create a representative graph of the wireless network as follows: Each wireless node is represented as a vertex in the graph. Two vertices are joined if the corresponding nodes are within a specified distance d of each other (i.e., within a 1-hop distance). Since the nodes are assigned any one of two genders (i.e. male or female), the connections can be viewed as a bipartite graph. Recall that in a bipartite graph (V, E), the set of vertices V consists of two disjoint groups V_1 and V_2, in the sense that each edge in E only connects a vertex in V_1 to another vertex in V_2 [11]. In the representative wireless graph, the nodes of each gender fall into a group. Then the connections between the nodes that can communicate with each other form a bipartite graph. It should also be noted that we want each node to connect to approximately half of its 1-hop neighbors.

Based on these observations, we now formulate our problem as follows. Let the representative graph of the wireless nodes be $G = (V, E)$. Assume genders are assigned (two colors) to the vertices. Therefore for each vertex v, its neighbors are either of opposite or the same gender. We denote $N_B(v)$ to be all the neighbors of different gender from v, and $N_C(v)$ to be all the neighbors of the same gender as v. *Our goal is to solve the following optimization problem: determining a gender assignment such that $|N_B(v)| \approx |N_C(v)|$ for each vertex $v \in V$* (where $|A|$ denotes number of elements in the set A). This will satisfy the requirement that each node has approximately equal number of neighbors from each gender.

Although this problem seems to be similar to extracting the maximal (in the sense that the number of vertices is maximized) bipartite subgraph from some graph, it has certain additional interesting aspects as discussed below. The problem of finding the maximal bipartite subgraph is NP-complete. Here our problem is different and more complicated, because our target is not finding the "largest" subgraph, but rather an almost equal division between the 1-hop neighbors of each node. Moreover, because the wireless connections of all the nodes may not be known globally, we have to do this assignment using local information, one vertex at a time.

We present our *heuristic distributed gender assignment algorithm* below which attempts to maintain a balance between the number of male (M) and female (F) nodes in each 1-hop neighborhood. Also note that the placement of the male and female nodes in the 1-hop neighborhood of any node, in terms of the direction from this node, should be almost evenly distributed. The advantage of our algorithm is that it "locally optimizes the solution", i.e., it makes sure that each node has almost equal number of neighbors with different genders in its 1-hop neighborhood.

Before presenting the pseudocodes of our algorithm, we make some notation conventions. For each node v, $g(v)$ denotes its gender assignment ($1 = $ M and $0 = $ F). Then we note that $1 - g(v)$ is the opposite gender assignment to v's. We also denote $deg_G(v)$ to be the number of 1-hop neighbors of v in the graph G. We call G_1 a complementary graph of G_2 in G, if G_1, G_2 have disjoint sets of vertices, disjoint sets of edges and $G = G_1 \cup G_2$. Moreover, if the set of vertices of G_2 is V_2, we denote $G_1 = G \backslash V_2$. Finally we use $\arg \max_{v \in V} deg_G(v)$ to denote the set of vertices in V having the largest number of neighbors in G (V is not necessarily the set of vertices of G). As a main result, the pseudocodes of our *heuristic distributed gender assignment algorithm* is given below:

Pseudocodes: Heuristic Distributed Gender Assignment Algorithm

1: **INPUT:** a visible subgraph $G = (V, E)$ of the global graph, with $V = \{v_1, \ldots, v_n\}$ and $E = \{e_1, \ldots, e_m\}$

2: *Step 1: First Seed Node Selection and its Gender Assignment.*

3: Randomly equally likely pick v_0 in V. Denote by N_0 the set of v_0's 1-hop neighbors in G.

4: Randomly equally likely pick v in $\arg \max\limits_{u \in N_0} deg_G(u)$; $g(v) \leftarrow 1$;

5: *Step 2: Gender Assignment in 1-hop Neighborhood of the First Seed.*

6: *Determine 1-hop neighborhood. Let $(v_1, \ldots, v_p)$ be all the neighbors of v in clockwise order, starting from the North Direction.*

7: $N \leftarrow (v_1, \ldots, v_p)$;

8: *Assign genders to 1-hop neighborhood of v.*

9: **for** k in $\{1, \ldots, p\}$, **do:**

10:     if k is odd, do: $g(v_k) \leftarrow 0$; else do: $g(v_k) \leftarrow 1$;

11: **end for**

12: *Step 3: Next Seed Node Selection.*

13: *Consider the complementary graph of unassigned nodes G'.*

14: $G' \leftarrow G \backslash N$; randomly equally likely pick v' in $\arg \max\limits_{u \in N \cup \{v\}} deg_{G'}(u)$;

15: *Step 4: General Gender Assignment in 1-hop Neighborhood (Phase 1).*

16: *Determine unassigned 1-hop neighborhood. Let $(v'_1, \ldots, v'_q)$ be all the unassigned neighbors of v' in clockwise order, starting from the North Direction.*

17: $N' \leftarrow (v'_1, \ldots, v'_q)$;

18: *Assign genders to unassigned 1-hop neighborhood of v'.*

19: **for** k in $\{1, \ldots, q\}$, **do:**

20:     if $|N_B(v')| - |N_C(v')| > 2$, do: $g(v'_k) \leftarrow g(v')$;

21:     else if $|N_C(v')| - |N_B(v')| > 2$, do: $g(v'_k) \leftarrow 1 - g(v')$;

22:     else do: $N'_k \leftarrow$ "1-hop neighborhood of v'_k";

23:         if $|\{u \in N'_k, \ g(u) = 1\}| > |\{u \in N'_k, \ g(u) = 0\}|$, do: $g(v'_k) \leftarrow 0$;

24:         else if $|\{u \in N'_k, \ g(u) = 1\}| < |\{u \in N'_k, \ g(u) = 0\}|$, do: $g(v'_k) \leftarrow 1$;

25:         else do: $g(v'_k) \leftarrow g(v')$;

26: **end for**

27: *Step 5: General Gender Assignment in 1-hop Neighborhood (Phase 2).*

28: **for** k in $\{1, \ldots, q\}$, **do:**

29:     if v'_k disconnects other 1-hop neighbors of v', do: $g(v'_k) \leftarrow 1 - g(v')$;

30: **end for**

31: *Step 6: Run Steps 3-5 with updated G until all the nodes are assigned genders.*

32: **OUTPUT:** Graph G with gender-assigned nodes.

The following example uses a connected graph with 16 nodes. We assume that the 1-hop neighbors of a node are known. The algorithm iteratively selects a seed node in each step, and assigns the gender to the seed node and its 1-hop neighboring nodes. The 1-hop neighbors of a node are shown connected with the node by a solid or broken link in the graphs in our examples.

1. *First Seed Node Selection:*

   The first seed node is selected from a group of nodes consisting of a randomly selected starting node and its 1-hop neighbors, as described below. The number of 1-hop neighbors (whose gender has not yet been assigned) of each node from the above group, is recorded and the node with the maximum number of unassigned 1-hop neighbors is selected as the first seed node. If more than one node has the maximum number of unassigned 1-hop neighbors, one of them is randomly selected as the first node. This seed node is assigned the "male" gender.

   For example in Fig. 2a, node 5 is selected as the starting node. It has four 1-hop neighbors (9, 10, 15, 14). Among these five nodes, node 15 is selected as the first seed and assigned the male gender, because it the maximum number of seven unassigned 1-hop neighbors.

2. *Gender Assignment in 1-hop Neighborhood of the First Seed:*

   Gender of the 1-hop nodes of first seed is assigned in an alternating order (e.g., M followed by F, or F followed by M) in the clockwise direction, starting from the node in the North direction, such that all the nodes in 1-hop neighborhood are equally (or nearly equally) divided in two genders. The first 1-hop node is assigned a gender opposite of the seed node. Two consecutive nodes in the clockwise order are assigned opposite genders. Note that our algorithm will work equally well if we select the anti-clockwise order and chose a node starting in a direction other than the North, as long as the same convention is followed.

   In Fig. 2a, Node 15 is the first seed node and its 1-hop neighbors in the clockwise order starting from North are 3, 1, 4, 14, 5, 10 and 7. The gender to

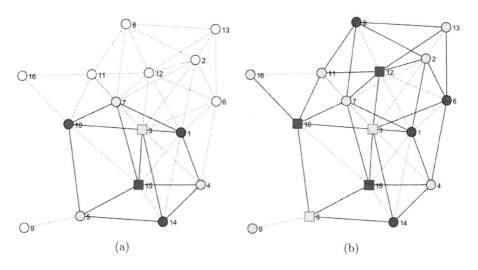

(a)   (b)

Fig. 2. (a) First seed selection for the neighborhood of node 5 and gender assignment in the neighborhood of node 15, (b) Gender assignment in phase 2 for the neighborhood of seed node 5. Node 9 will change to a male gender to keep it connected.

first node (node 3) is assigned as F (which is opposite of seed node) and the remaining nodes are assigned gender in an alternating order in the clockwise direction. The result of this assignment is shown in Fig. 2a.

3. *Next Seed Selection:*
The next seed is selected from the list of all the assigned nodes in the graph. The node which has the maximum number of the unassigned nodes in its 1-hop neighborhood is the candidate seed node. If more than one such nodes exist, the seed is randomly selected from the list of candidate seed nodes.

In Fig. 2a, node 3 is selected as the next seed (after seed 15) since it has the maximum number of four unassigned neighbors with nodes [12, 2, 6, 11].

4. *General Gender Assignment in 1-hop Neighborhood (Phase 1):*
A gender to each unassigned 1-hop neighbor node of the current seed is assigned in the clockwise order, by considering the following two criteria: (*i*) The difference in the number of male and female nodes in the 1-hop neighborhood of the seed should not exceed a threshold (we use a threshold of 2 here), and (*ii*) The difference in the number of already-assigned male and female nodes in its 1-hop neighborhood, including the gender of the current node is minimized.

For example, in Fig. 2a, node 3 is the seed with nodes [12, 2, 6, 11] as its unassigned 1-hop neighbors in the clockwise order. For deciding the gender of the first node 12, we first check the difference between the male and female nodes in the 1-hop neighborhood of seed node 3, which is one. In the 1-hop neighborhood of node 12, the nodes 3 and 7 are female and nodes 1 is the only male. Therefore, its gender is assigned as male which satisfies both the above criteria. For the next node 2, again we first check the difference between the male and female nodes in the 1-hop neighborhood of seed node 3 which in this case is three; so node 2 is assigned the female gender to minimize the difference between the male and female nodes in the neighborhood of seed node based on criteria (i).
Exception: If there are equal number of already-assigned male and female nodes in the 1-hop neighborhood of a given node, its gender is assigned as the opposite gender of the current seed.

5. *General Gender Assignment in 1-hop Neighborhood (Phase 2):*
After assigning gender to all the 1-hop neighbors of the current seed node in Phase 1, the connectivity of each of these nodes is checked. If any node is left unconnected, its gender is changed to the opposite gender of the current seed. For example, the node 9 in the neighborhood of seed node 5 in Fig. 2b is left unconnected in Phase 1. Therefore, its gender will change to male, which is opposite gender of seed. Thus, node 9 will have one link available for communicating with its one hop neighbor (node 5).

6. The next seed is selected as discussed in Step 3 and gender assignment performed in its 1-hop neighborhood until all the nodes are assigned a gender.

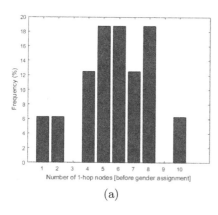

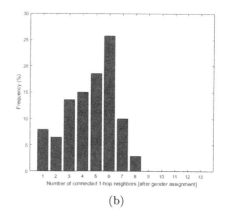

(a)   (b)

Fig. 3. (a) Histogram of the number of 1-hop neighbors before the gender assignment, (b) Histogram of the number of connected 1-hop neighbors after the gender assignment, for the random graph of 16 nodes shown in Fig. 2b.

3   Performance Evaluation

In this section, performance of the proposed gender assignment algorithm is evaluated for two different graphs. The first example is a random graph with 16 nodes (Fig. 2b) and the second example is a 5×5 grid graph. The results are average for 1000 runs with different starting nodes. The 1-hop neighbors of a node are determined such that their Euclidean distance from the node is below a threshold. A histogram representing the frequency of occurrence of the number of 1-hop neighbors for each node in the random graph of 16 nodes (before gender assignment) is shown in Fig. 3a. For example, an average of 12 % of nodes have 4 nodes in their 1-hop neighborhood (before gender assignment). A node can communicate with other nodes of opposite gender in its 1-hop neighborhood. We call these nodes as the "connected 1-hop neighbors". Figure 3b shows the histogram representing the frequency of occurrence of the connected 1-hop neighbors for each node in the graph, after the gender assignment. We observe that more than 90% nodes have at least 2 connected 1-hop neighbors, and about 85% nodes have 3 or more connected 1-hop neighbors. Figure 4a shows the number of male and female nodes in the 1-hop neighborhood of each node of the graph. We observe that the 1-hop neighbors of a node are almost equally divided among the male and female genders for all the 16 nodes of the graph. A histogram of the difference between the number of male and female nodes in 1-hop neighborhood of each node of the graph is shown in Fig. 4b. This plot further verifies that the maximum difference between the number of male and female nodes is less than 2.2 in 1-hop neighborhood of each node.

A sample of the gender assignment output for a 5×5 grid graph is shown in Fig. 5a. Before the gender assignment in this graph, nine nodes had eight 1-hop neighbors each, 12 edge nodes had five 1-hop neighbors each, and the four corner nodes had 3 neighbors each. After the gender assignment (see Fig. 5b), we

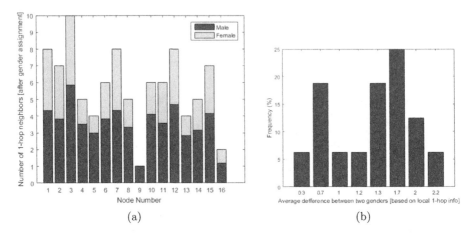

(a)   (b)

Fig. 4. (a) Gender distribution in the 1-hop neighborhood of each node, (b) Histogram of the difference between two genders among 1-hop neighbors, for the random graph of 16 nodes shown in Fig. 2b.

observe that all nodes have at least two connected 1-hop neighbor, and 80% nodes have three or more connected 1-hop neighbors. Figure 6a shows that the 1-hop neighbors of every node are almost equally divided among the male and female genders. Figure 6b shows that the maximum difference between the number of male and female nodes in the 1-hop neighborhood of each node of the graph is less than 2.3.

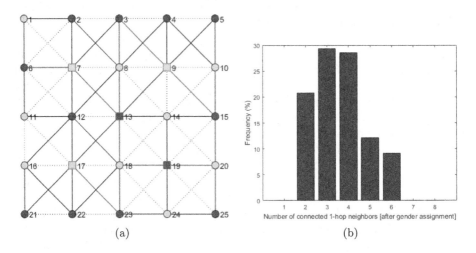

(a)   (b)

Fig. 5. (a) A sample gender assignment result, for the 5 × 5 grid graph, (b) Histogram of the number of connected 1-hop neighbors after gender assignment.

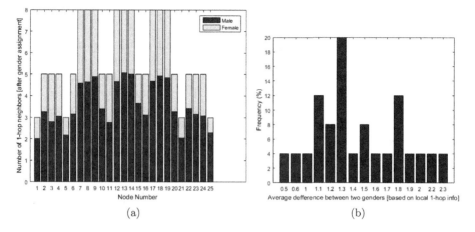

(a)   (b)

Fig. 6. (a) Gender distribution in the 1-hop neighborhood of each node, (b) Histogram of the difference between the nodes of two genders among 1-hop neighbors.

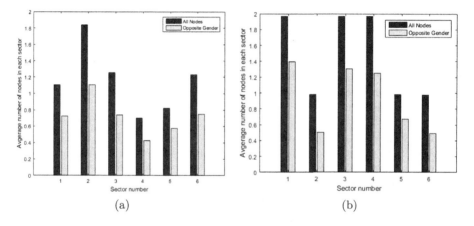

(a)   (b)

Fig. 7. Average distribution of genders in 1-hop neighborhood of all the seed nodes: (a) for the random graph in Fig. 2b, (b) for the grid graph in Fig. 5a. The results are average of 1000 runs.

Average distribution of nodes in different directions for both graphs is illustrated in Fig. 7. We have used six sectors of 60° each in these plots, which cover all the directions. We observe that our gender assignment algorithm assigns nearly half the 1-hop nodes to opposite gender in each sector. The Y-axis of these plots shows fractional number because we have taken an average for all the seeds over 1000 runs.

Our algorithm thus assigns equal (or nearly equal) number of colors (genders) in every neighborhood for both graphs. After the gender assignment, there isn't any unconnected node in either graph and each node can communicate with approximately half of its neighbors in different directions.

4   Conclusion and Future Work

We considered a multihop wireless network consisting of the directional FDD nodes equipped with MBDA, operating in full-duplex mode. Such networks offer the advantages of higher throughput, better link reliability, and superior interference mitigation. However, the multihop FD-FDD communication partitions the nodes in two classes (or genders) wherein the nodes of the same class (or gender) cannot communicate with each other. This can seriously impact the communication between neighboring nodes. Therefore, the gender (or frequency channel) of these nodes should be selected such that each node is able to simultaneously communicate with its multiple 1-hop neighbors, located in different directions.

A heuristic, distributed gender assignment algorithm was presented to assign the appropriate gender to these directional, FD-FDD nodes in a multi-hop network. Our simulation results demonstrated that our algorithm successfully assigned the gender to all the nodes in the network, such that each node could establish communication links with approximately half of its 1-hop neighbors in different directions. This arrangement can also help in establishing the robust and multi-path routes which pass through nodes in different parts of the network while avoiding the congested network nodes and regions. To the best of our knowledge, this is the first algorithm to extract a balanced bipartite graph, where the neighbors of the bipartite graph is roughly equal to the remaining neighbors in the graph. In addition to being an interesting and complex problem, this problem is solved using only local node information, and not the global knowledge of the network.

Acknowledgments and Disclaimer. This work was supported by U.S. Department of Defense under Grant No. FA8750-14-1-0075. Approved for public release; Distribution Unlimited: 88ABW-2016-3523, 18 July 2016. Any opinions, findings and conclusions or recommendations expressed in this material are those of the author(s) and do not necessarily reflect the views of DOD or U.S. government. The authors also acknowledge Drs. John Boyd and Kevin Chang, Cubic Corp. for technical discussions.

References

1. Otyakmaz, A., Walke, B.H.: Concurrent operation of half- and full-duplex terminals in future multi-hop FDD based cellular networks. In: International Conference on Wireless Communications Network and Mobile Computing, October 2008. 10.1109/WiCom.2008.72
2. Pabst, R., et al.: Relay-based deployment concepts for wireless and mobile broadband radio. IEEE Commun. Mag. **44**, 80–89 (2004)
3. Schoenen, R., et al.: On PHY and MAC performance of 3G-LTE in a multi-hop cellular environment. In: WiCom 2007, September 2007
4. Chan, P.W.C., Lo, E.S., et al.: The evolution path of 4G networks: FDD or TDD. IEEE Commun. Mag. **44**, 42–50 (2006)
5. ETSI. GSM recommendations, 04 May1993, Data Link Layer - General aspects (1993)

6. 3GPP, TS 25 211, Physical Channels and Mapping of Transport Channels onto Physical Channels (FDD) (2001)
7. IEEE Std, I. 802.16-2004, IEEE Standard for Local, Metropolitan Area Networks, Part 16: Air Interface for Fixed Broadband Wireless Access Systems, October 2004
8. Pedersen, K.I., et al.: A flexible 5G frame structure design for frequency-division duplex cases. IEEE Comm. Mag. **54**, 53–59 (2016)
9. Ramanathan, R.: On the performance of Ad hoc networks with beamforming antennas. In: Proceedings of ACM MobiHoc, pp. 95–105 (2001)
10. Jain, V., Gupta, A., Agrawal, D.P.: On-demand medium access in multihop wireless networks with multiple beam smart antennas. IEEE Trans. Parallel Distrib. Syst. **19**(4), 489–502 (2008)
11. Feige, U., Kogan, S.: Hardness of approximation of the Balanced Complete Bipartite Subgraph problem. Department of Computer Science and Applied Mathematics, Weizmann Institute of Science, Rehovot, Israel, Technival Report MCS04 2004 (2004)

A Massive MIMO Panel Array at Ka-Band with Flexible Patterns and Beam Steering Performance

Satish K. Sharma[(⊠)] and Sandhya Krishna

Antenna and Microwave Lab (AML),
Department of Electrical and Computer Engineering,
San Diego State University, 5500 Campanile Drive,
San Diego, CA 92182-1309, USA
ssharma@mail.sdsu.edu,
sandhyakrish2014@gmail.com

Abstract. In this paper, we are presenting a software defined based massive multiple input multiple output (MIMO) panel array antenna with flexible radiation patterns and beam steering application. This design involves a Ka-band (31.7 GHz to 33.46 GHz) 4×4 microstrip patch planar array with dual linear polarizations and good isolation which is required for a massive MIMO antenna system. In this case, by controlling input excitations of the selected radiating elements in the array, we generate radiation patterns with variable beamwidths or directivity at the broadside angle. Further, by applying variable amplitude and phase excitations to the partial or full array, we generate beam steering radiation patterns. Additional analysis and measured performance results will be presented during the conference.

Keywords: Software defined phased array · Massive MIMO panel · Microstrip patch · Dual linear polarization · Millimeter wave

1 Introduction

High gain directional antennas are highly desired in long range communication scenarios such as multipoint communications, airborne communication networks, airborne to ground communications, satellite to ground terminals and satellite to airborne platforms, such as shown in Fig. 1. The antenna design becomes more challenging when the separation between the transmitter and receiver system is variable such as an airborne communication system communicating with another such airborne communication system. Similarly, if the same communication system is involved with other communication platforms then we would need multiple radiating beams. These can be implemented using a software defined or digital beam forming (DBF) approach where array beams, both in transmit or receive mode, are constructed in virtual domain.

To accomplish above discussed antenna system design, we need an array antenna aperture with individual input excitations such as in a massive multiple input multiple output (MIMO) communication system. This array aperture can be reconfigured by selective excitations of the radiating elements so that we can generate radiated beams

© ICST Institute for Computer Sciences, Social Informatics and Telecommunications Engineering 2017
Y. Zhou and T. Kunz (Eds.): ADHOCNETS 2016, LNICST 184, pp. 402–413, 2017.
DOI: 10.1007/978-3-319-51204-4_33

with flexible beamwidths or directivity for a chosen polarization and matching band-width. Similarly, by simultaneous excitation of a group of radiating elements with proper amplitude and phase excitations, we can generate multiple radiated beams in the digital domain. Beams can be steered by applying variable amplitude and progressive phase shifts, therefore a lot can be achieved with one array antenna aperture with the help of software defined approach contrary to an analog beam forming network (BFN) based array antenna where flexibility is limited.

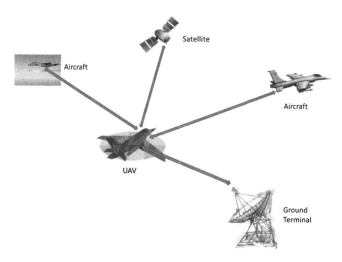

Fig. 1. Scenario showing the airborne (UAV) communication platform communicating with other communication platforms.

There are some research efforts towards the introduction of a massive MIMO antenna for beamforming which are documented in [1–3]. The design challenge involves a large size array starting from several radiating elements to a 100th of radiating elements in an array. Some conventional 2 elements MIMO antennas for several handheld communication platforms are presented in [4–6]. Different phased arrays antennas and beam forming algorithms can be found in [7–10]. However, in this paper, design of a massive MIMO panel array (4 × 4 subarray) with dual linear polarization at millimeter wave frequency is presented which can be used as a building block for realizing a large size massive MIMO antenna system. Results include impedance matching, isolation and radiation patterns for different selective excitations of the array radiating elements.

2 Massive MIMO Panel Array Geometry

Figure 2 shows the proposed massive MIMO panel array geometry where each of the radiating elements are dual linear polarized (Fig. 2(a)). The patch is fed using micro-strip transmission lines from below the ground plane using via connections to the patch, where via refers to a wire. The 4 × 4 subarray or panel array is shown in Fig. 2(b) where patch to patch inter-element spacing is maintained around 0.585λ (free space wavelength (λ) is at the design frequency of 32.6 GHz). Overall subarray or panel size is 23.2 mm × 23.2 mm. Since feedlines are placed below the ground plane, any spurious radiation from these would be eliminated. Additionally, it would be easy to connect software defined radios from this arrangement than when feed ports are placed on the same side as the radiating elements. In the fabricated version of this array, we plan to employ stripline instead of the microstrip feedlines so that the 50Ω K-connectors can be placed.

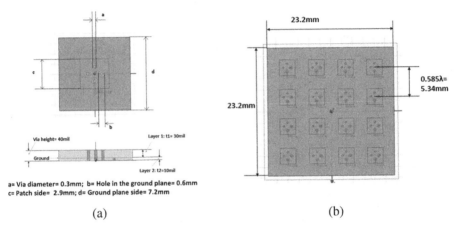

(a)   (b)

Fig. 2. Proposed millimeter wave (Ka-band) (a) Single microstrip patch antenna with dual linear polarizations and (b) a 4 × 4 massive MIMO panel array with dimensions and inter-element spacing.

3   Impedance Matching and Isolation Performance

The impedance matching (S_{ii}/S_{jj}) for both polarizations (X-linear and Y-linear) are shown in Fig. 3(a). From this figure, it can be seen that, impedance matching of $S_{11} \leq -10$ dB is maintained between 32.20 GHz to 34.20 GHz for all the X-polarized feed ports. The matching performance for Y-polarized feed ports is shown in Fig. 3(b) which has the same bandwidth of 32.20 GHz to 34.20 GHz.

Figure 4 shows isolation between (i) the two linear ports in each of the patches and (ii) the adjacent patches for both X-polarized and Y-polarized feed excitations. The good isolation between the two linear ports of the individual patches is a desired feature for a massive MIMO antenna system which is shown in Fig. 4(a). From Fig. 4(a), it can be seen that both polarizations in each of the patches are maintaining port isolation of better than 22 dB therefore, the two ports will not couple energy between them. From Fig. 4(b) for X-polarization case, it can be seen that isolation is better than 15 dB throughout the matching band of 32.20 GHz to 34.20 GHz, therefore two adjacent patches do not couple energy that can be of concern. Similarly, Fig. 4(c) for Y-polarization shows that isolation of better than 20 dB is maintained throughout the matching band and therefore, would offer almost no coupling between the patches. Thus, the isolation study confirms that we have good isolation between linear polarizations of the individual patches as well as between the adjacent patches for array implementations, all of which are desired feature of a massive MIMO antenna design.

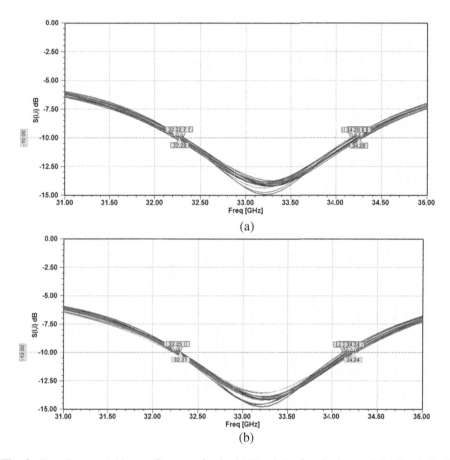

Fig. 3. Impedance matching performance for the (a) X-polarized excitations and (b) Y-polarized excitations.

4 Generation of Flexible Radiation Patterns

Since the array has individual feed port excitations for both X- and Y-polarization cases, hence by exciting the ports with proper amplitude and phase coefficients, we can generate flexible radiation patterns such as discussed in this section. We should make it clear that for simplicity, we have kept amplitude and phase values constant for all the radiating elements which are excited and chosen for the pattern generation.

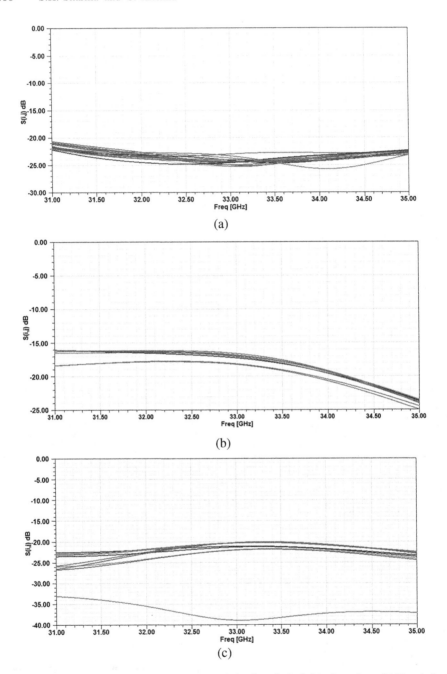

Fig. 4. Isolation between the (a) two linear polarizations in individual patches, (b) X-polarized feed excitations adjacent patches in the array, and (c) Y-polarized feed excitations between the adjacent patches in the array.

4.1   X-Polarization Based Flexible Radiation Patterns

This subsection shows flexible radiation patterns when X-polarization ports in the array are excited while Y-polarization ports are matched terminated. Figure 5 shows a case when we excite single element, and linear arrays of 1 × 2, 1 × 3 and 1 × 4 combinations for the two principal cut planes (Phi = 0 deg and 90 deg). Figure 5(a) shows Phi = 0 deg cut plane patterns where it can be seen that pattern's broadside gain is increasing as array size is increasing hence better directivity. Similarly, Fig. 5(b) shows radiation pattern performance for the Phi = 90 deg cut plane and along the array axis hence the array factor contribution is seen. The patterns show higher gain or directivity and consequently, narrow beamwidths as array size increases. We can do the similar exercise for other linear arrays as part of this planar array, however for the sake of brevity these results are not shown here. Thus, it can be seen that by selective excitation of the radiating elements, radiation patterns with variable gain or beamwidths can be generated.

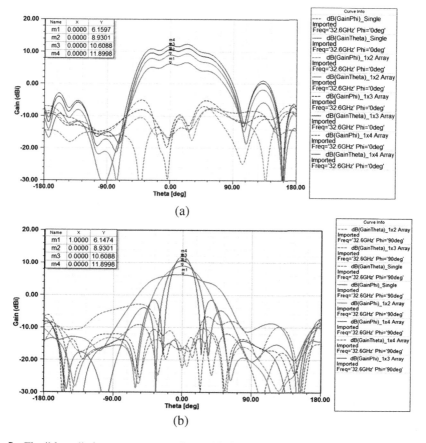

Fig. 5. Flexible radiation pattern generation at 32.6 GHz from the 4 × 4 subarray by selective excitation of the 1 × 4 linear radiating elements for the X-polarization case (a) along one radiating element and (b) along the array axis (single, 1 × 2, 1 × 3, 1 × 4 linear cases).

Figure 6 shows flexible radiation pattern generation for the two principal cut planes (Phi = 0 deg and 90 deg) when we selectively excite radiating elements in planar array fashion, i.e., single, 2 × 2, 3 × 3 and 4 × 4 array cases. From Fig. 6(a), it can be seen that, pattern beamwidth gets narrower as we increase the number of radiating elements. In other words, the broadside gain increases to 17.35 dBi for 4 × 4 array case compared to 6.16 dBi that of a single radiating element. Similar is the observation for Fig. 6(b) which shows the patterns for Phi = 90 deg cut plane. Finally, it should be noted that, there are several other possible linear and planar radiating element combinations which are not included here due to limited paper size.

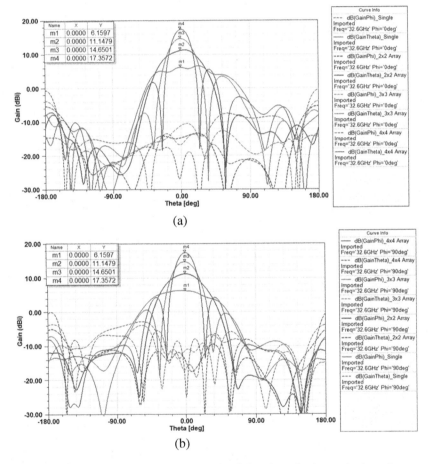

(a)

(b)

Fig. 6. Flexible radiation pattern generation at 32.6 GHz by selective excitation of a group of radiating elements in planar array configuration for the X-polarization (single, 2 × 2, 3 × 3 and 4 × 4 array) (a) Phi = 0 deg cut plane and (b) Phi = 90 deg cut plane.

4.2   Y-Polarization Based Flexible Radiation Patterns

This subsection shows flexible radiation patterns when Y-polarization ports in the array are excited while X-polarization ports are matched terminated. Figure 7 shows a case when we excite single element, and linear arrays of 1 × 2, 1 × 3 and 1 × 4 combinations for the two principal cut planes (Phi = 0 deg and 90 deg). Figure 7(a) shows Phi = 0 deg cut plane patterns from which it can be seen that the pattern broadside gain is increasing as array size is increasing hence better directivity. Similarly, Fig. 7(b) shows radiation pattern performance for the Phi = 90 deg cut plane and along the array axis hence array factor contribution is seen. The patterns show higher gain or directivity and consequently, narrow beamwidths as array size increases. We can do the similar exercise for other linear arrays as part of this planar array, however for the sake of brevity these results are not shown here. Thus, it can be seen that by selective excitation of the radiating elements, radiation patterns with variable gain or beamwidths can be generated.

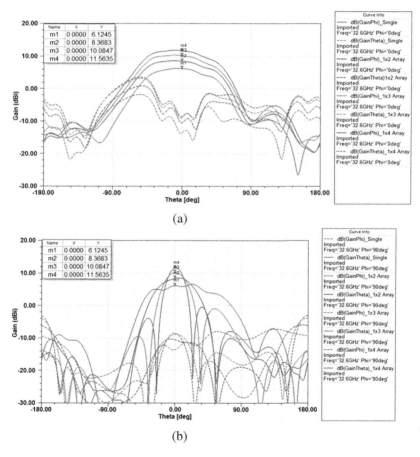

Fig. 7. Flexible radiation pattern generation at 32.6 GHz from the 4 × 4 subarray by selective excitation of the 1 × 4 linear radiating elements for the Y-polarization case (a) along one radiating element and (b) along the array axis (single, 1 × 2, 1 × 3, 1 × 4 linear cases).

Figure 8 shows flexible radiation pattern generation for the two principal cut planes (Phi = 0 deg and 90 deg) when we selectively excite radiating elements in planar fashion, i.e., single, 2 × 2, 3 × 3 and 4 × 4 array cases. From Fig. 8(a), it can be seen that, the pattern beamwidth gets narrower as we increase the number of radiating elements. In other words, the broadside gain increases to 17.37 dBi for 4 × 4 array case compared to 6.12 dBi that of a single radiating element. Similar observation can be drawn for Fig. 8(b) which shows the patterns for Phi = 90 deg cut plane. Finally, it should be noted that, there are several other such linear and planar radiating element combinations which are not included here.

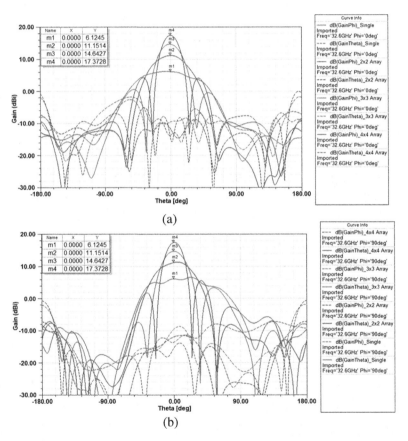

Fig. 8. Flexible radiation pattern generation at 32.6 GHz by selective excitation of a group of radiating elements in planar configuration for the Y-polarization (single, 2 × 2, 3 × 3 and 4 × 4 array) (a) Phi = 0 deg cut plane and (b) Phi = 90 deg cut plane.

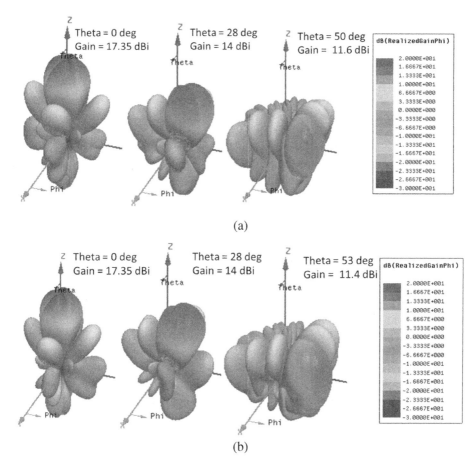

Fig. 9. Beam steering performance of the 4 × 4 planar array configuration for 32.6 GHz by varying progressive phase shifts for the (a) X-polarization and (b) Y-polarization.

5   Generation of Beam Steering Radiation Patterns

Since the proposed massive MIMO panel array (Fig. 2(b)) has individual feed port excitations for both X- and Y-polarization cases, hence by exciting the ports with equal amplitude and progressive phase shifts, we can also generate steered radiation patterns such as shown in this section. We have kept amplitudes constant for simplicity while phase shifts have been varied. In real implementations, we would compute proper amplitude and phase coefficients for generating an arbitrary beam scan radiation pattern by employing beam forming algorithms such as in [7–10] although it is not discussed here.

Figure 9 shows 3D radiation patterns for the beam scan cases for both X- and Y-polarizations for the positive half of the elevation angles only. The beam scan performance for the negative half of the elevation angles is almost the same as in case of the positive half of the elevation angles, hence for the sake of brevity is not shown here. From Fig. 9(a), it can be seen that as the beam scans towards elevation angle of theta = 50 deg, gain of the main beam drops to 11.6 dBi from 17.35 dBi that of the broadside beam, hence there is gain drop of almost 6 dB as beam scans. This gain drop is a feature of any beam steering array. Similarly, from Fig. 9(b) for the Y-polarization, it can be seen that as the beam scans towards elevation angle of theta = 53 deg, gain of the main beam drops to 11.4 dBi from 17.35 dBi that of the broadside beam, hence there is gain drop of almost 6 dB as well. Comparison of the two polarization cases shows that both polarizations almost perform similar which is a desired feature of a dual polarization beam steering array antenna. Once again we should note that, we can generate beam scan performance for the different linear combinations of the array also. Further, by applying variable amplitude in addition to the phase shifts, additional radiation pattern features can be generated.

6 Fabrication and Implementation Approach

The full array will be fabricated using the multilayer printed circuit board (PCB) approach with metal plated vias. The amplitude and phase excitations of the patch elements would depend on the computed amplitude and phase values based on the selected beamforming algorithms. For receive mode of the array, each of the radiating elements will be backed by low noise amplifiers (LNAs), variable gain attenuators (VGAs) and phase shifters as shown in [11]. Consequently, for the transmit mode of the array, each of the radiating elements will be backed by power amplifiers (PAs), in addition to the VGAs and phase shifters. There will be a diplexer to switch between the transmit and receive mode of the operation. Finally, to control the amplitude and phase of the array elements, we would employ microcontrollers such as done in [11].

7 Conclusions and Future Study

This paper presented simulation study results of a 4 × 4 microstrip planar massive MIMO panel array with dual polarization capability in Ka-band frequency range. In such an array, we have control of all the feed ports (in this case 32 ports, 16 ports each for the X-polarization and Y-polarization). By applying variable amplitude and phase coefficients to the selected feed ports in a selected polarization, we can obtain flexible radiation patterns for the fixed beam and steered beam cases. For simplicity, in this paper, we assumed amplitude to be constant while we varied phase values for a group of radiating elements. In practical realization, we will fabricate array antenna aperture in addition to RF input ports and beam forming network for the transmit and/or receive mode. The beamforming algorithm controlled through the microcontroller would be applied to the array feed ports which is a software defined based approach in reference to this array antenna. Further, the panel size can be increased and/or a large massive MIMO antenna can be implemented using the proposed massive MIMO panel array antenna of 4 × 4 size. During the conference, additional analysis and measured results will be presented.

References

1. Chin, M., Leung, V., Lai, R.: IEEE access special section editorial: 5G wireless technologies: perspectives on the next generation of mobile communications and networking. In: IEEE ACESS, January 30 (2015)
2. Choudhary, D.: 5G wireless and millimeter wave technology evolution: an overview. In: IEEE IMS 2015 (2015)
3. Talwar, S., Choudhury, D., Dimou, K., Aryafar, E., Bangerter, B., Stewart, K.: Enabling technologies and architectures for 5G wireless. In: IEEE IMS 2014 (2014)
4. Fernandez, S., Sharma, S.K.: Multi-band printed meandered loop antennas with MIMO implementations for wireless routers. IEEE Antennas Wirel. Propag. Lett. **12**, 96–99 (2013)
5. Choukiker, Y., Sharma, S.K., Behera, S.K.: Hybrid fractal shape planar monopole antenna covering multiband wireless communications with MIMO implementation for handheld mobile devices. IEEE Trans. Antennas Propag. **62**(3), 1483–1488 (2014)
6. Kulkarni, A., Sharma, S.K.: Frequency reconfigurable microstrip loop antenna covering LTE bands with MIMO implementation and wideband microstrip slot antenna all for portable wireless DTV media player. IEEE Trans. Antennas Propag. **61**(2), 964–968 (2013)
7. Balanis, C.A.: Antenna Theory: Analysis and Design, 4th edn. Wiley, Hoboken (2016)
8. Labadie, N., Sharma, S.K., Rebeiz, G.: Investigations on the use of multiple unique radiating modes for 2D beam steering. IEEE Trans. Antennas Propag. **64**(11), 4659 (2016). Accepted Feb 2016
9. Labadie, N., Sharma, S.K., Rebeiz, G.: A novel approach to beam steering using arrays composed of multiple unique radiating modes. IEEE Trans. Antennas Propag. **63**(7), 2932–2945 (2015)
10. Shafai, L., Sharma, S.K., Shafai, L., Daneshmand, M., Mousavi, P.: Phase shift bandwidth and scan range in microstrip arrays by the element frequency tuning. IEEE Trans. Antennas Propag. **54**(5), 1467–1473 (2006)
11. Babakhani, B., Sharma, S.K., Labadie, N.: A frequency agile microstrip patch phased array antenna with polarization reconfiguration. IEEE Trans. Antennas Propag. Accepted July 2016

Author Index

Printed in the United States
By Bookmasters